ANALYSIS AND DIFFERENTIAL EQUATIONS

Second Edition

Other World Scientific Titles by the Author

Functional Estimation for Density, Regression Models and Processes
ISBN: 978-981-4343-73-2

Inequalities in Analysis and Probability
Third Edition
ISBN: 978-981-123-134-6

Probability and Stochastic Processes: Work Examples
ISBN: 978-981-121-352-6
ISBN: 978-981-121-446-2 (pbk)

Orthonormal Series Estimators
ISBN: 978-981-121-068-6

Estimations and Tests in Change-Point Models
ISBN: 978-981-3231-76-4

Inequalities in Analysis and Probability
Second Edition
ISBN: 978-981-3143-98-2

Analysis and Differential Equations
ISBN: 978-981-4635-95-0

Statistical Tests of Nonparametric Hypotheses: Asymptotic Theory
ISBN: 978-981-4531-74-0

Inequalities in Analysis and Probability
ISBN: 978-981-4412-57-5

ANALYSIS AND DIFFERENTIAL EQUATIONS

Second Edition

Odile Pons

French National Institute for Agronomical Research, France

NEW JERSEY • LONDON • SINGAPORE • BEIJING • SHANGHAI • HONG KONG • TAIPEI • CHENNAI

Published by

World Scientific Publishing Co. Pte. Ltd.
5 Toh Tuck Link, Singapore 596224
USA office: 27 Warren Street, Suite 401-402, Hackensack, NJ 07601
UK office: 57 Shelton Street, Covent Garden, London WC2H 9HE

Library of Congress Cataloging-in-Publication Data
Names: Pons, Odile, author.
Title: Analysis and differential equations / Odile Pons,
French National Institute for Agronomical Research, France.
Description: Second edition. | New Jersey : World Scientific Publishing Co. Pte. Ltd., [2023] |
Includes bibliographical references and index.
Identifiers: LCCN 2022047680 | ISBN 9789811268564 (hardcover) |
ISBN 9789811268571 (ebook for institutions) | ISBN 9789811268588 (ebook for individuals)
Subjects: LCSH: Differential equations--Textbooks. | Calculus--Textbooks.
Classification: LCC QA371 .P575 2023 | DDC 515/.35--dc23/eng/20221013
LC record available at https://lccn.loc.gov/2022047680

British Library Cataloguing-in-Publication Data
A catalogue record for this book is available from the British Library.

For any available supplementary material, please visit
https://www.worldscientific.com/worldscibooks/10.1142/13206#t=suppl

Printed in Singapore

Preface

The methods developed by the mathematicians during the past centuries for the calculus of distances and trajectories aimed to establish maps or astronomical laws for an improvement of the calendar and predictions. They have been the origin of the most important advances in mathematics. Large systems of partial differential equations were developed for dynamics models in physics, they have provided finer constants in physics from new measurements made jointly in several places.

The classical form of the theory of differential equations dates from the 18th century with d'Alembert, Fourier, Laplace, Cauchy and Lagrange. The expansions in series and the projections on orthonormal functions have a large part in their resolution. D'Alembert, Leibnitz, Taylor and Mclaurin and Cauchy expansions of functions of a real variable have been extended to complex functions of complex variables and to higher dimensions under Cauchy's necessary conditions. They enable to solve differential equations for the time and spatial variations of manifolds with Fourier series. The classical formalism and many examples has remained nearly unchanged in this approach. A large amount of literature devoted to numerical methods is not so important as it seems. The finite differences and the approximations of solutions by linearization lead to approximations of the differential equations and to Hilbert operators.

The theory of elliptic and hyperbolic equations has been much developed since the early 19th century. Though simple physical and biological systems of differential equations with constant coefficients have most often solutions that may be calculated with the usual functions, explicit solutions of the differential equations are not always possible and implicit solutions are obtained by integration. The graphical methods have won in accuracy and have been fastened by the approximations of implicit solutions when there is no explicit solution. Many classes of differential equations that could not be explicitly solved have led to the definition of new mathematical functions, mainly for second order differential equations. The primitive functions such as Bessel's functions and the orthonormal polynomials have been widely studied since the XIXth century, they will be presented in Chapters 2 and 7. Chapters 2, 3, 4 and 7 present several new primitive functions and their properties.

This book is also an introduction to the ordinary and partial differential equations based on the classical theory of the differential calculus and its generalizations. Chapters 4 and the followings can be read directly after Chapter 1 by students more interested in this part, after a course on differential and integral calculus. They gather techniques for the resolution of the differential equations with new results and examples of classes of solutions. It is organized according to the dimension of the problems, the order of the derivatives and of the classes of differential equations. The explicit resolution of examples of differential equations and the equivalence of several classes of differential equations by reparametrizations should enable the reader to solve new cases.

Odile M.-T. Pons
May 2014

Preface of the second edition

The main changes of the second edition are the addition of theoretical sections in Chapters 4, 5 and 6. They prove of the existence and unicity of the solutions for linear differential equations on real and complex spaces and for nonlinear differential equations defined by locally Lipschitz functions of the derivatives.

Approximations of nonlinear parabolic, elliptic and hyperbolic equations with locally differentiable operators enable to prove the existence and unicity of their solutions.

The behavior of the solutions of differential equations under small perturbations of the initial condition or of the differential operators is studied.

Finally several models of nonlinear partial differential equations have been added.

The text is throughout revised and completed by more examples and exercises.

Odile M.-T. Pons
September 2022

Preface of the second edition

Contents

Chapter 1

Introduction

This chapter introduces the ordinary and partial differential equations formaly and through working examples. We present the implicit equations related to differential equations and solutions of second order differential equations and system of partial differential equations, with the extensions developed later.

1.1 Differential equations

Let $(E_k, \|.\|)$ be metric spaces of real functions, for $k = 0, \ldots, n$, and let f be a continuous function in an open convex set of $\mathbb{R} \times E_0 \times \ldots \times E_n$. A differential equation of order n in $\mathbb{R}$ is defined for a function u of E_1, with derivatives $u^{(k)}$ belonging to E_k, $k = 1, \ldots, n$, as

$$f(x, u_x, u'_x, \ldots, u_x^{(n)}) = 0 \tag{1.1}$$

with a function f depending on u and its first n derivatives. Denoting the vector of u_x and its derivatives by $U_{x,n} = (u_x, u'_x, \ldots, u_x^{(n)})^t$, the primitive equation of (1.1) is a differentiable functional F of $\mathbb{R} \times E_0 \times \ldots \times E_{n-1}$ such that $F(x, U_{x,n-1}) - \alpha = 0$, with an arbitrary constant α defined by initial values of x and $U_{x,n-1}$. Its derivative with respect to x is a sum of partial derivatives

$$f(x, U_{x,n}) = F'_1(x, U_{x,n}) + \sum_{k=1}^{n} u_x^{(k)} F'_{k+1}(x, U_{x,n-1})$$

where F'_j is the derivative of F with respect to its jth component, for $j = 1, \ldots, n+1$.

A linear differential equation of order n in $\mathbb{R}$ is defined by a linear function f of $U_{x,n-1}$ and such that

$$u_x^{(n)} + f(x, u_x, u'_x, \ldots, u_x^{(n-1)}) = 0. \tag{1.2}$$

Its primitive equation is a linear and differentiable functional

$$F(x, U_{x,n-1}) - F(x_0, U_{x_0,n-1}) = 0$$

with an arbitrary initial value x_0 and its derivative with respect to x is

$$\begin{aligned}\frac{d}{dx}F(x, U_{x,n-1}) &= F_1'(x, U_{x,n-1}) + \sum_{k=1}^{n} u_x^{(k)} F_{k+1}'(x, U_{x,n-1}),\\ &= u_x^{(n)} + f(x, U_{x,n-1}).\end{aligned}$$

This implies

$$f(x, U_{x,n-1}) = F_1'(x, U_{x,n-1}) + \sum_{k=1}^{n} u_x^{(k)} F_{k+1}'(x, U_{x,n-1}),$$

with $F_{n+1}'(x, U_{x,n-1}) = 1$.

A nth order differential operator L_n is linear if for all constants c_1 and c_2 and for all functions u_1 and u_2 of $C_n(\mathbb{R})$, $L_n(c_1u_1 + c_2u_2) = c_1L_nu_1 + c_2L_nu_2$. Inhomogeneous linear differential equations of order n in $C_n(\mathbb{R})$ have the form $L_n(x)\, u_x = g_x$ with a linear operator

$$L_n(x) = \sum_{k=0}^{n} A_{k,x} \frac{d^k}{dx^k}$$

depending on real functions $A_{0,x}, \ldots, A_{n,x}$. Their solutions depend on initial conditions for the first $(n-1)$th derivatives at $(n-1)$ values of x.

Let Ω be an open and bounded subset of $\mathbb{R}^p$, the functional spaces $L^\alpha(\Omega) = \{u; \int_\Omega u(x)\, dx < \infty\}$ and

$$H_\alpha^\beta(\Omega) = \{u; u^{(s)} \in L^\alpha(\Omega), s \leq \beta\}$$

are metric spaces. Let L be a linear differential operator of order β on H_α^β, the equation $Lu = f$, with f in $L^\alpha(\Omega)$, is equivalent to

$$\int vLu\, dx = \int fv\, dx$$

for every v in $L^\alpha(\Omega)$. The problem is to give exact solutions of differential equations satisfying boundary conditions on the frontier of Ω.

A nth order differential equation $L_nu = f$ has n distinct solutions if it can be solved. Let u be a solution of the nth order differential equation (1.1) then $y_x = u_xv_x$ is also a solution if v_x is solution of a $n-1$th order equation

$$L_{2,n-1}v = u_xv_x^{(n)} + h(x, U_{x,n-1}, V_{x,n-1}) = 0$$

with a function h depending on the first $(n-1)$th derivatives of u_x and v_x. Repeating this argument with a solution of the k-th order differential equation $L_{k,n-k}v_k = 0$, with $k = 1, \ldots, n-2$, solves the differential equation.

1.2 Second order differential equations

The equation of an ellipse centered at (x_0, y_0) in the plane is

$$\left(\frac{x-x_0}{a}\right)^2 + \left(\frac{y-y_0}{b}\right)^2 = 1,$$

with constants $a > 0$ and $b > 0$. Let $y = y_x$, the ellipse is defined by the first or second order nonlinear differential equations

$$a^2(y_x - y_0)y'_x + b^2(x - x_0) = 0,$$
$$(y_x - y_0)y''_x + y'^2_x + a^{-2}b^2 = 0,$$

y having the value y_0 at x_0. The motion on the ellipse is therefore described by a first order differential equation

$$a^2(y_x - y_0)v_x + b^2(x - x_0) = 0,$$

where the velocity v_x is the first derivative of the function y_x

$$v_x = y'_x = -\frac{b^2(x-x_0)}{a^2(y_x - y_0)}.$$

The equations are also written as $a^2(y_x - y_0)\,dy_x + b^2(x - x_0)\,dx$ and $(y_x - y_0)\,dv_x + v_x^2\,dx + a^{-2}b^2\,dx = 0$.

The constants x_0 and y_0 are suppressed from the third differential equation of the ellipse which is equivalent to

$$y_x^{(3)}\{a^2y'^2_x + b^2\} - 3a^2y'_xy''^2_x = 0,$$

using the second differential equation.

The observations of the planetes and comets have led to the description of their motion based on difficult trigonometric calculus during the 18th and the 19th centuries. The elliptic curve of a comet turning around the sun may be defined by adding lines with a parametric slope to the equation of the half-circle

$$x^2 + y^2 = 1 \pm (1-k)x, \ y > 0, \tag{1.3}$$

with a constant k in $]0, 1[$. Its second order differential equation is the same as the circle, $x^2 + y^2 = k_1x + k_2$ and the constants of (1.3) are

$$k_1 = \pm(k^{-1} - 1), \qquad k_2 = k^{-1}.$$

The curves coming towards and away from the circle belong to a periodic curve larger than the circle. According to the slope $1-k$, they belong to an elliptic curve or there is a cross-over.

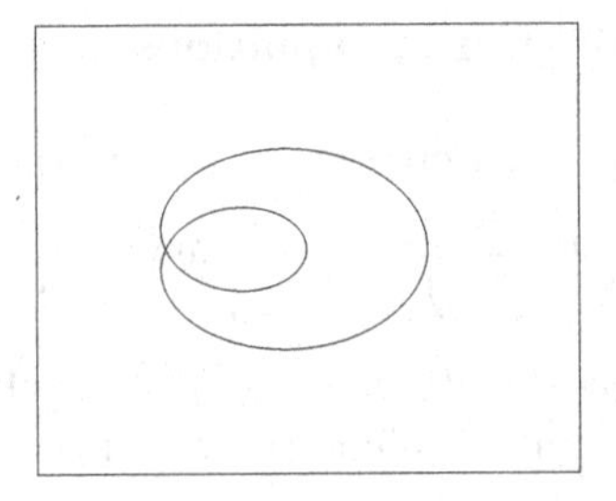

Fig. 1.1 Graph of the complex polynom $(z - .3i)^2$.

Differential equations for trigonometric functions may be defined as equations for their coordinates or for real or complex functions. Equations of complex circles present singularities as shown by Figure (1.1) for a complex polynom on the circle $|z| = 1$, for $z = x + iy$

$$\begin{aligned} p(z) &= (z - ki)^2 \\ &= 1 - 2y(y - k) - k^2 + 2ix(y - k), \end{aligned}$$

with a constant k in $]0, 1[$. It is described by the algebraic equation in $\mathbb{R}^2$

$$y^2(k - k^{-1}x^2) + y(1 + 2x^2) - k^2x^2 = 0.$$

For all real numbers $a > 1$ and x, the real conjugate roots of the polynom $P(x) = x^2 - 2ax + 1$ are $\zeta = a + \sqrt{a^2 - 1}$ and $\bar{\zeta} = a - \sqrt{a^2 - 1}$. For every integer m, the polynom P_m with real conjugate roots ζ^m and $\bar{\zeta}^m$ is

$$\begin{aligned} P_m(x) &= \{x - \zeta^m\}\{x - \bar{\zeta}^m\} \\ &= x^2 - 2(\zeta^m + \bar{\zeta}^m)x + 1 \\ &= x^2 + 1 - 2x\Big\{a^m + \sum_{k=1}^{[\frac{m}{2}]} \binom{m}{2k} a^{m-2k}(a^2 - 1)^k\Big\} \end{aligned} \tag{1.4}$$

where $[\frac{m}{2}]$ is the integer part of $\frac{m}{2}$.

The existence of real conjugate roots enables the factorization of trigonometric polynomials such as

$$\begin{aligned} Q_m(x) &= x^2 - a\{\cos(m\theta) + \sin(m\theta)\}x + \frac{1}{2}a^2\cos(2m\theta) \\ &= \{x - a\cos(m\theta)\}\{x - a\sin(m\theta)\}. \end{aligned}$$

In the complex space $\mathbb{C}$, for all complex conjugate numbers $\zeta = a + ib$ and $\bar{\zeta} = a - ib$ defined by real numbers a and b, the polynom P_1 of (1.4) is written as

$$\begin{aligned} P(z) &= \{z - \zeta\}\{z - \bar{\zeta}\} = z^2 - 2az + a^2 + b^2 \\ &= x^2 - y^2 + 2i(x - a)y - 2ax + a^2 + b^2 = (z - a)^2 + b^2. \quad (1.5) \end{aligned}$$

In the trigonometric parametrization, let $\zeta = ae^{i\theta} = a(\cos\theta + i\sin\theta)$ and let $\bar{\zeta} = a(\cos\theta - i\sin\theta) = -ae^{-i\theta}$, with $a = \|\zeta\|_2$, $2\cos\theta = e^{i\theta} + e^{-i\theta}$ and $2\sin\theta = e^{i\theta} - e^{-i\theta}$. For every integer m, the complex polynom P_m defined by (1.4) for every z complex is written as

$$P_m(z) = z^2 - 2a^m z\cos(m\theta) + a^{2m}.$$

The first derivative at z_0 of an analytic function f on an open subset of $\mathbb{C}$ is defined as

$$f'(z) = \lim_{\|z - z_0\| \to 0} \frac{f(z) - f(z_0)}{z - z_0},$$

the first derivative of P_m is therefore

$$P'_m(z) = 2z - (\zeta^m + \bar{\zeta}^m) = 2\{z - a^m\cos(m\theta)\}$$

and P_m is solution of the same differential equation for real and complex variables

$$P_m(z) = zP'_m(z) - z^2 + a^{2m}.$$

Considering P_m as a function of θ, its second derivative with respect to θ is $P''_{m,\theta}(z,\theta) = 2m^2 a^m z\cos(m\theta)$, it satisfies the inhomogeneous second order differential equation

$$P''_{m,\theta}(z,\theta) + m^2 P_{m,\theta}(z,\theta) - m^2(z^2 + a^{2m}) = 0.$$

For every non null real x different from $a\cos(m\theta)$ and $a\sin(m\theta)$, and for every θ in $]0,\pi[$, an expansion of the inverse of the function

$$Q_m(x) = \{x - a\cos(m\theta)\}\{x - a\sin(m\theta)\}$$

is written as as a trigonometric series

$$Q_m^{-1}(x) = x^{-2} \sum_{j,k \geq 0} \left\{\frac{a\cos(m\theta)}{x}\right\}^j \left\{\frac{a\sin(m\theta)}{x}\right\}^k$$

it develops as

$$Q_m^{-1}(x) = x^{-2} \sum_{j,k \geq 0} \frac{a^{j+k}}{x^{j+k}} \Big\{ \frac{\cos(2m\theta)}{2} \Big\}^{j \wedge k}$$

$$\cdot \{ \cos^{j-k}(m\theta) 1_{\{j>k\}} + \sin^{k-j}(m\theta) 1_{\{k>j\}} \},$$

$$= x^{-2} \sum_{j,k \geq 0} \frac{a^{j+k}}{x^{j+k}} \Big[\Big\{ \frac{\cos(2m\theta)}{2\sin(m\theta)} \Big\}^j \sin^k(m\theta) 1_{\{k \geq j\}}$$

$$+ \Big\{ \frac{\cos(2m\theta)}{2\cos(m\theta)} \Big\}^k \cos^j(m\theta) 1_{\{j>k\}} \}\Big].$$

The trigonometric expansions have been much used to solve the differential equations, so Chebychev's polynomials constitute an orthonormal set of trigonometric functions. Legendre's polynomials are defined as the coefficients of z^m for every integer $m \geq 0$, in the expansion of the first derivative of the complex function $(1 - 2\cos\theta z + z^2)^{-\frac{1}{2}}$ in a series.

The solutions of the homogeneous second order differential equations $f(u_x, u''_x) = 0$ with constant coefficients are linear combinations of exponential functions under constraints on their parameters. Thus, the differential equation

$$y''_x + \alpha y_x = 0,$$

with a constant $\alpha > 0$ and an initial value $y_0 \neq 0$, is the linear differential equation of an oscillator. Its solutions are sine and cosine functions as it may be verified denoting $y_x = e^{ax+b} = y_0 e^{ax}$ with the initial condition $y_0 = e^b$ with a constant a such that $a^2 + \alpha = 0$. Multiplying the equation by $v_x = y'_x$, it becomes

$$v_x \, dv_x + \alpha y_x y'_x \, dx = 0$$

and its primitive is $y'^2_x + \alpha y^2_x = k$ with a constant $k > 0$. This equation is satisfied as v_x is a sine and cosine functions of x and with the constant $k = \alpha$, then either $y_x = \sin(\sqrt{\alpha}x + \omega)$ and $v_x = \sqrt{\alpha}\cos(\sqrt{\alpha}x + \omega)$ or $y_x = \cos(\sqrt{\alpha}x + \omega)$ and $v_x = -\sqrt{\alpha}\sin(\sqrt{\alpha}x + \omega)$. The solutions of the inhomogeneous equation $y''_x + \alpha_x y_x = f_x$, with functions α_x and f_x, are functionals of α_x and f_x and they are studied hereafter.

The solutions of the homogeneous second order differential equations $f(x, u_x, u''_x) = 0$ with non constant coefficients are not combinations of exponential functions. As examples, let us consider the linear second order differential equations satisfied by the power functions. The coefficient of

u_x in the differential equations cannot be a square number. The function $y_x = kx^a$ is solution of the homogeneous differential equations

$$x^2 y_x'' - a(a-1)y_x = 0, \quad \text{if } a > 0,$$
$$x^2 y_x'' - a(a+1)y_x = 0, \quad \text{if } a < 0.$$

With $a = 2$, the function

$$y_x = x^2\Big(\frac{y_0}{x_0^2} + \int_{x_0}^{x} s^{-4} F_s\,ds\Big),$$

with initial value y_0 at an initial value $x_0 > 0$, is solution of the inhomogeneous first order differential equation

$$x^2 y_x' - 2xy_x = F_x = F_0 + \int_{x_0}^{x} f_s\,ds$$

and of the second order differential equation

$$x^2 y_x'' - 2y_x = f_x.$$

With $a = -2$, the function

$$y_x = x^{-2}\Big(y_0 x_0^2 + \int_{x_0}^{x} F_s\,ds\Big),$$

with initial value y_0 at $x_0 > 0$, is solution of the inhomogeneous first order differential equation

$$x^2 y_x' + 2xy_x = F_x$$

and of the second order equation

$$x^2 y_x'' + 4xy_x' + 2y_x = f_x.$$

These examples are sufficient to prove that the solutions of second order differential equations are not always convolutions of the second member with a solution of the homogeneous equation. The solutions are not always uniquely defined and methods for the construction of other solutions of these linear differential equations are developped in Chapter 4.

The method of the variation of the constant for such second order differential equations consists in the search of solutions in the form $y_x = x^a \alpha_x$ with a function α_x. For a constant $a > 0$, a solution of the differential equation

$$x^2 y_x'' + (2-a)xy_x' - ay_x = f_x$$

with initial value y_0 at $x_0 > 0$ is

$$y_x = x^a\Big(\frac{y_0}{x_0^a} + \int_{x_0}^{x} s^{-(a+2)} F_s\,ds\Big).$$

Under the same initial condition, the function

$$y_x = x^{-a}\Big(y_0 x_0^a + \int_{x_0}^x s^{a-2} F_s\, ds\Big)$$

is solution of the differential equation

$$x^2 y_x'' + (2+a)xy_x' + ay_x = f_x,$$

and the function

$$y_x = x^a\Big\{\frac{y_0}{x_0^a} + \frac{x-x_0}{x_0^a}\Big(y_0' - \frac{ay_0}{x_0}\Big) + x\int_{x_0}^x s^{-a}\, dF_s - \int_{x_0}^x s^{1-a}\, dF_s\Big\}$$

is solution of the differential equation

$$x^2 y_x'' + 2axy_x' + a(a+1)y_x = f_x$$

with initial values $y_{x_0} = y_0$ and $y'_{x_0} = y'_0$.

The nonlinear equation

$$y''_{xx} + f(x)y'_x + g(x)y'^2_x = 0,$$

is equivalent to

$$z'_x + f(x)z + g(x)z^2 = 0$$

and its primitives with values y_1 at x_1 and y_0 at x_0 satisfy the implicit equations

$$y'_x = y'_1 \exp\Big\{\int_x^{x_1} f(s)\, ds\Big\} \exp\Big\{\int_x^{x_1} y_s g(s)\, ds\Big\},$$

$$\int_{x_0}^x \exp\Big\{-\int_\xi^{x_1} y_s g(s)\, ds\Big\}\, dy_\xi = y'_1 \int_{x_0}^x \exp\Big\{\int_\xi^{x_1} f(s)\, ds\Big\}\, d\xi.$$

This equation is explicitly solved in Chapter 4. If the coefficients are constant, the solution is the well known logistic function. Higher order equations that involve a small number of terms may be solved by a change of variable or by reducing the degree of the equation.

It may be difficult to find out the exact solutions of differential equations and their properties. For exemple, the solution of the Airy differential equation $xy''_x - y_x = 0$ is an unknown function which is not oscillatory (Chapter 4). The functions defined as solutions of differential equations cannot always be expressed in a closed form with algebraic functions, they may be solutions of implicit equations or limits of convergent series deduced from an expansion of the solution as a series.

1.3 Differential equations for functions on $\mathbb{R}^p$

A first order differential equation for a real function u_x of $C_1(\mathbb{R}^p)$ is a system of partial differential equations, for $k = 1, \ldots, p$

$$u'_{k,x} := \frac{\partial u_x}{\partial x_k} = f_k(x, u_x)$$

in a subset of $\mathbb{R}^p$, with initial conditions on its frontier. It is linear if f_k depends linearly on the components of u_x. The primitive equation $F(x, u_x)$ has the partial derivatives

$$\begin{aligned}\frac{\partial}{\partial x_k} F(x, u_x) &= F'_k(x, u_x) + u'_{k,x} F'_{1+p}(x, u_x) \\ &= F'_k(x, u_x) + f_k(x, u_x) F'_{1+p}(x, u_x),\ k = 1, \ldots, p. \quad (1.6)\end{aligned}$$

Let x by a time indexed function of $C_n(\mathbb{R}_+$ with values in a subset Ω of $\mathbb{R}^p$. The time derivative of the primitive $F(x, u_x)$ of a system of first order partial differential equations is the nonlinear equation

$$\frac{d}{dt} F(x, u_x) = \sum_{k=1}^{p} x'_{k,t} \{F'_k(x, u_x) + u'_{k,x} F'_{1+p}(x, u_x)\}$$

and a differential equation in $C_1(\mathbb{R}^p)$ has the form

$$\sum_{k=1}^{p} \alpha_k(x_t, x'_t) \{F'_k(x, u_x) + u'_{k,x} F'_{1+p}(x, u_x)\} = h(x). \quad (1.7)$$

It is equivalent to the system (1.6) of p equations with

$$x'_{k,t} = \alpha_k(x_t, x'_t),$$

with (1.6), for $k = 1, \ldots, p$. However the system (1.6) or the differential equation (1.7) may be solved without the restriction of differential equations for x as a time function.

Let $u(x, y)$ be a real function in $C_1(\mathbb{R}^2)$ with partial derivatives such that $ayu'_x - bxu'_y = 0$, with constants a and b. Since

$$\frac{d}{dt} u(x, y) = x'_t u'_x + y'_t u'_y,$$

a solution of the differential equation for u is solution of $x'_t = ay_t$ and $y'_t = -bx_t$ hence $x''_t = -abx_t$ and $y''_t = -aby_t$ for every t. The solutions x_t and y_t are determined by the sign of ab as sinusoid or hyperbolic functions and the function $u(x, y) = ax^2 + by^2$ is solution of the differential equation $ax^{-1}u'_x(x, y) = by^{-1}u'_y(x, y)$.

The function $u(x,y)$ solution of the differential equation $xu'_x - yu'_y = 0$ has the solution $u(x,y) = u_0 e^{axy}$, for every constant a. Adding initial conditions such as $u(0,y) = v_2(y)$ for every $y > 0$ and $u(x,0) = v_1(x)$ for every $x > 0$, a function of the form $u(x,y) = v_1(x)v_2(y)e^{axy}$, with non constant functions v_1 and v_2 is solution of the differential equation $xu'_x - yu'_y = 0$ if and only if $x = y$ $v_1(x) = e^{cx} = v_2(y)$, then the solution of the differential equation satisfies $u(x,x) = e^{ax^2+2cx}$.

A nth order differential equation for a real function u_x of $C_n(\mathbb{R}^p)$ is a system of partial differential equations, for $k = 1,\ldots,p$. The primitive equation $F(x, u_x, u_x^{(1)}, \ldots, u_x^{(n-1)})$ has the partial derivatives

$$\frac{\partial}{\partial x_k} F(x, u_x) = F'_k + u'_{k,x} F'_{1+p}$$

A second order linear differential equation for a real function u_x of $C_1(\mathbb{R}^p)$ is defined for $k = 1,\ldots,p$ by a second order differential operator

$$L_x = \sum_{j,k=1}^{p} A_{jk,x} \frac{\partial}{\partial x_k} \frac{\partial}{\partial x_j}$$

depending linearly on the second order partial differentiation, with functions $A_{jk,x}$ on $\mathbb{R}^p$. The equations are classified according to the questions they solve. Laplace's second order differential operator of a function u of $C_2(\mathbb{R}^p)$ is

$$\Delta u(x) = \sum_{k=1}^{p} \frac{\partial^2 u(x)}{\partial x_k^2}.$$

In $\mathbb{R} \times \mathbb{R}_+$, Poisson's equation

$$\Delta_x u(x,t) = f_{x,t}$$

is a linear second order differential equation for u forced by the second member f. Laplace's differential equation for the stable temperature of an isolated thin plate is $\Delta_x u(x,t) = 0$.

A function u depending the planar coordinates (x,y) through the Euclidean norm $r(x,y) = (\sum_{k=1}^{p} x_k^2)^{\frac{1}{2}}$ has the derivatives

$$u'_k(x) = \frac{x_k}{r} u'(r),$$

$$u''_k(x) = \frac{1}{r} u'(r) + \frac{x_k^2}{r^2} u''(r), \tag{1.8}$$

$$\Delta u = \frac{p}{r} u'(r) + u''(r). \tag{1.9}$$

For instance, the Poisson equation $\Delta u = \alpha^2 r^{-(\alpha+2)}$ with initial conditions $u_{r_0} = u_0$ has the solution

$$u(r) = u_0 r_0^{\alpha} r^{-\alpha},$$

in $\mathbb{R}^p$, this function u_r is solution of the differential equation

$$\Delta u = \alpha(\alpha - p + 1) r^{-(\alpha+2)}.$$

The potential of attraction due to a wire circle of radius c and mass M on a point at a distance r of the center of the circle is

$$V(r) = M(c^2 + r^2)^{-\frac{1}{2}}$$

according to the classical laws in physics. The potential function

$$V(\theta) = (r^2 - 2rr_1 \cos\theta + r_1^2)^{-\frac{1}{2}}$$

is solution of Laplace's equation

$$r^2 \Delta_r(rV) + \frac{1}{\sin\theta} \frac{d}{d\theta}\Big(\sin\theta \frac{d}{d\theta} V\Big) = 0.$$

The function $V(\theta)$ has the form

$$V(\theta) = \frac{1}{r} v(r^{-1} r_1, \theta) = \frac{1}{r_1} v(r_1^{-1} r, \theta),$$

where

$$v(z, \theta) = (1 - 2z\cos\theta + z^2)^{-\frac{1}{2}},$$

and it is finite for every z different from $cos\theta \pm i \sin\theta$ where it is infinite. Within a complex disk of radius 1 centered at the origin, the function $(r^2 - 2rr_1 \cos\theta + r_1^2)^{-\frac{1}{2}}$ has an expansion as a series according to the powers of z

$$\begin{aligned}\frac{1}{(r - 2rr_1 z + r_1^2)^{\frac{1}{2}}} &= \frac{1}{r} \sum_{m\geq 0} \Big(\frac{r_1}{r}\Big)^m P_m(\cos\theta), \quad \text{if } r_1 < r,\\ &= \frac{1}{r} \sum_{m\geq 0} \Big(\frac{r}{r_1}\Big)^m P_m(\cos\theta), \quad \text{if } r_1 > r,\end{aligned}$$

and it is integrable. The integral of $v(z, x) = (1 - 2xz + z^2)^{-\frac{1}{2}}$ with respect to z in $[-1, 1]$ and at $x = \cos\theta$ is calculated as $I = F_x(1) - F_x(-1)$ with the function $F_x(z)$ calculated in Exercise (1.6.17)

$$F_{\cos\theta}(z) = \arg\cosh \frac{(1 - 2\cos\theta z + z^2)^{\frac{1}{2}}}{\sin\theta}.$$

Fourier's heat equations (1822) in a thin object Ω of $\mathbb{R}^2$ or $\mathbb{R}^3$ with length l is defined for the variation of the temperature u of the object through the operator of derivatives

$$L_h = \Delta - \alpha$$

by $L_h u = 0$, with a constant $\alpha = 2h(kl)^{-1}$, where h and k are the inner and outer constants of conductibility of the object. The time varying temperature function u is a function in $\mathbb{R}^3 \times \mathbb{R}_+$, such that $u(\cdot, t)$ belongs to $C_2(\mathbb{R}^3)$ for every t in an interval $[0, T]$ and $u(x, \cdot)$ belongs to $C_1(\mathbb{R}_+)$ for every x in Ω. Fourier's heat conduction equation in a solid is defined by the partial derivatives operator

$$L_h(x,t) = \Delta - \alpha \frac{\partial}{\partial t}.$$

The norm of $u_0 = u(\cdot, 0)$ in $H_2^2(\Omega)$ is bounded by the norm of u in

$$\begin{aligned} H^2_{2,0}(\Omega, [0,T]) = \{u(x,t); u^{(s)}(\cdot,t) \in L^2(\Omega), s \leq 2, t \leq T; \\ u(x,\cdot) \in L^2([0,T]), x \in \Omega\}. \end{aligned}$$

A function u defined in an open bounded set Ω is extended by continuity on its frontier $\partial\Omega$ and Fourier's problem is

$$\begin{aligned} L_h(x,t)u(x,t) &= 0, \\ u(x,0) &= u_0(x), \quad \text{for every } x \text{ in } \Omega, \\ u(x,t) &= 0, \quad \text{for all } x \text{ in } \partial\Omega, t \text{ in } [0,T]. \end{aligned}$$

The temperature is stationary in time as $u'_t = 0$ and Fourier's equation reduces to Poisson's equation with a null second member. For a product function $u(x,t) = h(x)k(t)$ in $\mathbb{R} \times \mathbb{R}_+$, solutions of the heat equation are solutions of the equation $h''(x)k(t) - \alpha h(x)k'(t) = 0$, with a constant $\alpha > 0$ and under the constraint of the initial conditions $u(x, 0) = u_0$ for every x in an interval $[x_0, x_1]$ and $u(x_0, t) = u(x_1, t) = 0$ for every t in $[0, T]$.

Solutions are deduced from marginal equations for the functions h and k. A particular solution has marginal functions solutions of the equations $h''(x) = \gamma h(x)$ in $[x_0, x_1]$, with a constant $\gamma > 0$ and under the initial conditions $h(x_0) = 0$, $h(x_1) = a$ and $k'(t) = -\alpha^{-1}\gamma k(t)$ in $[0, T]$ under the initial condition $k(t_0) = k_0$. It follows that for every constant $\gamma > 0$

$$u_\gamma(x,t) = u_1 \sin\{\sqrt{\gamma}(x - x_0)\} e^{-\alpha^{-1}\gamma(t-t_0)} \tag{1.10}$$

is solution of the equation $\Delta u - \alpha u'_t = 0$, with u_1 depending on k_0, a and x_1. Under the conditions, the function $u(x,t)$ may be a combination of such functions depending on constants γ_j for the sine functions and $\alpha^{-1}\gamma_j$ for the exponential functions, it is stationary in time at t_0.

The heat conduction in a solid is not necessarily a wave function. If the constant γ is strictly negative, the exponential is increasing and the sine functions are replaced by hyperbolic sine or cosine functions. Chapter 5 presents other solutions for homogeneous heat differential equations that do not split as a product of time and space functions and solutions of inhomogeneous heat differential equations. The case of non constant coeficient α is also considered. The solutions of heat equations change according to the geometry of the objects in $\mathbb{R}^3$.

A function u in $C_2(\mathbb{R}^2 \times \mathbb{R}_+)$ satisfies a wave equation in a domain Ω of $\mathbb{R}^2$ and in an interval $[0, T]$ if there exists a constant $\alpha > 0$ such that

$$\Delta u(x,t) - \alpha u_t''(t,t) = f(x,t),$$

with initial conditions at $t_0 = 0$ and on the frontier $\partial\Omega$. The homogeneous wave equation is defined as

$$\begin{aligned} \Delta u(x,t) - \nu^{-2} u_t''(t,t) &= 0, \qquad (1.11)\\ u(x_0,t) &= \varphi(t), \quad x_0 \in \partial\Omega, t > 0,\\ u(x,0) &= \psi(x), \quad x \in \partial\Omega, \end{aligned}$$

and $\varphi(0) = \psi(x_0) = u_0$.

Figure (1.2) is the graph of the function $u(x,t) = \sin(\omega x) + \sin(\nu\omega t)$, this is solution of the wave equation on $\mathbb{R} \times \mathbb{R}_+$

$$\Delta u(x,t) + \nu^{-2} u_t''(t,t) + 2\omega^2 u(x,t) = 0. \qquad (1.12)$$

Writing the solution of (1.12) as the product of a decreasing exponential function $h(t) = e^{-\gamma t}$ and a function $k(x)$, the differential equation for the function k is $h'' + (\nu^{-2}\gamma^2 - \omega^2)h = 0$ and its solution is a trigonometric or an hyperbolic function according to the sine of $\nu^{-2}\gamma^2 - \omega^2$.

Figure (1.3) is the graph of the time varying wave function

$$f(t) = \sum_{j=0}^{m} (-1)^k (2k+1)^{-1} \sin\{(2k+1)t\},$$

this is a combination of sine functions solutions of the differential equations with varying parameters $u_{tt}'' + (2k+1)u = 0$. It may be viewed as a model for the oceanic tide due to the variations of the temperature in the equatorial sea according to the sunshine. The duration at a stable level is

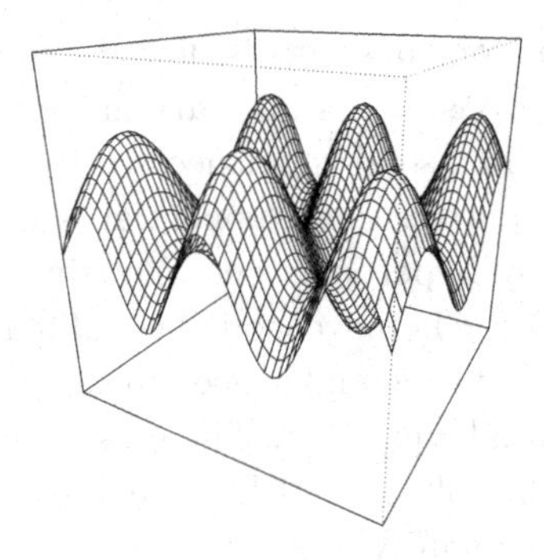

Fig. 1.2 Additive wave function $u(x,t)$.

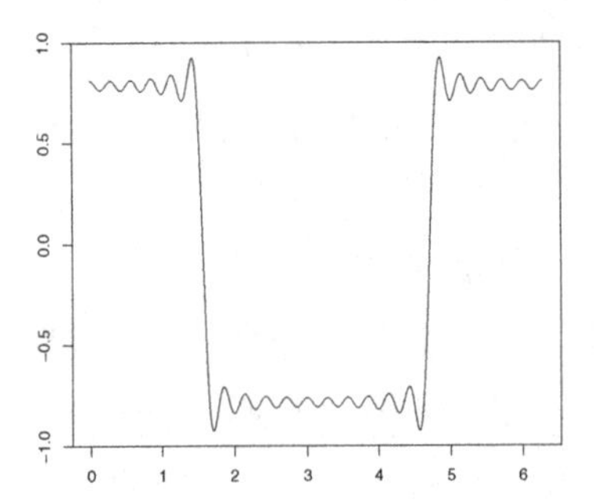

Fig. 1.3 Fourier's stationary wave function.

about 12 hours. The level has a fast change during a short time period and the sinusoïdal variations are not modified by this motion.

The wave equations are generalized by replacing the constant α with a function. For a multiplicative function $u(x,t) = \varphi(r(x))\psi(t)$ of $C_2(\mathbb{R}^2 \times \mathbb{R}_+)$ and a continuous function $\alpha(x) = \alpha(r)$ depending both on x through the $l_{2,2}$-norm $r(x)$ of x, a wave equation is written as

$$\left\{\frac{2}{r}\varphi'(r) + \varphi''(r)\right\}\psi(t) - \alpha(r)\varphi(r)\psi''(t) = f(r,t).$$

In $C_2(\mathbb{R} \times \mathbb{R}_+)$, a wave equation along a straight line has the form

$$u''_x(x,t) - \alpha(x,t)u''_t(x,t) = f(x,t).$$

For an equation having a multiplicative second member $f(x,t) = kx^{-3}e^{at}$, with a constant k in $[0,2[$, the function $u_1(x,t) = x^{-1}e^{at}$ is solution of the wave equation with the constant $a = \{\alpha^{-1}(2-k)\}^{\frac{1}{2}}$ and the function $\alpha(x) = \alpha x^{-2}$.

The function $u_2(x,t) = u_0 \exp(-2t^2(\lambda x)^{-1}\}$, for all $x > 0$ and $\lambda > 0$, is solution of the equation

$$u''_x(x,t) - \frac{4x^2}{t^2} u''_t(x,t) = \frac{4t^2}{\lambda x^3}(4-x),$$

with initial values $u(x,0) = u_0$ and $u'_t(x,0) = 0$. After a fast decay, they are slowly varying functions.

The solutions of the waves and heat equations depend on the domain Ω and on the functional spaces of u and f and on $\Omega \times [0,T]$. Several equations and domains will be considered in Chapter 3.

1.4 Multidimensional differential equations

Systems of coupled differential equations describe the evolution of functions in time or in a spatio-temporal space. In $C_1(\mathbb{R}_+)$, they are indexed by a time variable, in $C_1(\mathbb{R}_+ \times \mathbb{R}^d)$ they are indexed by time and space variables. They define the motion of an object along a linear curve, in a plane or in the space according to the equations

$$f(t, x, u, u'_t, u'_x, \ldots, u_t^{(n)}, u_{t,x,\ldots,x}^{(n)}, \ldots, u_{t,\ldots,t,x}^{(n)}, u_x^{(n)}) = 0, \tag{1.13}$$

with t in $\mathbb{R}_+$, x in $\mathbb{R}^d$ and u in $\mathbb{R}^p$, $d \geq 1$ and $p \geq 1$.

In the time evolution of populations, the components of the system may be several classes of the population. The birth-and-death equation is a system of two time-dependent linear differential equations

$$\begin{aligned} x'_t &= a_{11}x_t - a_{12}y_t, \\ y'_t &= -a_{21}x_t + a_{22}y_t, \end{aligned} \tag{1.14}$$

where dx_t and dy_t are the variations of birth and the death numbers of a populations at t, with initial values x_0 and y_0 at t_0. The function u_t with components x_t and y_t satisfies the first order differential equation $u'_t = Au_t$. Its solution (x_t, y_t) satisfy an implicit exponential equation obtained by integrating (1.14)

$$x_t = e^{a_{11}(t-t_0)} \Big\{ x_0 - a_{12} \int_{t_0}^t e^{-a_{11}(s-t_0)} y_s \, ds \Big\},$$

$$y_t = e^{a_{22}(t-t_0)} \Big\{ y_0 a_{21} \int_{t_0}^t e^{-a_{22}(s-t_0)} x_s \, ds \Big\}.$$

The function

$$u(x,y) = \frac{dy_t}{dx_t}$$

obtained from Equations (1.14) is

$$\begin{aligned} u(x,y) &= = \frac{-a_{21}x + a_{22}y}{a_{11}x - a_{12}y} \\ &= -\frac{a_{22}}{a_{12}} - \frac{D}{a_{12}} \frac{x}{a_{11}x - a_{12}y} := A + \frac{Bx}{a_{11}x - a_{12}y}, \\ D &= a_{11}a_{22} - a_{12}a_{21}. \end{aligned}$$

Its primitive provides an implicit solution y of the differential equation $(y'_x - A)(a_{11}x - a_{12}y) = Bx$

$$a_{11} \int_{x_0}^{x} s\,dy_s - \frac{a_{12}y^2}{2} + Aa_{12} \int_{x_0}^{x} y_s\,ds = \frac{(B + a_{11}A)x^2}{2} + C,$$

with a constant C depending on the initial conditions. The coefficients may be estimated from observations of the numbers of individuals in each state (alive, dead) or from their numbers of transitions between consecutive states, larger models include migrations. The time-varying functions x and y are not suitable in populations showing variations according to the age of the individuals. The population may be divided in classes according to the ages or x and y may be indexed by the age and the calendar time. Replacing the constant coefficients by time variables requires other methods for the resolution of the differential equations.

Biological models with functions depending on a delay explain the variations of a function from its past values, with a fixed delay. The nonlinear differential equation

$$u'_t = u(t-\tau)\{1 - f(u(t-\tau))\},\ t > \tau$$

leads to the prediction formula

$$u_t = u(t-\tau) + u_0 e^{\int_t^{\tau} \{1 - f(u(s-\tau))\}\,ds}.$$

Differential equations of the function u_t describing the sizes of interacting populations of prey and predators are nonlinear, they are studied in Chapter 6. Replacing the constant coefficients by time-varying functions, the equations reflect time-varying environmental conditions and their solutions are modified.

Trigonometric functions $u_t = (x_t, y_t)$ of $C_2([0, 2\pi])$ with values in $\mathbb{R}^2$ define several classes of geometric objects, among them the spirals, the astroïds and the cycloïd. They are generalized to spatial geometric objects.

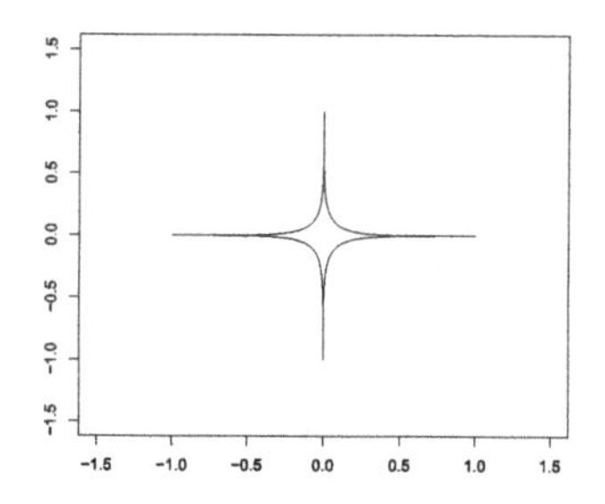

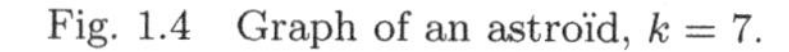
Fig. 1.4 Graph of an astroïd, $k = 7$.

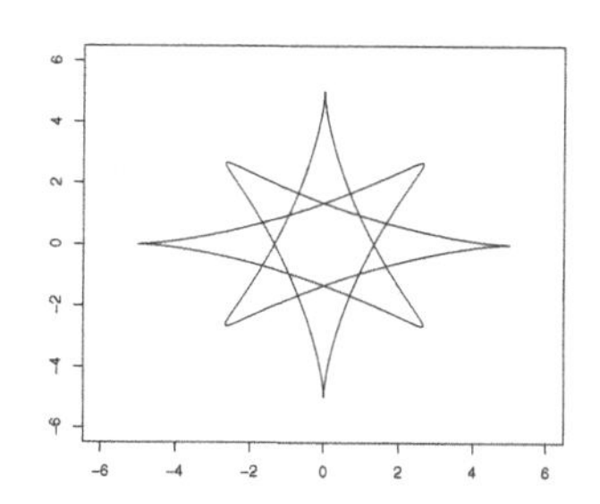

Fig. 1.5 Graph of a star, $k = 5$.

The equation of the astroïds is

$$x_t = \cos^k t, \quad y_t = \sin^k t, \tag{1.15}$$

with an odd parameter $k \geq 3$.

Their cartesian equation is

$$x_t^{\frac{2}{k}} + y_t^{\frac{2}{k}} = 1.$$

Figure (1.4) is the graph of this equation with parameter $k = 7$. Other astroïds are defined by the second derivatives

$$\begin{aligned} x_t'' &= -k \cos^{k-1} t + k(k-1) \sin^2 t \cos^{k-2} t, \\ y_t'' &= -k \sin^{k-1} t + k(k-1) \sin^{k-2} t \cos^2 t, \end{aligned} \tag{1.16}$$

with an integer parameter $k = 5, 6, 7$ or an odd k between 5 and 17. The star with unequal eight arms of Figure (1.5) has the same equations with parameter $k = 5$.

Using the parameters $k = 4$ and 6 in (1.15), the graphs of Figures (1.6) and (1.7) are possible trajectories of comets.

The hypocycloïd curves are circular but not smooth, they have turn back points. A cycloïd rolling along a line has the equations

$$x_t = a + t + \cos t, \quad y_t = b + ct + \sin t,$$

they are the equations of circles centered along the line $(a + t, b + ct)$ and moving with a point of the circle. Its graph presents halph-circles along this line. The webs are cycloïds rolling along a circle, their equations are

$$x_t = k \cos t + \cos(kt), \quad y_t = k \sin t - \sin(kt),$$

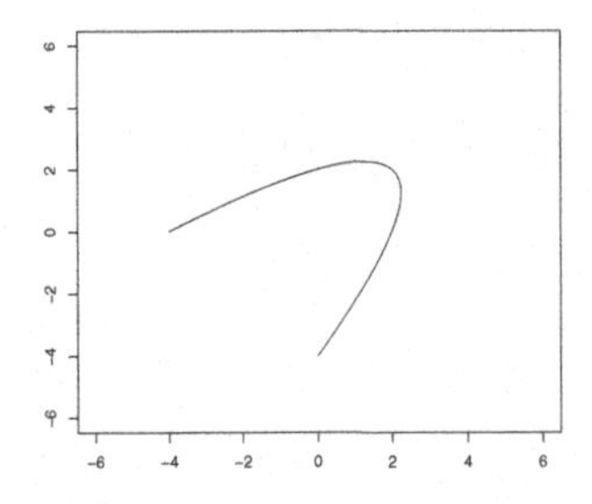

Fig. 1.6 Graph of a star, $k = 4$.

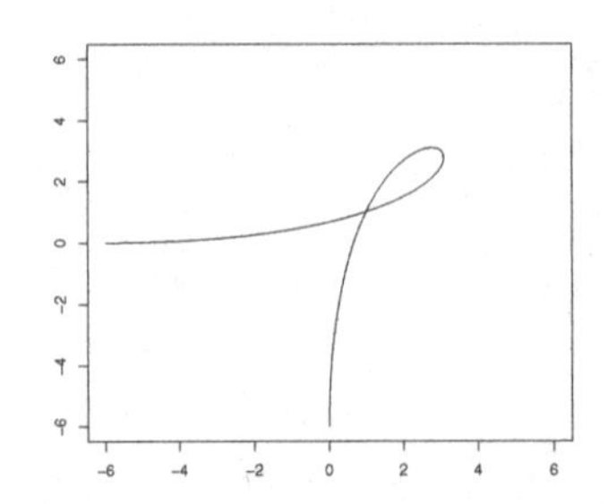

Fig. 1.7 Graph of a star, $k = 6$.

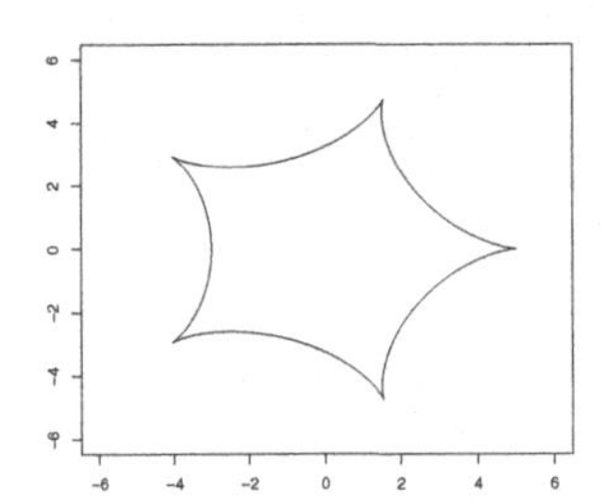

Fig. 1.8 Graph of a web with $k = 4$.

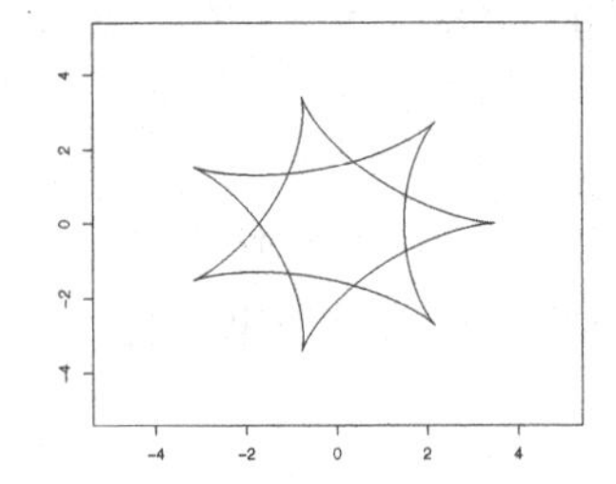

Fig. 1.9 Graph of a regular web-star.

hence $\{x_t - k\cos(t)\}^2 + \{y_t^2 - k\sin(t)\} = 1$, this is the equation of a circle with radius 1 centered moving along a circle with center zero, radius k and a period $(k\cos(t), k\sin(t))$, multiple of the period of the circle with radius 1. As t increases, the circle with radius 1 moving inside the circle with radius k draws $k+1$ arcs of circles. The period k is a multiple of the period of the rolling circles and a web with parameter k has $k+1$ peaks.

For instance, Euler's deltoïd equations is the derivative of a cycloïd with parameter $k = 2$ and the curve of Figure (1.8) has the same equation with $k = 4$. If the period k of the larger circle is not a multiple of the period of the rolling circles, several webs overlap and their graph is a regular star. In Figure (1.9), the parameter is $k = 2.5$ and the graph of shows $2(k+1) = 7$ arms, $2k$ being a multiple of the period of the small circles.

1.5 Overview

The next chapter presents methods for the approximation of functions by classes of orthonormal polynomials. They have been defined to solve the differential equations of potential and heat equations from expansions of their solutions in series by projections on the polynomials. We present the conditions for the convergence of expansions, analytical properties of the polynomials and we several generalizations. Bilinear polynomials provide models for the waves of an elastic surface and several classes of bivariate orthogonal functions for spatial representations combining several scales.

Chapter 3 generalizes the optimization theory and the Euler-Lagrange conditions for the differentiability of integrals. It is illustrated by examples concerning the calculus of integrals and optimal volumes. The partial derivative equations are applied to elliptic differential functions and to the equations of curves in a plane or objects in polar coordinates. We prove Fourier's nonlinear differential equations for the first and second order derivatives of functions in a ball, a sphere or in ellipsoïds.

Chapter 4 deals with the first and second order linear differential equation in $\mathbb{R}$. Conditions for the existence of solutions are given and the method of the variation of the constant provides a solution for the first order equations with functional coefficients and a second member. This is not an universal method and the form of the solutions cannot uniquely defined. The solutions of inhomogeneous second order equations are defined from those of the homogeneous equations according to the parameters and implicit equations are given when the solutions cannot explicited.

Several classes of equations are considered, with non standard cases including special functions defined by differential equations and solutions of differential equations based on reparametrizations. We present general solutions of the Sturm-Liouville equations and the solutions of examples based on expansions in series are detailed. The methods for the search of linearly independent solutions of the higher order linear differential equations are presented.

Chapter 5 concerns the second order linear differential equations for time and space varying functions, with boundary conditions. Several problems of Poisson's equation, potential equations, heat conduction differential equations and wave equations for multiplicative functions $u_1(x)u_2(t)$ and non multiplicative functions are presented. The solutions of the spatio-temporal homogeneous equations are not unique, we give solutions of several different multiplicative and additive forms. They are not distinguished by the boundary conditions. Controlled solutions of inhomogeneous equations are also studied.

Chapter 6 is devoted to the solutions of partial differential equations in several spaces and systems of multidimensional differential equations for time variables by separation of their components. Several models of population dynamics are considered as applications of the methods. Conditions for the existence of solutions of nonlinear first and second order equations are examined and several cases are solved. Algorithms for the calculus of the solutions are deduced from implicit equations equivalent to the differential equations.

The special functions studied in Chapter 7 are Euler and Legendre integral functions with two or three real variables, and the Airy, Bessel, Hermite, Laguerre and other functions solutions of second order differential equations.

1.6 Exercises

1.6.1. (Frenet) Write a differential equation that does not depend on the functions φ and ψ for the function

$$u(x,y) = x^n \varphi\Big(\frac{y}{x}\Big) + y^n \psi\Big(\frac{y}{x}\Big).$$

1.6.2. Write a differential equation that does not depend on φ for

$$u(x,y,z) = \varphi\Big(\frac{x}{y}, \frac{y}{z}, \frac{z}{x}\Big)$$

with independent variables x, y and z.

1.6.3. (Frenet) Write a differential equations for $u(x,y) = \varphi(ay + bx)\psi(ay - bx)$.

1.6.4. (Frenet) Write a partial differential equation that does not depend φ and ψ for $u(x,y,z) = \varphi(ax + cz) = \psi(ax - by)$.

1.6.5. (Frenet) Write a partial differential equation for the equation $u^2(x,y) = xy + \varphi(x^{-1}y)$.

1.6.6. Solve the equation $(x + y)\,dx + (x + z)\,dy + (x + y)\,dz = 0$.

1.6.7. Solve the equation $y'_x = y(1 - y)$.

1.6.8. Solve the equations $y'_x - ay_x^2 = a$, $y'_x - ay_x^2 + by_x = a + \frac{b^2}{4a}$, $xy'_x - y^2 + y = 1$.

1.6.9. Solve the equation $y' + y = \cos x$.

1.6.10. Solve the equation $y'_x + \frac{xy}{1-x^2} = xy^{\frac{1}{2}}$.

1.6.11. Solve $(x+a)y''_x + \{2 - (x+a)^{\frac{1}{2}}\}y'_x = 0$, for all strictly positive and distinct real x and a.

1.6.12. Solve $(ax^{-n} + b)y'_x = ax^{-1}y$, with n integer larger than 1.

1.6.13. Solve $x^2 + x^{-2} + (2x)^{-1}y^2 - yy'_x = 0$.

1.6.14. Primitive of $f(x) = \arcsin(ax)$.

1.6.15. Primitive of $f(x) = \arctan(ax)$.

1.6.16. Primitive of $f(x) = \cos x\,(a^2\cos^2 x + b^2\sin^2 x)^{-1}$.

1.6.17. Primitive of $f(x) = e^{ax}\cos(nx)$, $a > 0$.

1.6.18. Primitive of $f(x) = e^{ax}\sin(nx)$, $a > 0$.

1.6.19. Primitive of $f(x) = (1 \pm 2xz + x^2)^{-\frac{1}{2}}$, for $|z| \leq 1$.

1.6.20. Primitive of $f(x) = (x^2 + 6x + 10)^{-\frac{1}{2}}$.

1.6.21. Write the derivatives of the astroïds equations (1.15) and draw their graphs with the parameter $k = 3$.

1.6.22. Prove $\int_0^x (a^2 + u^2)^{-\frac{1}{2}}\,ds = \log\{a^{-1}x + (1 + a^{-2}x^2)^{\frac{1}{2}}\}$, for $x > 0$ and $a > 0$.

1.6.23. Primitive of $f(x) = x^{-1}(a^2 + x^2)^{\frac{1}{2}}$.

1.6.24. Primitive of $f(x) = (1 - x^2)^{\frac{1}{2}}$.

1.6.25. Primitive of $f(x) = (1 + x^2)^{\frac{1}{2}}$.

1.6.26. Primitive of $f(x) = x^2(a^2 + x^2)^{-\frac{1}{2}}$.

1.6.27. Primitive of $f(x) = (1 - x^2)^{-\frac{5}{2}}$.

1.6.28. Primitive of $f_{\frac{7}{2}}(x) = (1 + x^2)^{-\frac{7}{2}}$.

1.6.29. Calculate $I = \int_0^1 (1 - x^2)^{-\frac{1}{2}} \log(x^{-1})\, dx$.

1.6.30. Solve the differential equation $y''_x + 2(y - y^3) = 0$.

1.6.31. Prove the convergence of the sequence $\gamma_n = \sum_{k=1}^{n} k^{-1} - \log n$.

1.6.32. Find the limit of the series $\sum_{k=0}^{\infty}(-1)^k\{k(k+1)\}^{-1}$ and $\sum_{k=1}^{\infty}(-1)^k k^{-2}$.

1.6.33. Write an integral expression of Euler's function $\zeta(s) = \sum_{k=1}^{\infty} k^{-s}$ depending on the logarithm, for every $s \geq 2$.

Chapter 2

Expansions with orthogonal polynomials

In this chapter, we present methods of expansions of functions by projections in bases of squared integrable functions with respect to real measures. We define Laguerre, Hermite and Legendre orthonormal polynomials and recursive methods for their calculus, with their graphs. The conditions for the convergence of the expansions of functions in polynomial series depend on the properties of the measures defining their orthogonality. Several generalizations to the real line and the plane and their differential equations are presented. They will be used in the following chapters to provide numerical approximations of the solutions of differential equations.

2.1 Introduction

Let $L^2(\mathbb{R}, \mu)$ be the space of square integrable functions with respect to the measure μ. The scalar product of functions f and g in $L^2(\mathbb{R}, \mu)$ is

$$< f, g >_\mu = \int_{\mathbb{R}} f(x)g(x)\, d\mu(x)$$

and the norm of a function f is $\|f\|_\mu = \{\int_{\mathbb{R}} f^2(x)\, d\mu(x)\}^{\frac{1}{2}}$. Let $(\psi_k)_{k\geq 0}$ be an orthonormal basis of functions in $L^2(\mathbb{R}, \mu)$ endowed with the norm $\|\cdot\|_\mu$. Every function f of $(L^2(\mathbb{R}), \mu)$ has an expansion as the infinite series

$$S_f(x) = \sum_{k\geq 0} a_k \psi_k(x)$$

with coefficients defined by the scalar product of f and ψ_k

$$a_k = < f, \psi_k >_\mu .$$

The norm of S_f is

$$\|S_f\|_\mu = \Big(\sum_{k\geq 0} a_k^2\Big)^{\frac{1}{2}}$$

and the series S_f converges in $L^2(\mathbb{R},\mu)$ to f if $(\sum_{k\geq 0} a_k^2)^{\frac{1}{2}}$ is finite and the inequalities of the Hilbert spaces apply to the basis $(\psi_k)_{k\geq 0}$. The squared approximation error of a function f of $(L^2(\mathbb{R}),\mu)$ by a finite sum $f_m(x) = \sum_{k=0}^m a_k\psi_k(x)$ is $\|f - f_m\|_\mu = \sum_{k\leq m+1} a_k^2$ and it converges to zero as m tends to infinity since $\|f\|_\mu$ is finite.

Let $\widehat{F}_n$ be the empirical distribution function of a n-sample of a variable X with density f with respect to μ and let $\nu_n = n^{\frac{1}{2}}(\widehat{F}_n - F)$ be the empirical process of the sample. The density f is approximated by an empirical estimator of the function f_m

$$\widehat{f}_{m,n}(x) = \sum_{k=0}^m \widehat{a}_{k,n}\psi_k(x), \tag{2.1}$$

where the coefficients of the series are estimated from the empirical distribution function $\widehat{F}_n$ by

$$\widehat{a}_{k,n} = \int_{\mathbb{R}} \psi_k(x)\, d\widehat{F}_n(x). \tag{2.2}$$

For every integer k, the estimator $\widehat{a}_{k,n}$ is $n^{\frac{1}{2}}$-consistent and $n^{\frac{1}{2}}(\widehat{a}_{k,n} - a_{k,n}) = \int_{\mathbb{R}} \psi_k(x)\, d\nu_n(x)$ converges weakly to the centered Gaussian variable $\int_{\mathbb{R}} \psi_k(x)\, dW_1 \circ F(x)$ defined by the standard Brownian bridge W_1, its variance is $\sigma_k^2 = \int_{\mathbb{R}} \psi_k^2(x)\, F(x) - \{\int_{\mathbb{R}} \psi_k(x)\, F(x)\}^2$. If F is continuous, the limit is also written $\int_0^1 \psi_k \circ F^{-1}\, dW_1$, with the quantile function F^{-1}.

The squared estimation error of f converges to zero in L^2 with $\|\widehat{f}_{m,n} - f_m\|_2^2 + \|f - f_m\|_\mu^2$. It is the sum of the approximation error $\|f - f_m\|_\mu^2$ which is the squared bias of the estimator and the squared estimation error of the approximating function f_m

$$\|\widehat{f}_{m,n} - f_m\|_2^2 = \sum_{k\leq m} (\widehat{a}_{k,n} - a_k)^2$$

the mean of which $E\|\widehat{f}_{m,n} - f_m\|_2^2$ equals the variance of $\widehat{f}_{m,n}$. The convergence rate of the norm the estimator is the rate of the sum of the norms $\|\widehat{f}_{m,n} - f_n\|_2 = O(n^{-\frac{1}{2}}m^{1/2})$ and $\|f_m - f\|_2 = (\sum_{i=m+1}^\infty a_k^2)^{\frac{1}{2}}$. As the sample size tends to infinity, the order m of the approximating function f_m must be chosen such that both errors have the same order hence $m = o(n)$ and m tends to infinity.

Proposition 2.1. *The optimal convergence rate of an estimator of a density f defined by (2.1) and (2.2) equals the optimal convergence rate of a kernel estimator for f in C^s, with $s \geq 2$. Then $m = O(n^{\frac{1}{2s+1}})$.*

Proof. The optimal convergence rate of a kernel estimator of a density f in C^s is $x_n = n^{-\frac{s}{2s+1}}$ and ithas the same order as the convergence rate $mn^{-\frac{1}{2}}$ of $\widehat{f}_{m,n} - f_n$ if $m = O(n^{\frac{1}{2}-\frac{s}{2s+1}}) = O(n^{\frac{1}{2s+1}})$. With this order, $mn^{-\frac{1}{2}} = O(x_n)$. □

The expansion $S_f = \sum_{k\geq 0} a_k(f)\psi_k$ of a function f of $L^2(\mathbb{R}, \mu)$ satisfies the following inequality due to the Hölder inequality. All coefficients $a_k(f)$ of the expansion H_f for a function f belonging to $L^2(\mu)$ have a norm $\|a_k(f)\|_{L^2(\mu)} \leq \|f\|_{L^2(\mu)}$. Moreover $\|f\|_{L^2(\mu)} = \|S_f\|_{L^2(\mu)}$ since $< f, \psi_k >=< S_f, \psi_k >$ for every $k \geq 0$.

Proposition 2.2. *Functions f and g of $L^2(\mathbb{R}, \mu)$ satisfy*

$$\int_{\mathbb{R}} S_f S_g \, d\mu = \sum_{j,k=0}^{\infty} a_j(f) a_k(g)$$
$$\leq \Big\{\sum_{k\geq 0}\Big(\int f\psi_k \, d\mu\Big)^2\Big\}^{\frac{1}{2}} \Big\{\sum_{k\geq 0}\Big(\int_{\mathbb{R}} g\psi_k \, d\mu\Big)^2\Big\}^{\frac{1}{2}}.$$

The derivatives of the estimator $\widehat{f}_{m,n}$ defined by (2.1) and (2.2) are

$$\widehat{f}^{(k)}_{m,n}(x) = \sum_{i=0}^{m} \widehat{a}_{k,n}\psi_i^{(k)}(x),$$

for $k \geq 1$. This is also an series estimator of the kth derivative of the density. If ψ_i is a polynomial of degree i, then $\psi_i^{(k)} = 0$ for every $i \leq k-1$ and $\widehat{f}^{(k)}_{m,n}(x) = \sum_{i=k}^{m} \widehat{a}_{k,n}\psi_i^{(k)}(x)$.

2.2 Laguerre's polynomials

Laguerre's orthonormal polynomials are defined by $L_0(x) = 1$ and the derivatives

$$L_n(x) = \frac{e^x}{n!}\frac{d^n}{dx^n}(e^{-x}x^n), \; n \geq 1.$$

Let $L_2(\mathbb{R}_+, \mu_{\mathcal{E}})$ be the space of square integrable functions with respect to the measure $\mu_{\mathcal{E}}$ defined by the exponential density with parameter 1 with respect to the Lebesgue measure in $\mathbb{R}_+$. In $L_2(\mathbb{R}_+, \mu_{\mathcal{E}})$, the scalar product of functions f and g is $< f, g >= \int_0^\infty f(x)g(x)e^{-x}\,dx$. The first Laguerre orthonomal polynomials are calculated, for every real $x \geq 0$, as

$$
\begin{aligned}
L_1(x) &= 1 - x, \\
L_2(x) &= 1 - 2x + \frac{1}{2}x^2, \\
L_3(x) &= 1 - 3x + \frac{3}{2}x^2 - \frac{1}{3!}x^3, \\
L_4(x) &= 1 - 4x + 3x^2 - \frac{2}{3}x^3 + \frac{1}{4!}x^4, \\
L_5(x) &= 1 - 5x + 5x^2 - \frac{5}{3}x^3 + \frac{5}{4!}x^4 - \frac{1}{5!}x^5, \\
L_6(x) &= 1 - 6x + \frac{15}{2}x^2 - \frac{11}{6}x^3 + \frac{15}{4!}x^4 - \frac{1}{20}x^5 + \frac{1}{6!}x^6.
\end{aligned}
$$

The sequence of the coefficients b_{kj} of x_j in L_k is increasing with k. For every integer n, $L_n(x)$ is equivalent to $1 - nx$, as x tends to zero. Several

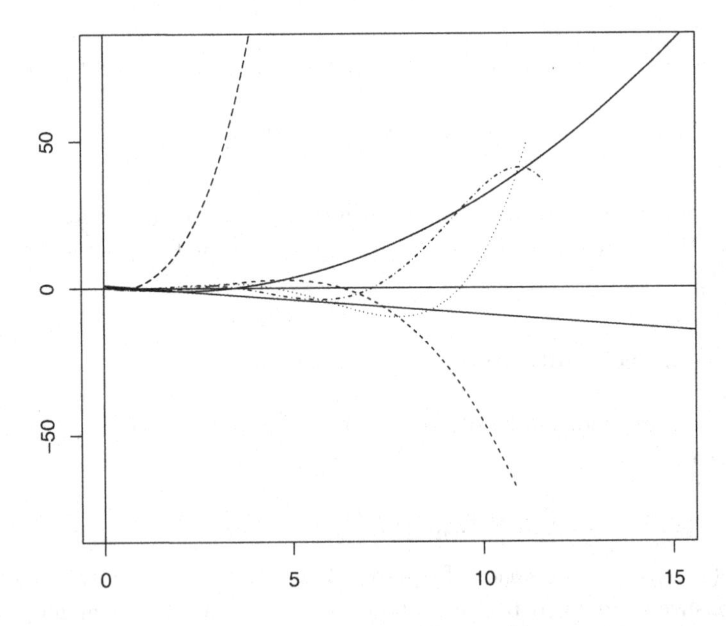

Fig. 2.1 Laguerre's polynomials.

recurrence formulas have been established

$$(n+1)L_{n+1}(x) - (x - 2n - 1)L_n(x) + nL_{n-1}(x) = 0, \ n \in \mathbb{N},$$

with the initial values $L_0(x) = 1$ and $L_1(x) = 1 - x$, and a differential recurence equation

$$\begin{aligned} &xL_n'(x) - nL_n(x) + nL_{n-1}(x) = 0, \\ &L_n(0) = 1, \ n \in \mathbb{N}. \end{aligned} \tag{2.3}$$

Proposition 2.3. *A solution of the recurrence equation (2.3) with initial value $L_n(a)$ at $a > 0$ is*

$$L_n(x) = L_n(a)a^{-n}x^n\Big\{1 - n\int_a^x y^{-(n+1)}L_{n-1}(y)\,dy\Big\}, \ n \in \mathbb{N}.$$

Proof. Equation (2.3) is linear, the solution of the homogeneous equation $xL_n'(x) = nL_n(x)$ is $L_n(x) = L_n(a)(a^{-1}x)^n$ for every $a > 0$. A solution of the non homogeneous equation (2.3) has the form $L_n(x) = u_n(x)x^n$ with a function u_n satisfying the differential equation $x^{n+1}u_n'(x) + nL_{n-1}(x) = 0$, the result follows. □

Since L_{n-1} is a polynomial of degree $n-1$, an approximation of the polynomials as x tends to infinity is written in the form $L_n(x) \sim c_nx^n + b_nx^{n-1}$, for every integer n, so Laguerre's polynomials diverge to infinity at infinity, with positive values if n is even and negative values if n is odd. The constants are calculated as $c_n = (-1)^n\frac{1}{n!}$ and $b_n = nc^{n-1}$ then, as x tends to infinity

$$L_n(x) \sim (-1)^n\Big\{\frac{1}{n!}x^n - \frac{n}{(n-1)!}x^{n-1}\Big\}, \ n \in \mathbb{N}.$$

The first derivatives of Laguerre's polynomials are deduced from the recurrence formula (2.3) as $L_n^{(1)}(x) = nx^{-1}\{L_n(x) - L_{n-1}(x)\}$, $x > 0$, then

$$\begin{aligned} L_n^{(2)}(x) &= \frac{n(n-1)}{x^2}\{L_n(x) - 2L_{n-1}(x) + L_{n-2}(x)\}, \\ L_n^{(3)}(x) &= \frac{n(n-1)}{x^3}\{(n-1)L_n(x) - (3n-5)L_{n-1}(x) \\ &\quad + (3n-5)L_{n-2}(x) - (n-2)L_{n-3}(x)\} \end{aligned}$$

and all derivatives are calculated by recurrence.

Let f be a function of $C^n(\mathbb{R}_+)$ such that $\lim_{x\to\infty} f(x)x^ke^{-x} = 0$ for every $k = 1, \ldots, n$, its scalar products with Laguerre's polynomials are calculated with integrations by parts

$$< f, L_n > = \frac{1}{n!} \int_{\mathbb{R}_+} f(x)(x^n e^{-x})^{(n)}\, dx \tag{2.4}$$
$$= \frac{(-1)^n}{n!} \int_{\mathbb{R}_+} f^{(n)}(x) x^n e^{-x}\, dx = \frac{(-1)^n}{n!} < f^{(n)}(x), x^n >,$$

Laguerre's polynomials L_n are therefore equivalent to differentiation operators for every $n \geq 1$.

From (2.4), the scalar product of real polynomials with the functions L_n are expressed in terms of the gamma functions $\Gamma_{n+1} = \int_{\mathbb{R}_+} x^n e^{-x}\, dx = n!$.

Proposition 2.4. *For every $x \geq 0$ and for all integers k and n*

$$< x^n, L_n(x) > = (-1)^n n! = (-1)^n \Gamma_{n+1},$$
$$< x^{k+n}, L_{k+n}(x) > = (-1)^k \frac{k!n!}{(k+n)!} < x^{k+n}, L_n(x) >$$
$$= (-1)^n \frac{(k+n)!}{k!} < x^k, L_k(x) >,$$
$$< x^{k+n}, L_n(x) > = (-1)^n \frac{(k+n)!^2}{k!n!},\ k > 0,$$
$$< x^k, L_{k+n}(x) > = 0,\ n > 0.$$

Example 1. The function $l = \sum_{n \geq 0} (-1)^n \frac{1}{\sqrt{n!}} L_n$ converges in $L^2(\mathbb{R}_+, \mu_\mathcal{E})$ and its norm is

$$< l(x), l(x) >= \sum_{n \geq 0} \frac{1}{n!} \int_{\mathbb{R}_+} L_n^2(x) e^{-x}\, dx = \sum_{n \geq 0} \frac{1}{n!} = e.$$

Example 2. Let f be a function of $C^\infty(\mathbb{R}_+)$, for every real $x \geq 0$, the series $\sum_{n \geq 0} \frac{1}{n!}(-x)^n f^{(n)}(x)$ and $\sum_{n \geq 0} < f, L_n > (x)$ converge by a Taylor expansion of $f(0)$ at x and they equal $f(0)$.

Example 3. The operator

$$T(x) = \sum_{n \geq 0} \frac{(-1)^n}{n!} L_n(x)$$

provides series of functions $Tf(x) = \sum_{n \geq 0} (-1)^n \frac{1}{n!} L_n(x) f(x)$, for every function f of $C^\infty(\mathbb{R}_+)$ having a finite Laplace transform. It satisfies

$$< Tf, 1 >= \sum_{n \geq 0} \frac{1}{n!} < f^{(n)}(x), x^n >= \int_{\mathbb{R}_+} f(2x) e^{-x}\, dx < \infty.$$

A Laguerre transform $L(x) = \sum_{k\geq 0} a_k L_k(x)$ converges if its norm $< L, L >= \sum_{k\geq 0} a_k^2$ is a convergent series. Let $\|f\|_2$ be the L^2-norm of a function f with respect to Lebesgue's measure.

Proposition 2.5. *Let f be a function of $C^\infty(\mathbb{R}_+)$ such that its derivatives satisfy $(2n+1)\|f^{(n+1)}\|_2 < \|f^{(n)}\|_2$ for every integer n, then f has a convergent Laguerre expansion.*

Proof. Integrating by part, the Laguerre coefficients of f are

$$\begin{aligned} a_{n,f} &= \int_0^\infty f(x)e^{-x}L_n(x)\,dx = \frac{1}{n!}\int_0^\infty f(x)(x^n e^{-x})^{(n)}\,dx \\ &= \frac{(-1)^n}{n!}\int_0^\infty f^{(n)}(x)x^n e^{-x}\,dx. \end{aligned}$$

The norm of the Laguerre expansion of f is

$$\sum_{n\geq 0} a_{n,f}^2 = \sum_{n\geq 0} \frac{1}{n!^2}\Big\{\int_0^\infty f^{(n)}(x)x^n e^{-x}\,dx\Big\}^2$$

and it is bounded using the Cauchy-Schwarz inequality as

$$\begin{aligned} \sum_{n\geq 0} a_{n,f}^2 &\leq \sum_{n\geq 0} \frac{1}{n!^2}\|f^{(n)}\|_{L^2}^2 \int_0^\infty x^{2n}e^{-2x}\,dx \\ &\leq \frac{(2n)!^2}{n!^2\, 2^{2n+1}}\|f^{(n)}\|_{L^2}^2, \end{aligned}$$

hence $\sum_{n\geq 0} a_{n,f}^2 \leq \sum_{n\geq 0} u_n^2 \|f^{(n)}\|_{L^2}^2$ with $u_{n+1}u_n^{-1} = 2n+1$. It converges as the derivatives satisfy the condition of decreasing norms. □

The polynomials of degree n have a finite expansion in the basis $(L_0, \ldots, L_n)$, we have $x = 1 - L_1(x)$, $x^2 = 2L_0(x) - 4L_1(x) + 2L_2(x)$, $x^3 = 30L_0(x) - 54L_1(x) + 18L_2(x) - 6L_3(x)$. The condition of Proposition 2.5 is not necessary, under the condition $\sum_{n\geq 0} a_n^2(f)$ finite, then the partial sum $S_n(f) = \sum_{k\leq n} a_k(f)L_k(x)$ converges to f if and only if the error $R_n(f) = f - S_n(f)$ in the estimation of f converges to zero, with the inequality

$$\|S_n(f)\|_{L^2(\mu_\mathcal{E})} < \|S(f)\|_{L^2(\mu_\mathcal{E})} = \|f\|_{L^2(\mu_\mathcal{E})}.$$

The exponential density satisfies the following expansions.

Proposition 2.6. *Laguerre's transform of the exponential density $f_\mathcal{E}$ with parameter 1 converges to the exponential density and its coefficients are $a_n = 2^{-(n+1)}$ for every $n \geq 0$.*

Proof. Laguerre's coefficients of $f_{\mathcal{E}}$ are $a_n = \int_0^\infty e^{-2x} L_n(x)\,dx$, for every $n \geq 0$. Integrating by parts yields

$$\begin{aligned} a_n &= \frac{1}{n!}\int_0^\infty (x^n e^{-x})^{(n)} e^{-x}\,dx = \frac{1}{n!}\int_0^\infty x^n e^{-2x}\,dx \\ &= \frac{1}{2^{n+1}n!}\Gamma_{n+1} = \frac{1}{2^{n+1}}. \end{aligned} \tag{2.5}$$

Then $\sum_{n\geq 0} a_n^2 = \sum_{n\geq 0} 4^{-(n+1)} = \frac{1}{12}$. □

The norm of the error $R_n(f_{\mathcal{E}})$ tends to zero and it satisfies $\|R_n(f_{\mathcal{E}})\|^2_{L^2(\mu_{\mathcal{E}})} = \sum_{k>n} a_k^2(f_{\mathcal{E}})$ with coefficients a_k given by Proposition 2.6, hence

$$\|R_n(f_{\mathcal{E}})\|^2_{L^2(\mu_{\mathcal{E}})} = \frac{1}{3}\frac{1}{4^{n+1}}.$$

2.3 Hermite's polynomials

Let $L^2(\mathbb{R}, \mu_{\mathcal{N}})$ be the space of square integrable functions with respect to the measure $\mu_{\mathcal{N}}$ having the normal density $f_{\mathcal{N}}$ with respect to Lebesgue's measure in $\mathbb{R}$. In $(L_2(\mathbb{R}), \mu_{\mathcal{N}})$, the scalar product of functions f and g is

$$< f, g >_{\mu_{\mathcal{N}}} = \int_{\mathbb{R}} f(x)g(x)\,d\mu_{\mathcal{N}}(x) = \int_{\mathbb{R}} f(x)g(x)f_{\mathcal{N}}(x)\,dx.$$

Hermite's polynomials $(H_n)_{n\in\mathbb{N}}$ is a basis of orthogonal real functions in $L^2(\mathbb{R}, \mu_{\mathcal{N}})$ defined from the nth derivatives of the normal density function $f_{\mathcal{N}}$ as

$$H_k(t) = (-1)^k \frac{d^k e^{-\frac{t^2}{2}}}{dt^k} e^{\frac{t^2}{2}} = \frac{1}{f_{\mathcal{N}}(x)}\frac{d^k f_{\mathcal{N}}(x)}{dx^k}, \tag{2.6}$$

the functions H_{2k} are symmetric at zero and the functions H_{2k+1} are odd, for every integer k.

Let $\varphi(t) = e^{\frac{-t^2}{2}}$ be Fourier's transform of the normal density, by the inversion formula

$$\frac{d^k f_{\mathcal{N}}(x)}{dx^k} = H_k(x)f_{\mathcal{N}}(x) = \frac{1}{2\pi}\int_{-\pi}^{\pi} (it)^k e^{itx}\varphi(t)\,dt, \tag{2.7}$$

or by the real part of this integral. Hermite's polynomials are recursively defined by $H_0 = 1$ and

$$H_{k+1}(x) = xH_k(x) - H_k'(x), \tag{2.8}$$

for every $k > 2$, hence H_k is a polynomial of degree k.

Proposition 2.7. *The function H_k solution of the differential equation (2.8) satisfies the implicit equation*

$$H_k(x) = e^{\frac{x^2}{2}}\Big\{\int_0^x H_k^{-1}(t)H_{k+1}(t)\,dt\Big\}.$$

Proof. The solution of the homogeneous equation $xH_k(x) = H_k'(x)$ with $H_k(0) = 1$ is $H_k(x) = e^{\frac{x^2}{2}}$. A solution of the non homogeneous equation (2.8) has the form $H_k(x) = a_k(x)e^{\frac{x^2}{2}}$ with a function $a_k(x) = e^{-\frac{x^2}{2}}H_k(x)$. Its derivative $a_k'(x) = e^{\frac{-x^2}{2}}H_k'(x) - xa_k(x)$ satisfies the differential equation

$$H_k'(x) = \{a_k'(x) + xa_k(x)\}e^{\frac{x^2}{2}} = \frac{a_k'}{a_k}(x)H_k(x) + xH_k(x),$$

since $H_k'(x) = xH_k(x) - H_{k+1}(x)$, this is equivalent to

$$\frac{a_k'}{a_k}(x) + \frac{H_{k+1}}{H_k}(x) = 0.$$

By integration $a_k(x) = a_{k0}\exp\{-\int_0^x H_k^{-1}H_{k+1}\}$ and the result follows from the initial value of H_k at zero, $a_k(0) = H_k(0)$. □

From (2.6), the $L^2(\mathbb{R}, \mu_{\mathcal{N}})$-norm of H_k is c_k such that

$$c_k^2 = \int_{\mathbb{R}} f_{\mathcal{N}}(x)H_k^2(x)\,dx = \sqrt{2\pi}\int_{\mathbb{R}} \{f_{\mathcal{N}}^{(k)}(x)\}^2 e^{\frac{t^2}{2}}\,dx = k!,$$

due to integrations by parts and it is calculated from the even moments of the normal distribution.

The normalized functions

$$h_k = \frac{H_k}{\sqrt{k!}}$$

are equivalent to $(k!)^{-\frac{1}{2}}x^k$ as x tends to infinity. At zero, the polynomials converge to finite limits

$$h_{2k+1}(x) \sim 0,$$
$$h_{2k}(x) \sim \frac{1}{2k}.$$

The coefficients of highest degrees x^k and x^{k-1} in h_k are 1 and, respectively, $\frac{1}{2}k(k-1)$. They have the form

$$h_{2k}(x) = \sum_{j=0}^{k-1} b_{2j}x^{2j} + x^{2k},$$

$$h_{2k+1}(x) = \sum_{j=0}^{k-1} b_{2j+1}x^{2j+1} + x^{2k+1}$$

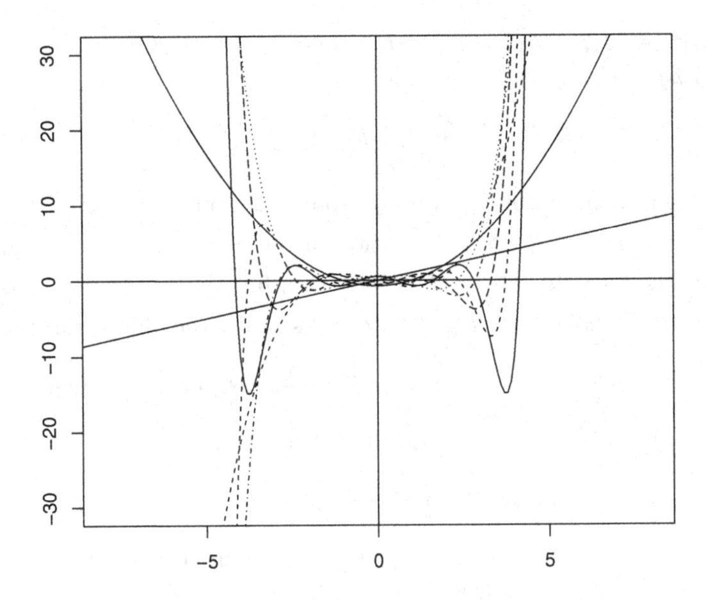

Fig. 2.2 Normalized Hermite's polynomials h_k, $k = 1, \ldots, 8$.

and all coefficients are calculated iteratively. The first normalized polynomials are calculated as

$$\begin{aligned}
h_1(x) &= x, \\
h_2(x) &= x^2 - 1, \\
h_3(x) &= x^3 - 3x, \\
h_4(x) &= x^4 - 6x^2 + 3, \\
h_5(x) &= x^5 - 10x^3 + 15x, \\
h_6(x) &= x^6 - 15x^4 + 45x^2 - 15, \\
h_7(x) &= x^7 - 21x^5 + 105x^3 - 105x, \\
h_8(x) &= x^8 - 28x^6 + 210x^4 - 420x^2 + 105, \text{ etc.}
\end{aligned}$$

The absolute value of the n_kth coefficients of the normalized polynomial h_k is increasing with k.

Proposition 2.8. *Let a_k^* be the largest zero of the normalized polynomial h_k, then $(a_k^*)_k$ is an increasing sequence.*

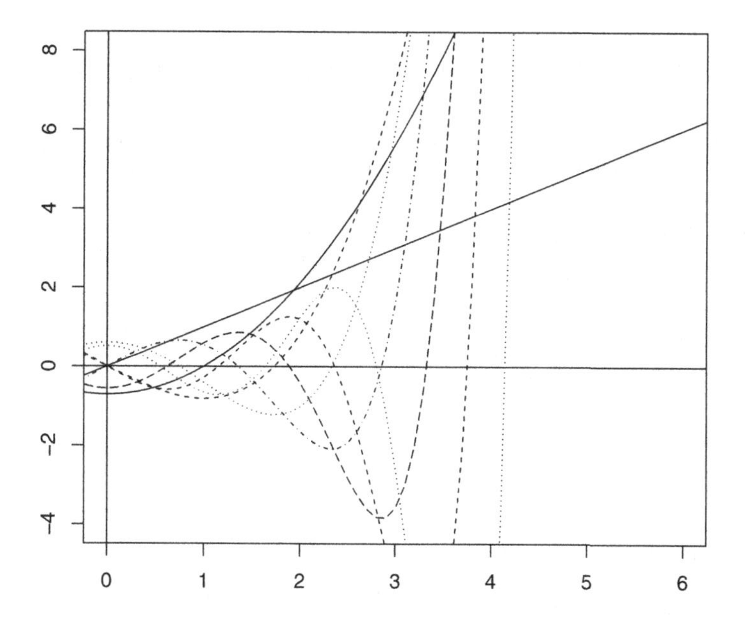

Fig. 2.3 Zeros and local maxima of Hermite's polynomials.

Proof. The assertion is proved recursively using (2.8). By symmetry of the zeros of h_k, $a_k^* > 0$. Since the coefficient of x^k in h_k is one, the polynomials are increasing at infinity and $h_k'(a_k^*) > 0$ for every $k > 1$. From (2.8), $h_k'(a_k^*) + h_{k+1}(a_k^*) = 0$ this implies $h_{k+1}(a_k^*) < 0$, equivalently $h_{k+1}(a_k^*) < h_{k+1}(a_{k+1}^*)$. The function h_k is increasing at infinity from a_{k+1}^* therefore $a_{k+1}^* > a_k^*$. □

Deriving (2.8), a_k^* satisfies the following equations

$$a_k^* = \frac{H_k^{(2)} - H_{k+1}^{(2)}}{H_k'}(a_k^*) = \frac{H_{k+1}^{(2)} \dot{-} H_k^{(2)}}{H_{k+1}}(a_k^*). \tag{2.9}$$

Figure (2.3) suggest the following property.

Conjecture 2.1. *The sequence $(a_k^* - a_{k-1}^*)_k$ is decreasing.*

Proposition 2.9. *The polynomial h_k has a local minimum at a_{k-1}^* and the largest local minimum of h_k is located at a_{k-1}^*, for every $k > 1$.*

Proof. For the first assertion, it is sufficient to prove that $h'_{k+1}(a_k^*) = 0$ and $h''_{k+1}(a_k^*) > 0$, then h_{k+1} is locally convex in a neighborhood of a_k^* and $h_{k+1}(x) \geq h_{k+1}(a_k^*)$ locally near a_k^*, with equality if and only if x equals a_k^*. This can be proved computionally with the explicit expressions of the polynomial coefficients, like the second assertion. □

Proposition 2.10. *The Hermite transform $H_f = \sum_{k\geq 0} a_k(f)H_k$ of a function f has the coefficients $a_k(f) = (-1)^k c_k^{-1} E_{\mu_\mathcal{N}} f^{(k)}(X)$.*

Proof. For every integer $k > 1$, if these integrals are finite

$$\begin{aligned} a_k(f) &= \frac{1}{c_k}\int_\mathbb{R} f(x)H_k(x)f_\mathcal{N}(x)\,dx = \frac{1}{c_k}\int_\mathbb{R} f(x)f_\mathcal{N}^{(k)}(x)\,dx \\ &= \frac{(-1)^k}{c_k}\int_\mathbb{R} f(x)^{(k)} f_\mathcal{N}(x)\,dx. \end{aligned}$$

□

The scalar product of the Hermite expansions of functions f and g in $(L^2(\mathbb{R}), \mu_\mathcal{N})$ is written as $E\{H_f(X)H_g(X)\}$, with a normal variable X, and it develops as $\int_\mathbb{R} H_f H_g \, d\mu_\mathcal{N} = \sum_{k\geq 0} k!\, a_k(f)a_k(g)$. A condition for the convergence of the transform H_f of a function f is the convergence of the series

$$\|H_f\|_2^2 = \sum_{k\geq 1} k!\, a_k^2(f).$$

Under this condition, a function f is equal to its normalized Hermite expansion H_f since $< f, H_k >=< H_f, H_k >$ for every $k \geq 0$ by orthogonality of the basis.

The first derivatives of Hermite's polynomials are deduced from the recurrence formula (2.8) as

$$\begin{aligned} H_k^{(1)}(x) &= xH_k(x) - H_{k+1}(x), \\ H_k^{(2)}(x) &= (1+x^2)H_k(x) - 2xH_{k+1}(x) - H_{k+2}(x), \\ H_k^{(3)}(x) &= x(3+x^2)H_k(x) - 3(1+x^2)H_{k+1}(x) + xH_{k+2}(x) + H_{k+3}(x), \\ H_k^{(4)}(x) &= (3+6x^2+x^4)H_k(x) - 4x(3+x^2)H_{k+1}(x) + 4(1+x^2)H_{k+2}(x) \\ &\quad - H_{k+4}(x), \end{aligned}$$

they have the form

$$H_k^{(n)}(x) = \sum_{i=0}^{n} \xi_{i,k}(x)H_{k+i}(x)$$

with polynomial functions $\xi_{i,k}$ of degree $k-i$. The explicit calculus of the integrals of functions of the derivatives $H_k^{(n)}$ must be performed using the numerical values of the coefficients of the polynoms H_k.

Proposition 2.11. *The second derivative of h_k is $h_k^{(2)} = k(k-1)h_{k-2}$, its jth derivative is $h_k^{(j)} = k!\{(k-j)!\}^{-1}h_{k-j}$.*

This property is proved computationally and it is easily checked for small degrees k. From this, it is possible to solve ordinary differential equations by searching solutions as a series $f = \sum_{k\geq 0} a_k(f)h_k$ or as its finite projection $f_N = \sum_{k=0}^N a_k(f)h_k$ satisfying the equation, the coefficients of x^m in h_k being given. The problem becomes the search of the roots of a rational polynomial of degree N. The error of the approximation $R_N(f) = \sum_{k>N} a_k^2(f)$ must be smaller than an arbitrary constant $\varepsilon > 0$.

Proposition 2.12. *For a function f of $L^2(\mathbb{R}, \mu_{\mathcal{N}})$ the second order differential equation $f^{(2)} - \alpha f = 0$, with boundary conditions $f(0)$ and $f^{(1)}(0)$, has an unique solution $f = \sum_{k\geq 0} a_k h_k$ defined by the coefficients*

$$a_{2k} = a_0 \frac{\alpha^k}{(2k)!}, \quad a_{2k+1} = a_1 \frac{\alpha^k}{(2k+1)!}$$

where the constants a_0 and a_1 are determined by the initial condtions

$$f_0 = a_0 \sum_{k\geq 0} \frac{\alpha^k}{(2k)!} h_{2k}(0), \quad f_0^{(1)} = a_1 \sum_{k\geq 1} \frac{\alpha^k}{(2k+1)!} h_{2k+1}^{(1)}(0).$$

The expansion of a function f is written as

$$f(x) = f_0 \frac{\sum_{k\geq 0} \frac{\alpha^k}{(2k)!} h_{2k}(x)}{\sum_{k\geq 0} \frac{\alpha^k}{(2k)!} h_{2k}(0)} + f_0^{(1)} \frac{\sum_{k\geq 1} \frac{\alpha^k}{(2k+1)!} h_{2k+1}^{(1)}(x)}{\sum_{k\geq 1} \frac{\alpha^k}{(2k+1)!} h_{2k+1}^{(1)}(0)}.$$

Direct solutions of the equation of Proposition 2.12 are linear combinations of the hyperbolic sine and cosine functions if α is positive, or trigonometric sine and cosine functions if α is negative, for instance

$$f(x) = f_0 e^{-x\sqrt{\alpha}} + (\sqrt{\alpha})^{-1} f_0^{(1)} \sinh(x\sqrt{\alpha}), \quad \text{if } \alpha > 0,$$
$$f(x) = f_0 e^{-ix\sqrt{\alpha}} + (\sqrt{\alpha})^{-1} f_0^{(1)} \sin(x\sqrt{\alpha}), \ \text{if } \alpha < 0.$$

These functions have then an expansions in Hermite's basis.

The generating functions related to the polynomials H_k provide translations of the normal density and therefore convolutions.

Proposition 2.13. *For every w such that $|w| < 1$*

$$f_{\mathcal{N}}(x+w) = \sum_{k\geq 0} H_k(x) f_{\mathcal{N}}(x) \frac{w^k}{k!},$$

$$\begin{aligned}\sum_{k\neq j\geq 0} H_k(t)H_j(t)e^{-t^2} \frac{w^{j+k}}{(j+k)!} &= \sum_{k\neq j\geq 0} \frac{d^{j+k}e^{-\frac{t^2}{2}}}{dt^{j+k}} \frac{w^{j+k}}{(j+k)!} \\ &= \sqrt{2\pi} f_{\mathcal{N}}(t+w).\end{aligned}$$

For every function g of $L^2(\mathbb{R}, \mu_{\mathcal{N}})$

$$\sum_{k\geq 0} \int_{\mathbb{R}} g(x) H_k(x) f_{\mathcal{N}}(x) \frac{w^k}{k!}\, dx = \int_{\mathbb{R}} g(x-w) f_{\mathcal{N}}(x)\, dx.$$

The derivatives of the Hermite transform of a function g of $(L^2(\mathbb{R}), \mu)$ are written as $g^{(j)} = \sum_{k\geq 0} a_k H_k^{(j)}$ where $H_k^{(1)}(x) = xH_k(x) - H_{k+1}(x)$ from (2.7) and the derivative of order j of H_k is a polynomial of degree $k+j$ recursively written in terms of $H_k, \ldots, H_{k+j}$ in the form

$$H_k^{(j)}(x) = \sum_{i=0}^{j} P_{ik}(x) H_{k+i}(x)$$

where P_{ik} is a polynomial of degree $j-i$.

Theorem 2.1. *Hermite's transform of the normal density is the series*

$$H_{f_{\mathcal{N}}}(x) = \sum_{k\geq 0} a_{2k}(f_{\mathcal{N}}) H_{2k}(x),$$

its coefficients are strictly positive

$$a_{2k}(f_{\mathcal{N}}) = \frac{1}{\sqrt{(2k)!}} \int_{\mathbb{R}} \{f_{\mathcal{N}}^{(2k)}(x)\}^2\, dx.$$

Corollary 2.1. *The norm of $H_{f_{\mathcal{N}}}$ is finite.*

The norm of $H_{f_{\mathcal{N}}}$ is expanded as

$$\|f_{\mathcal{N}}\|_{L^2(\mu_{\mathcal{N}})} = \sum_{k\geq 1} c_{2k}^2 a_{2k}^2(f_{\mathcal{N}}) = \sum_{k\geq 1} \int_{\mathbb{R}} \{f_{\mathcal{N}}^{(k)}(x)\}^2\, dx.$$

The integrals $\sum_{k\geq 1} \int_{|x|>A} \{f_{\mathcal{N}}^{(k)}(x)\}^2\, dx$ and $\int_{|x|>A} (1-x^2)^{-1} e^{-x^2}\, dx$ are equivalent as A tends to infinity and they converge to a finite limit.

The expansion of a function f in Hermite's basis is denoted by h_f for every function f. The error $R_n(f) = f - S_n(f)$ in estimating a function f by the sum $S_n(f) = \sum_{k \leq n} a_k(f) H_k(x)$ of the first n terms of its expansion has the norm $\|R_n(f)\|_{L^2(\mu_{\mathcal{N}})} = \sum_{k>n} k! a_k^2(f)$. Let us consider the norms in $L^p(\mu_{\mathcal{N}})$ of the error of the partial sum $S_n(f; h) = \sum_{k \leq n} a_k(f) h_k(x)$.

Lemma 2.1. *The coefficient a_k of the expansion h_f for a function f belonging to $L^2(\mu_{\mathcal{N}})$ has a norm $\|a_k\|_{L^2(\mu_{\mathcal{N}})} \leq \|f\|_{L^2(\mu_{\mathcal{N}})}$ where the equality is satisfied if and only if $f = h_k$.*

Proposition 2.14. *The partial sums $S_n(f; h)$ of the Hermite expansion h_f in the orthonormal basis $(h_k)_{k \geq 0}$ satisfy*

$$\|S_n(f; h)\|_{L^2(\mu_{\mathcal{N}})} \leq \|f\|_{L^2(\mu_{\mathcal{N}})}$$

and for all conjugate integers $p > 1$ and p'

$$\|n^{-1} S_n(f; h)\|_{L^2(\mu_{\mathcal{N}})} < \|f\|_{L^p(\mu_{\mathcal{N}})} n^{-1} \sum_{k>n} \|h_k\|_{L^{p'}(\mu_{\mathcal{N}})}.$$

Proof. The first inequality is a consequence of the equality of $\|S(f; h)\|_{L^2(\mu_{\mathcal{N}})}$ and $\|f\|_{L^2(\mu_{\mathcal{N}})}$ and it may be an equality like for polynomials. Hermite's expansion h_f in the orthonormal basis of polynomials satisfies $\|h_f\|^2_{L^2(\mu_{\mathcal{N}})} = \sum_{k \geq 0} \{E(f h_k)(X)\}^2$ and for every $k \geq 1$, Hölder's inequality implies the second inequality. □

Proposition 2.15. *For every function f of $L^2(\mu_{\mathcal{N}})$, the partial sum $S_n(f; H)$ has an error*

$$\|R_n(f)\|_{L^2(\mu_{\mathcal{N}})} \leq \|f\|_{L^2(\mu_{\mathcal{N}})} \|R_n(f_{\mathcal{N}})\|_{L^2(\mu_{\mathcal{N}})}$$

and it tends to zero as n tends to infinity.

Proof. The squared norm of $R_n(f)$ is $\|R_n(f)\|^2_{L^2(\mu_{\mathcal{N}})} = \sum_{k>n} c_k^2(f) a_k^2(f)$ with

$$a_k(f) = \frac{1}{c_k(f)} \int_{\mathbb{R}} f(x) f_{\mathcal{N}}^{(k)}(x)\, dx,$$
$$c_k^2 a_k^2(f) \leq \left\{ \int_{\mathbb{R}} f^2(x)\, dx \right\} \left\{ \int_{\mathbb{R}} f_{\mathcal{N}}^{(k)2}(x)\, dx \right\} = c_{2k} a_{2k}(f_{\mathcal{N}}) \|f\|^2 L^2(\mu_{\mathcal{N}}),$$

by Theorem 2.1. From Corollary 2.1, the Hermite transform of the normal density converges, it follows that $R_n(f_{\mathcal{N}}) = \sum_{k>n} c_{2k} a_{2k}(f_{\mathcal{N}})$ and therefore $\|R_n(f_{\mathcal{N}})\|_{L^2(\mu_{\mathcal{N}})}$ converge to zero as n tends to infinity. □

2.4 Legendre's polynomials

Legendre's polynomial $P_m(x)$, $m \geq 0$, is defined as the coefficient of z^m in the expansion of the first derivative of $(1-2xz+z^2)^{-\frac{1}{2}}$ in series, for every x in the interval $[-1,1]$. The even polynomials are symmetric and the odd polynomials are odd, then $P_{2m+1}(0)=0$. From this expansion, they are also defined as solutions of Legendre's differential equations

$$(1-x^2)\frac{d^2}{dx^2}P_m(x) - 2x\frac{d}{dx}P_m(x) + m(m+1)P_m(x) = 0 \tag{2.10}$$

with the initial value $P_0(x)=1$. In the space $L_2([-1,1])$ of the square functions with respect to the Lebesgue measure, the functions $P_m(x)$ satisfying (2.10) and having the norm 1 are defined by the normalized derivatives

$$P_m(x) = \frac{1}{2^m\, m!}\frac{d^m}{dx^m}\{(x^2-1)^m\},\ m \geq 1. \tag{2.11}$$

The first polynomials are calculated from (2.11), for every real x, as

$$\begin{aligned}
P_1(x) &= x,\\
P_2(x) &= \frac{1}{2}(3x^2-1),\\
P_3(x) &= \frac{1}{2}(5x^3-3x),\\
P_4(x) &= \frac{1}{8}(35x^4-30x^2+3),\\
P_5(x) &= \frac{1}{8}(63x^5-70x^3+15x),\\
P_6(x) &= \frac{1}{16}(231x^6-315x^4+105x^2-5),\\
P_7(x) &= \frac{1}{16}(429x^7-693x^5+315x^3-35x),\\
P_8(x) &= \frac{1}{128}(6435x^8-12012x^6+6930x^4-1260x^2+35).
\end{aligned}$$

At $x=1$, the expansion $(1-z)^{-1} = \sum_{m\geq 0} z^m$ implies $P_m(1)=1$, for every integer m.

The values of the odd polynomials at zero are zero and those of the even polynomials are

$$\begin{aligned}
&P_2(0) = -\frac{1}{2},\ P_4(0) = \frac{3}{2^3},\ P_6(0) = -\frac{5}{2^4},\\
&P_8(0) = \frac{35}{2^7},\ P_{10}(0) = \frac{63}{2^8},\ P_{12}(0) = \frac{231}{2^{10}},\\
&P_{14}(0) = \frac{117}{2^9},\ P_{16}(0) = \frac{945}{2^{12}}.
\end{aligned}$$

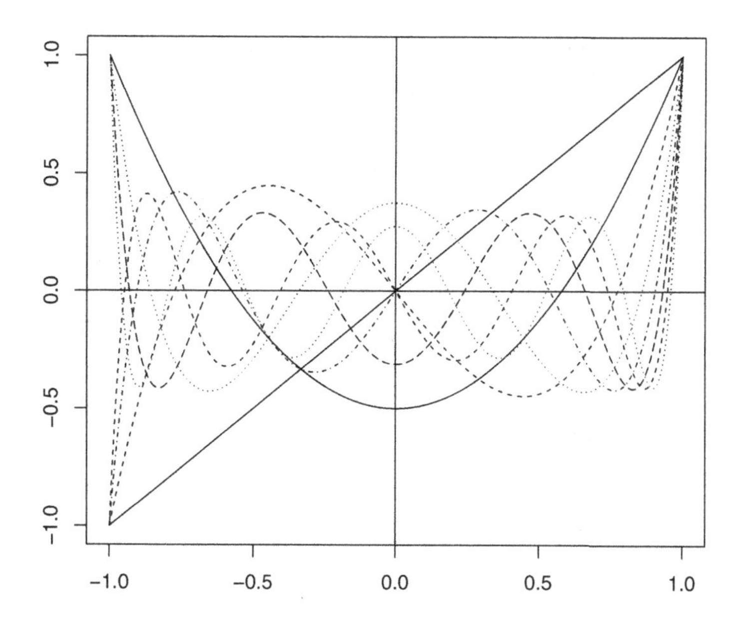

Fig. 2.4 Legendre's polynomials P_1 to P_8.

The function P_m is a polynomial of degree m obtained from the two previous functions by the formula

$$(m+1)P_{m+1}(x) - (2m+1)xP_m(x) + mP_{m-1}(x) = 0, \tag{2.12}$$

with the initial conditions $P_0 = 1$ and $P_1(x) = x$. The second derivative of $2^m m! P_m$ is $\{(x^2-1)^m\}^{(m+2)} = 2m\{x(x^2-1)^{m-1}\}^{(m+1)} = 2m\{(x^2-1)^{m-1}2(m-1)x^2(x^2-1)^{m-2}\}^{(m)} = 2m\{(2m-1)(x^2-1)^{m-1}+2(m-1)(x^2-1)^{m-2}\}^{(m)}$, therefore

$$P''_m(x) = (2m-1)P'_{m-1}(x) + P''_{m-2}(x), \tag{2.13}$$

for every integer $m \geq 5$. Integrating this expression implies

$$P'_m(x) = (2m-1)P_{m-1}(x) + P'_{m-2}(x),\ m \geq 5. \tag{2.14}$$

Proposition 2.16. *The functions P_m solutions of the differential equation (2.10) satisfy the implicit equations*

$$P_m(x) = P_m(0) + P_m^{(1)}(0)\int_0^x \frac{ds}{1-s^2} - m(m+1)\int_0^x \frac{\int_0^s P_m}{1-s^2}\,ds$$

for all x in $[-1,1]$ and $m \geq 1$.

Proof. Legendre's differential equation (2.10) is also written

$$\frac{d}{dx}\{(1-x^2)P_m^{(1)}(x)\} + m(m+1)P_m(x) = 0.$$

Let $y = (1-x^2)P_m^{(1)}(x)$, its derivative is $-m(m+1)P_m(x)$ which implies

$$P_m(x) = A + \int_0^x \frac{y_s}{1-s^2}\,ds, \text{ with}$$
$$y_x = C - m(m+1)\int_0^x P_m,$$

and with the constants $C = P_m^{(1)}(0)$ and $A = P_m(0)$, therefore

$$P_m(x) = P_m(0) + P_m^{(1)}(0)\int_0^x \frac{ds}{1-s^2} - m(m+1)\int_0^x \frac{\int_0^s P_m}{1-s^2}\,ds. \qquad \square$$

The second derivative $P_m^{(2)}$ is expressed fom (2.10) in terms of P_m and its first derivative. From Proposition (2.16), it is expressed as a nonlinear functional of P_m only.

They constitute an orthogonal basis in the space $L_2([-1,1])$ of the square functions with respect to the Lebesgue measure on $[-1,1]$. Their norm in $L_2([0,1])$ is

$$\|P_m\|_{L_2([0,1])} = \Big(\frac{2}{2m+1}\Big)^{\frac{1}{2}}.$$

The expansion of a symmetric function f has its odd coefficients equal to zero and its coefficient in the projection on P_{2m} is

$$a_{2m} = (4m+1)\int_0^1 f(x)P_{2m}(x)\,dx.$$

For an odd function f, all even coefficients of the projection of f on Legendre's basis are zero and its odd coefficients are

$$a_{2m+1} = (4m+3)\int_0^1 f(x)P_{2m+1}(x)\,dx.$$

Integrating by parts the functions P_m defined as (2.11), we obtain the following expression of the coefficients $a_m(f) = \int_{-1}^1 fP_m\,dx$.

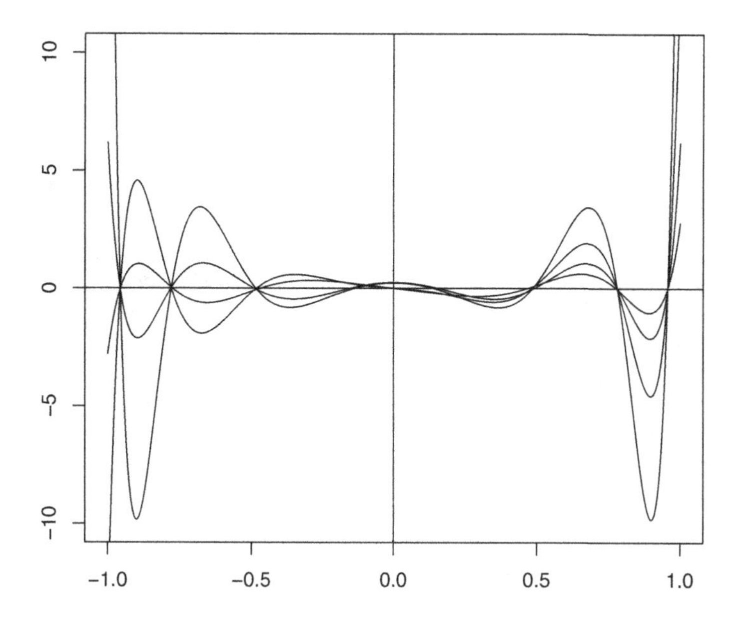

Fig. 2.5 Legendre's polynomials P_9 to P_{12}.

Proposition 2.17. *The Legendre transform* $H_f = \sum_{m\geq 0} a_m(f)P_m$ *of a function* f *of* $C_\infty([-1,1])$ *with derivatives vanishing at* ± 1 *has the coefficients*

$$a_m(f) = \frac{1}{m!\,2^m} a_m \int_{-1}^{1} (x^2-1)^m f^{(m)}(x)\,dx.$$

The functions have an increasing number of oscillations on decreasing sub-intervals on the edges of $[-1,1]$ and the maximal sizes of the functions on each period constitue a decreasing sequence with the order of the polynom near zero and an increasing sequence near the edges of the interval as we can see in Figure (2.5).

Proposition 2.18. *The sequences of zeros of the polynomials* P_m *converge to fixed points as* m *tends to infinity.*

Proof. Let a_m be a zero of the polynomial P_m in a subset of $[-1,1]$. From the recurrence equation (2.12), $(m+1)P_{m+1}(a_m) + mP_{m-1}(a_m) = 0$. If $P_{m-1}(a_m)$ then $P_{m+1}(a_m) = 0$. If $P_{m-1}(a_m)$ is not zero, it follows that $P_{m-1}^{-1}P_{m+1}$ tends to -1 at a_m as m tends to infinity. However P_{m-1} and P_{m+1} are both odd or even polynomials and this limit is impossible. $\square$

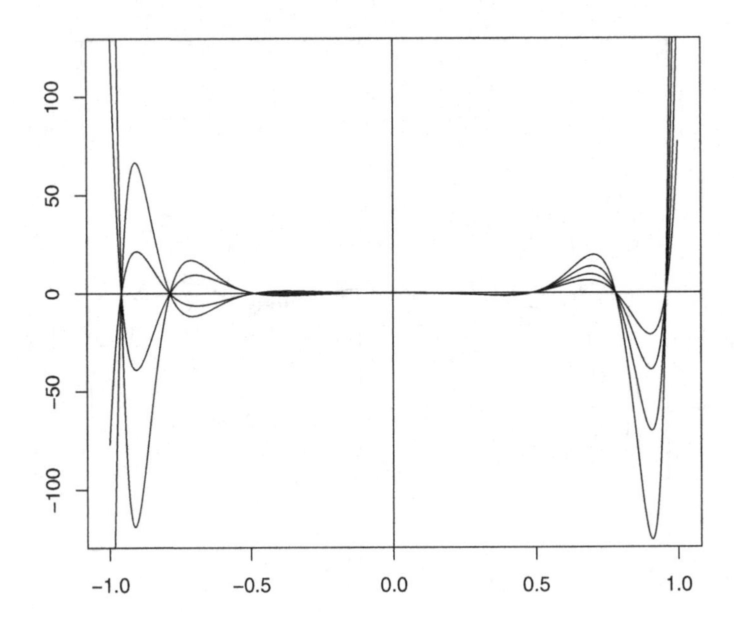

Fig. 2.6 Legendre's polynomials P_{13} to P_{16}.

2.5 Generalizations

The polynomials cannot be compared since their support and their range differ. At infinity in $\mathbb{R}_+$, Laguerre's even functions L_n have values in $\mathbb{R}_+$ and the odd functions L_n have values in $\mathbb{R}$ whereas all Hermite's polynomials H_k are positive. The divergence rate of Laguerre's polynomials is slower than for Hermite's polynomials as Figures (2.1) and (2.2) show. Near zero, all functions have an oscillatory behavior, the period and the size of the waves increase in both series.

Their support is modified by changing the location and the scale of the derivatives that define them. Parametric families of polynomials are defined in $\mathbb{R}$ from Hermite's polynomials as

$$H_k(t,\alpha,\lambda) = (-1)^k \frac{d^k e^{-\frac{(t-\alpha)^2}{2\lambda}}}{dt^k} e^{\frac{(t-\alpha)^2}{2\lambda}} = \frac{1}{f_{\mathcal{N}_{\alpha,\lambda}}(t)} \frac{d^k f_{\mathcal{N}_{\alpha,\lambda}}(t)}{dt^k}, \tag{2.15}$$

where the Gaussian density $f_{\mathcal{N}_{\alpha,\lambda}}$ has the mean α and the variance λ.

The coefficients of the polynomial functions generated by this parametrization are modified in consequence but not the form of the functions. They are polynomials of $\lambda^{-1}(t-\alpha)$ defined by the formula

$$H_{k+1}\Big(\frac{t-\alpha}{\lambda}\Big) = \frac{t-\alpha}{\lambda} H_k\Big(\frac{t-\alpha}{\lambda}\Big) - H_k'\Big(\frac{t-\alpha}{\lambda}\Big),$$

for every $k \geq 1$, with the initial function $H_0 = 1$.

Another generalization is obtained by variation of the exponent in their definition, with a real parameter $\alpha > 0$

$$G_k(t,\alpha) = (-1)^k e^{\frac{t^\alpha}{\alpha}} \frac{d^k e^{-\frac{t^\alpha}{\alpha}}}{dt^k}$$

with $G_0 = 1$. Then G_{k+1} is defined from G_k by

$$G_{k+1}(t,\alpha) = -G_k'(t,\alpha) + x^{\alpha-1} G_k(t,\alpha)$$

or equivalently

$$G_k(x) = -e^{\frac{-x^\alpha}{\alpha}} \int_0^x e^{\frac{-t^\alpha}{\alpha}} G_{k+1}(t)\,dt + G_k(0).$$

From Laguerre's polynomials we define a family of polynoms for $x \geq \alpha$ by the derivatives

$$L_n\Big(\frac{x-\alpha}{\lambda}\Big) = \frac{e^{\frac{x-\alpha}{\lambda}}}{n!} \frac{d^n}{dx^n}\Big\{e^{-\frac{x-\alpha}{\lambda}}\Big(\frac{x-\alpha}{\lambda}\Big)^n\Big\},\ n \geq 1,$$

with the initial condition $L_n(\alpha) = 1$ and with real parameters α and $\lambda > 0$. They are polynoms in $\lambda^{-1}(x-\alpha)$ and all their coefficients are modified.

The relationship between L_{n-1} and L_n is

$$\frac{x}{\lambda} L_n'\Big(\frac{x-\alpha}{\lambda}\Big) - nL_n\Big(\frac{x-\alpha}{\lambda}) + nL_{n-1}\Big(\frac{x-\alpha}{\lambda}\Big) = 0$$

and, with the initial condition its solution, is

$$L_n(x) = -n\Big(\frac{(x-\alpha)^n}{\lambda^{n+1}}\Big)^n \int_\alpha^x \Big(\frac{y-\alpha}{\lambda}\Big)^{-(n+1)} L_{n-1}\Big(\frac{y-\alpha}{\lambda}\Big)\,dy + 1,\ n \in \mathbb{N}.$$

A similar reparametrization for a change of location and scales is performed for Laguerre and Legendre polynomials. These polynomials are expressed in the global form

$$Q_n(x) = \frac{1}{\alpha_n f(x)} \frac{d^n}{dx^n}\{f(x)G^n(x)\},\ n \geq 1 \tag{2.16}$$

and Q_0 constant in a sub-interval I of $\mathbb{R}$. They may be generalized in this form with all functions f and G defined in I such that $(Q_n)_{n\geq 0}$ is an orthonormal basis of $L^2(I, \mu_f)$ endowed with the measure μ_f having the density f with respect to Lebesgue's measure. Conditions for orthogonal polynoms are $\int_I Q_k(x)f(x)\,dx = 0$ and

$$\int_I Q_k(x)Q_n(x)f(x)\,dx = \int_I \{f(x)G^k(x)\}^{(k)}\{f(x)G^n(x)\}^{(n)}\frac{dx}{f(x)} = 0$$

for all distinct integers n and k. If the functions f and G satisfy the conditions $\lim_{x\to\infty}(fG^n)^{(j)} = 0$, $\lim_{x\to 0}(fG^n)^{(j)} = 0$, for $j = 1,\ldots,n-1$, integrating by parts, we get

$$\begin{aligned}\int_I Q_k(x)Q_n(x)f(x)\,dx &= -\int_{I^2} Q_k'(x)1_{\{y\le x\}}Q_n(y)f(y)\,dy\,dx\\ &= -\frac{1}{\alpha_n}\int_I Q_k'(x)\{f(x)G^n(x)\}^{(n-1)}\,dx,\\ &= \frac{(-1)^n}{\alpha_n}\int_I Q_k^{(n)} f(x)G^n(x)\,dx.\end{aligned}$$

By the same arguments, their norm equals

$$\int_I Q_k^2(x)f(x)\,dx = \frac{1}{\alpha_n}\int_I Q_n^{(n)}(x)f(x)G^n(x)\,dx$$

and the scalar product Q_n with a function φ of $C_n(I)$ such that φ and $\varphi^{(j)}$, $j = 1\ldots,n$, belong to $L^2(I,\mu_f)$ is

$$\begin{aligned}<\varphi, Q_n>_f &= \int_I \varphi(x)f(x)Q_n(x)\,dx\\ &= (-1)^n <\varphi^{(n)}, G^n>_f .\end{aligned}$$

The convolution property of Proposition 2.13 is preserved, for every function φ of $L^2(I,\mu_f)$

$$\sum_{k=0}^{\infty}\alpha_k\frac{z^k}{k!}\int_I \varphi(x)Q_k(x)f(x)\,dx = \sum_{k=0}^{\infty}\alpha_k\int_I \varphi(x-z)Q_k(x)f(x)\,dx.$$

2.6 Bilinear functions

Let $f(x,y)$ and $G(x,y)$ by homogeneous functions of x and y in a domain I^2 of $\mathbb{R}^2$. The polynomials that generate bilinear form for wave equations in I^2 are written like (2.16) in the global form

$$Q_{m,n}(x,y) = \frac{1}{\alpha_{m,n}f(x,y)}\frac{d^{m+n}}{dx^m dy^n}\{f(x,y)G^{m+n}(x,y)\},\ n\ge 1 \qquad (2.17)$$

and Q_0 is constant in I^2. The functions $(Q_{m,n})_{n\ge 0}$ is an orthonormal basis of $L^2(I,\mu_f)$ endowed with the measure μ_f absolutely continuous with respect to Lebesgue's measure, with density f.

$$Q_{1,0}(x,y) = \frac{1}{\alpha_{1,0}}\Big\{G'_x(x,y) + \frac{f'_x(x,y)}{f(x,y)}G(x,y)\Big\},$$
$$Q_{2,0}(x,y) = \frac{1}{\alpha_{2,0}}\Big\{G''_{x,x}(x,y) + \frac{f''_{x,x}(x,y)}{f(x,y)}G(x,y) + 2\frac{f'_x(x,y)}{f(x,y)}G'_x(x,y)\Big\},$$
$$Q_{1,1}(x,y) = \frac{1}{\alpha_{1,1}}\Big\{G''_{x,y}(x,y) + \frac{f''_{x,y}(x,y)}{f(x,y)}G(x,y)$$
$$+ \frac{f'_x(x,y)}{f(x,y)}G'_y(x,y) + \frac{f'_y(x,y)}{f(x,y)}G'_x(x,y)\Big\},$$

and $Q_{0,2}$ is similar to $Q_{2,0}$, the higher order derivatives follow the binomial formula for the expansion of the crossed-order derivatives. Conditions for orthogonal polynoms are $\int_D Q_k(x,y)f(x,y)\,dx\,dy = 0$ and

$$\int_{D^2} Q_k(x)Q_j(x)f(x)\,dx = 0$$

for all distinct integers j and $k \geq 1$, their $L^2(D,f)$-norm is 1. Under such conditions, every function of $L^2(D,f)$ has an unique expansion in the orthonormal basis $(Q_{m,n})_{m\geq 0, n\geq 0}$. Integrating by parts the expansion yield properties for the integral of the derivatives of the function with respect to partial derivatives of fG^{m+n}.

The first two derivatives of $Q_{m,n}$ are calculated from the derivatives

$$\frac{d^{m+n+1}}{dx^{m+1}dy^n}\{fG^{m+n}\} = \frac{d^{m+n}}{dx^m dy^n}\{f'_x G^{m+n} + (m+n)fG'_x G^{m+n}\},$$
$$\frac{d^{m+n+2}}{dx^{m+2}dy^n}\{fG^{m+n}\} = \frac{d^{m+n}}{dx^m dy^n}\{f''_{x,x}G^{m+n} + 2(m+n)f'_x G'_x G^{m+n-1}$$
$$+ (m+n)fG''_{x,x}G^{m+n-1}$$
$$+ (m+n)(m+n-1)G'^2_x G^{m+n-2}\},$$
$$\frac{d^{m+n+2}}{dx^{m+1}dy^{n+1}}\{fG^{m+n}\} = \frac{d^{m+n}}{dx^m dy^n}[f''_{x,y}G^{m+n} + (m+n)\{(f'_x G'_y$$
$$+ f'_y G'_x)G^{m+n-1} + fG''_{x,y}G^{m+n-1}$$
$$+ (m+n-1)G'_x G'_y G^{m+n-2}\}],$$

and the derivatives with respect to dy^{n+1} are similar.

Proposition 2.19. *The functions defined in* $\mathbb{R}^2_+$ *by*

$$Q_{m,n} = \frac{e^{x+y}}{\alpha_{m,n}}\frac{d^{m+n}}{dx^m dy^n}\{e^{-(x+y)}(x+y)^{m+n}\}$$

satisfy the second order differential equation

$$\Delta Q_{m,n} = 2\frac{d^2}{dxdy}Q_{m,n}.$$

Proof. The polynomial are defined by the functions $f(x,y) = e^{-(x+y)}$ and $G(x,y) = (x+y)$ with derivatives $f'_x = f'_y = -f$ and $G'_x = G'_y = 1$, its derivatives satisfy

$$\begin{aligned}\frac{d}{dx}Q_{m,n} &= \frac{d}{dy}Q_{m,n} = \frac{1}{(m+n)!}\{(m+n)G^{m+n-1} - G^{m+n}\},\\ \frac{d^2}{dx^2}Q_{m,n} &= \frac{d^2}{dy^2}Q_{m,n} = \frac{1}{(m+n)!}\{G^{m+n} - 2(m+n)G^{m+n-1}\\ &\quad + (m+n)(m+n-1)G^{m+n-2}\},\end{aligned}$$

□

With the exponential function $f(x,y) = e^{-(x+y)}$, the first order polynomials are

$$\begin{aligned}Q_{1,0} &= \frac{2}{\alpha_{1,0}}(G'_x - G),\\ Q_{2,0}(x,y) &= \frac{2}{\alpha_{2,0}}\Big(GG''_{x,x} + GG'^2_x - GG'_x + \frac{1}{2}G^2\Big),\\ Q_{1,1} &= \frac{2}{\alpha_{1,1}}\Big(GG''_{x,y} + G'_yG'_x - GG'_y - GG'_x + \frac{1}{2}G^2\Big)\end{aligned}$$

and $Q_{0,0} = 1$. Conditions for the orthogonality of the functions $Q_{0,0}$, $Q_{1,0}$ and $Q_{0,1}$ are

$$\begin{aligned}\int_{\mathbb{R}^2_+} G_x(x,y)e^{-(x+y)}\,dx\,dy &= \int_{\mathbb{R}^2_+} G'_x(x,y)e^{-(x+y)}\,dx\,dy\\ &= \int_{\mathbb{R}^2_+} G'_y(x,y)e^{-(x+y)}\,dx\,dy,\\ 0 &= \int_{\mathbb{R}^2_+} \{(G'_x - G)(G'_y - G)\}(x,y)e^{-(x+y)}\,dx\,dy,\end{aligned}$$

the last equation and the orthogonality conditions for the second order functions imply

$$\begin{aligned}0 &= \int_{\mathbb{R}^2_+} \Big(GG''_{x,y} - \frac{1}{2}G^2\Big)e^{-(x+y)}\,dx\,dy,\\ 0 &= \int_{\mathbb{R}^2_+} \Big(GG''_{x,x} + GG'^2_x - GG'_x + \frac{1}{2}G^2\Big)e^{-(x+y)}\,dx\,dy,\\ 0 &= \int_{\mathbb{R}^2_+} \Big(GG''_{y,y} + GG'^2_y - GG'_y + \frac{1}{2}G^2\Big)e^{-(x+y)}\,dx\,dy.\end{aligned}$$

In particular, the functions defined in $\mathbb{R}_+^2$ by $f(x,y) = e^{-(x+y)}$ and $G(x,y) = (x+y)$ like Laguerre's polynomials in $\mathbb{R}_+$ do not satisfy the orthogonality properties.

2.7 Hermite's polynomials in $\mathbb{R}^2$

Let $L^2(\mathbb{R}^2, \mu_{\mathcal{N}})$ be the space of square integrable functions with respect to the measure $\mu_{\mathcal{N}}$ having the normal density with respect to Lebesgue's measure in $\mathbb{R}^2$, with density $f_{\mathcal{N}}(x,y) = (2\pi)^{-1} exp\{-\frac{x^2+y^2}{2}\}$. Let (X,Y) be a normal variable in $\mathbb{R}^2$ with density $f_{\mathcal{N}}$. The tensor product $(h_k)_{k\geq 0}^{\otimes 2}$ of Hermite's polynomials basis defines an orthonormal basis in $L^2(\mathbb{R}^2, \mu_{\mathcal{N}})$ where every function f is expanded in series

$$f(x,y) = \sum_{k\geq 0} \sum_{m\geq 0} a_{k,m} h_k(x) h_m(y)$$

with the coefficients

$$\begin{aligned} a_{k,m} &= \int_{\mathbb{R}^2} f(x,y) h_k(x) h_m(y) f_{\mathcal{N}}(x,y)\, dx\, dy \\ &= (c_k c_m)^{-1} E\{f(X,Y) H_k(X) H_m(Y)\}. \end{aligned}$$

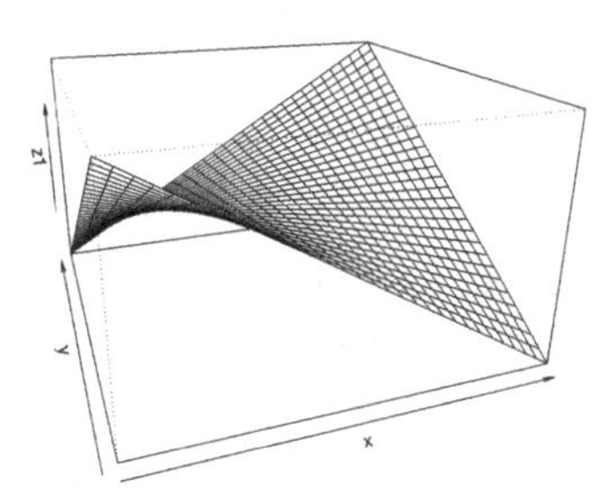

Fig. 2.7 Polynomial $H_{1,1}$.

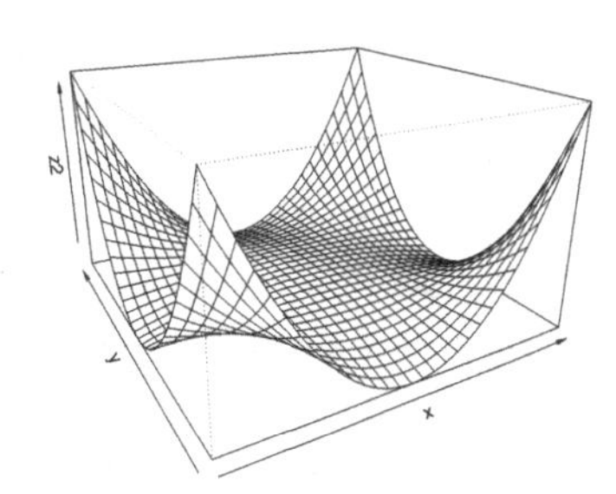

Fig. 2.8 Polynomial $H_{2,2}$.

Its norm in $L^2(\mathbb{R}^2, \mu_{\mathcal{N}})$ is

$$\|f\|_{L^2_{\mathcal{N}}}^2 = \sum_{k,m\geq 0} a_{k,m}^2 \|h_k\|_{L^2_{\mathcal{N}}}^2 \|h_m\|_{L^2_{\mathcal{N}}}^2 = \sum_{k,m\geq 0} a_{k,m}^2.$$

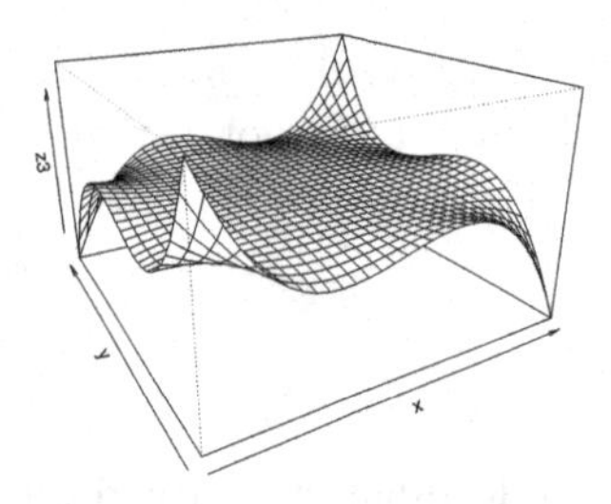

Fig. 2.9 Polynomial $H_{3,3}$.

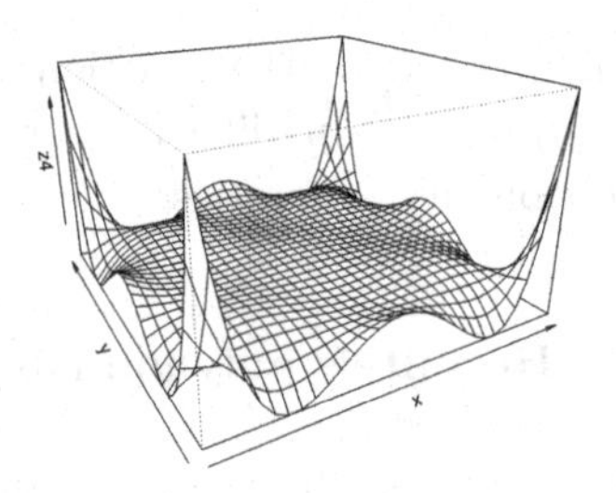

Fig. 2.10 Polynomial $H_{4,4}$.

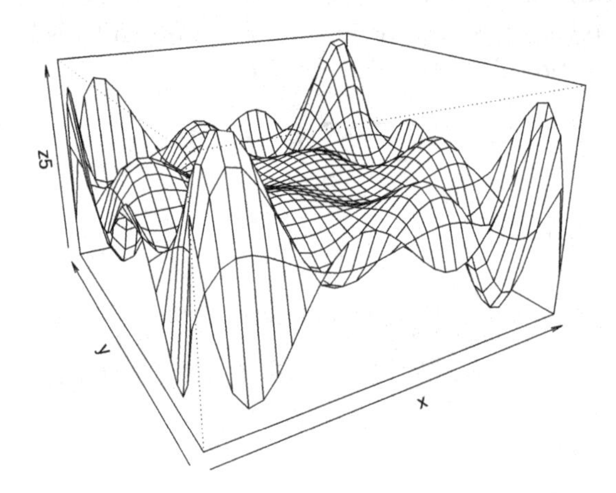

Fig. 2.11 Polynomial $H_{5,5}$.

Let $C^{(k,m)}(\mathbb{R}^2)$ be the space of functions $f(x,y)$ in $\mathbb{R}^2$ having a derivative $f^{(k,m)}$ of orders k with respect to x and m with respect to y.

Proposition 2.20. *Every function f of $L^2(\mathbb{R}^2, \mu_{\mathcal{N}}) \cap C^{(k,m)}(\mathbb{R}^2)$ satisfies*

$$a_{k,m} = (c_k c_m)^{-1} E f^{(k,m)}(X,Y),$$

$$f(X - w, Y - \zeta) = \sum_{k,m \geq 0} \frac{w^k}{k!} \frac{\zeta^m}{m!} f^{(k,m)}(X,Y) h_k(X) h_m(Y),$$

$$Ef(X - w, Y - \zeta) = \sum_{k,m \geq 0} c_k c_m a_{k,m} \frac{w^k}{k!} \frac{\zeta^m}{m!}.$$

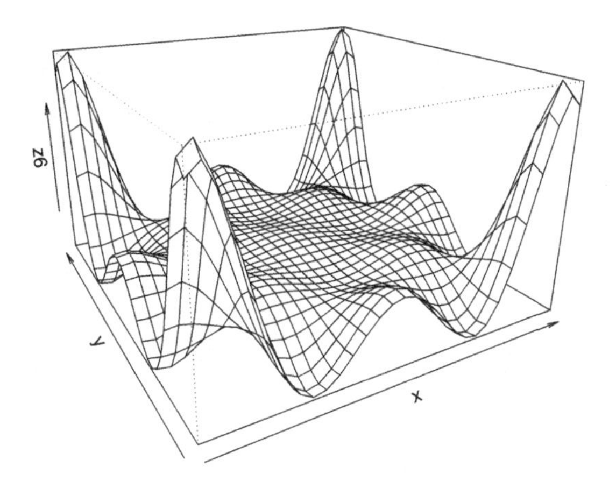

Fig. 2.12 Polynomial $H_{6,6}$.

Proof. By definition and integrating by parts

$$\begin{aligned} a_{k,m} &= \frac{(-1)^{k+m}}{c_k c_m} \int_{\mathbb{R}^2} f(x,y) f_{\mathcal{N}}(x) f_{\mathcal{N}}^{(k)}(x) f_{\mathcal{N}}^{(m)}(y)\, dx\, dy \\ &= (c_k c_m)^{-1} \int_{\mathbb{R}^2} f(x,y)^{(k,m)} f_{\mathcal{N}}(x,y)\, dx\, dy. \end{aligned}$$

By a Taylor expansion of $f(X-w, Y-\zeta)$ in (w,ζ) and applying this formula

$$\begin{aligned} Ef(X-w, Y-\zeta) &= \sum_{k,m\geq 0} \frac{w^k}{k!} \frac{\zeta^m}{m!} \int_{\mathbb{R}^2} f(x,y)^{(k,m)} f_{\mathcal{N}}(x,y)\, dx\, dy, \\ &= \sum_{k,m\geq 0} \frac{w^k}{k!} \frac{\zeta^m}{m!} c_k c_m a_{k,m}(f). \end{aligned}$$

□

In $L^2(\mathbb{R}^2, \mu_{\mathcal{N}})$, the functions $h_{k,m}(x,y) = h_k x) h_m(y)$ form an orthonormal basis of polynomials of degrees k with respect to x and m with respect to y, with $H_{0,0} \equiv 0$ and such that

$$H_{k+1,m+1}(x,y) = \{xH_k(x) - H_k'(x)\}\{yH_m(y) - H_m'(y)\},\ k \geq 1,\ m \geq 1.$$

The bivariate Hermite polynomials present an increasing number of pikes and holes that spread regularly in the plane.

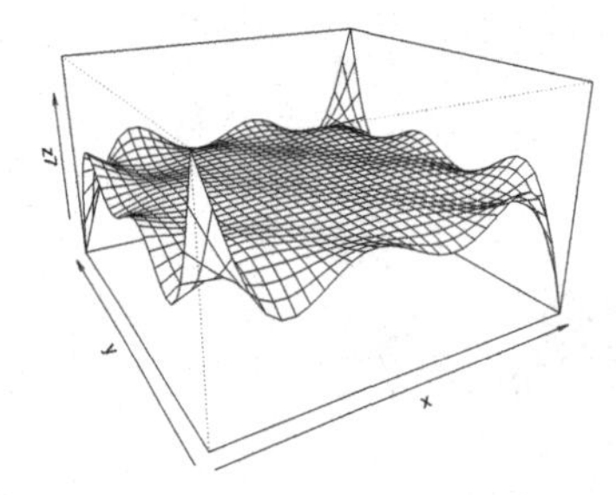

Fig. 2.13 Polynomial $H_{7,7}$.

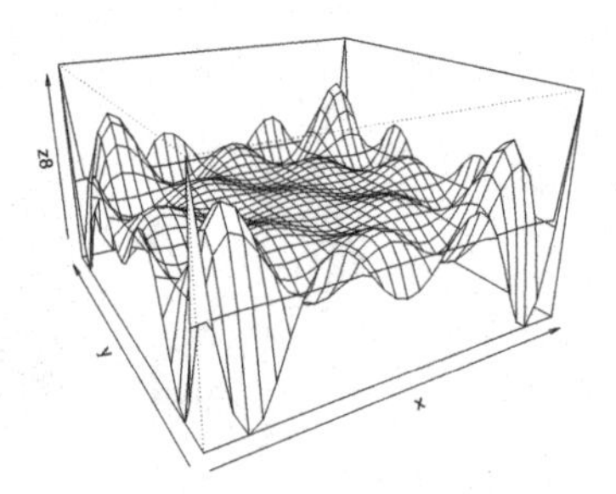

Fig. 2.14 Polynomial $H_{8,8}$.

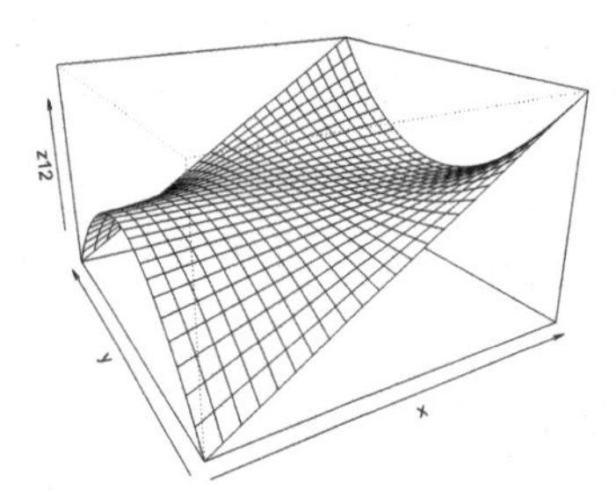

Fig. 2.15 Polynomial $H_{1,2}$.

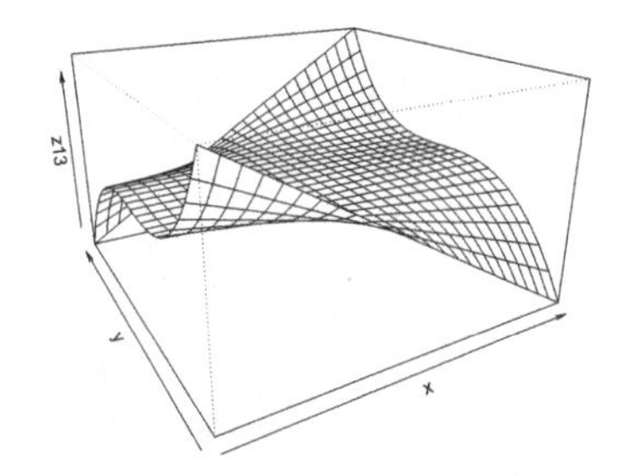

Fig. 2.16 Polynomial $H_{1,3}$.

2.8 Legendre's polynomials in $\mathbb{R}^2$

The equations are models for the interference of two sets of waves. The expressions of the first Hermite and Legendre polynomials are nearly identical so the similarity of their figures is not surprising. As their degree increases, the positive masses add up and create successive huge waves and flat fields like mixtures of distributions. They grow up fastly on the edges with a change of scale from low to higher dimensions. The waves $L_{2k,2k}$ are symmetric with respect to the axis of the coordinates though this does not appear in the figures due to their orientation, this is the reason of the interferences. The waves $L_{2k+1,2k+1}$ have no symmetry and they look more irregular (cf. Figure 2.5). They are not very different from Hermite's polynomials though they are flatter.

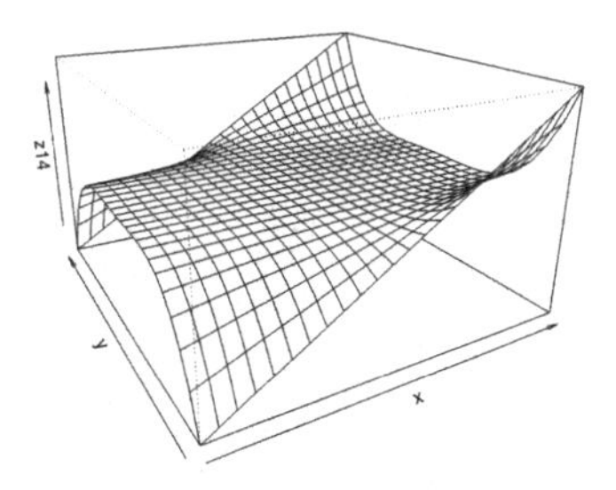

Fig. 2.17 Polynomial $H_{1,4}$.

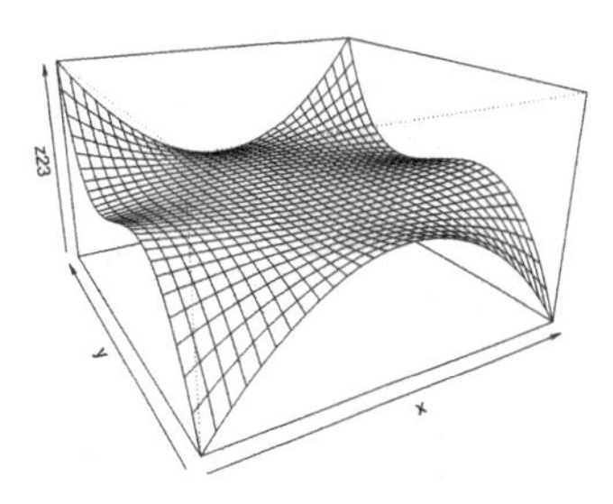

Fig. 2.18 Polynomial $H_{2,3}$.

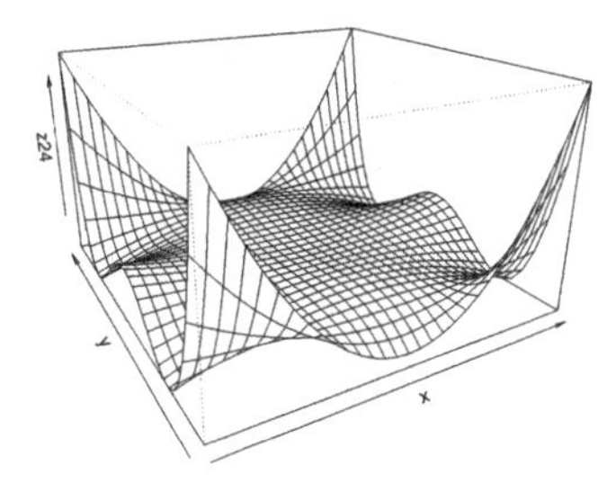

Fig. 2.19 Polynomial $H_{2,4}$.

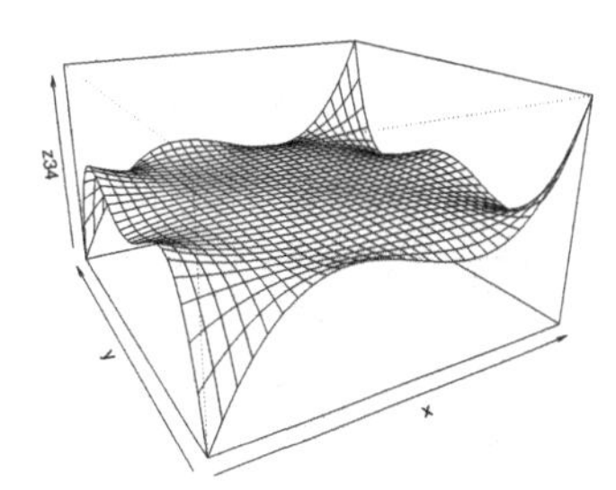

Fig. 2.20 Polynomial $H_{3,4}$.

Hermite generalized Legendre's polynomials as the coefficient of $a^m b^n$ in the expansion

$$\{(1 - ax - by)^2 - (a^2 + b^2)(x^2 + y^2 - 1)\}^{-\frac{1}{2}} = \sum_{m=0}^{\infty} \sum_{n=0}^{\infty} a^n b^m U_{m,n},$$

for all x and y in the disk $D = \{(x, y), x^2 + y^2 \leq 1\}$, where

$$U_{m,n} = \frac{1}{m!n!2^{m+n}} \frac{d^{m+n}}{x^m dy^n} \{(x^2 + y^2 - 1)^{m+n}\}.$$

The functions $U_{m,n}$ and $U_{\mu,\nu}(x, y)$ are orthogonal in $L^2(D, \mu)$ endowed with Lebesgue's measure μ, for all (m, n) and (μ, ν) such that $m + n$ differs from $\mu + \nu$. Another set of functions $V_{m,n}$ was defined from the expansion

$$(1 - 2ax - 2by + a^2 + b^2)^{-1} = \sum_{m=0}^{\infty} \sum_{n=0}^{\infty} a^n b^m V_{m,n}. \tag{2.18}$$

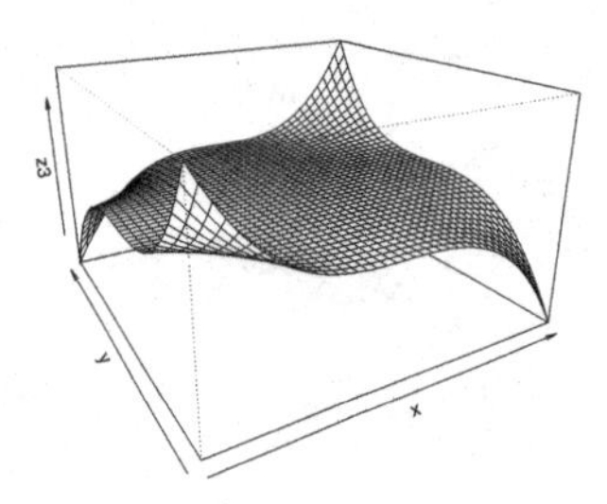

Fig. 2.21 Polynomial $L_{3,3}$.

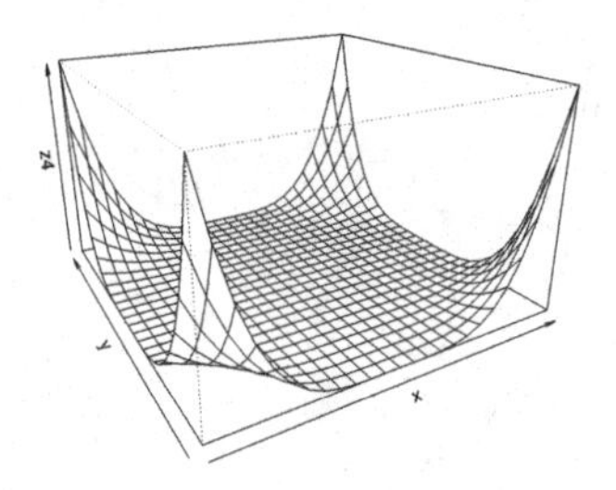

Fig. 2.22 Polynomial $L_{4,4}$.

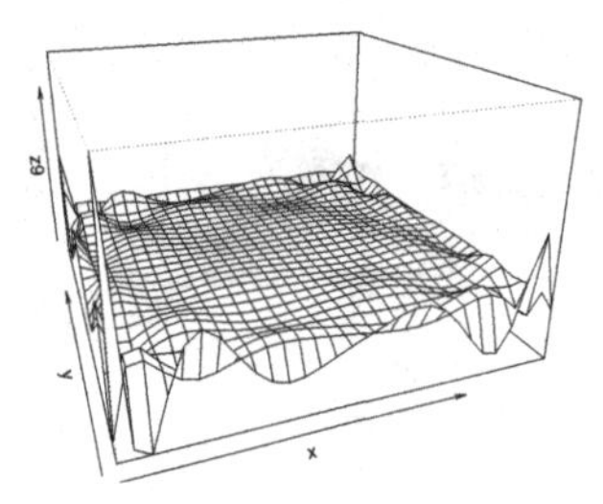

Fig. 2.23 Polynomial $L_{9,9}$.

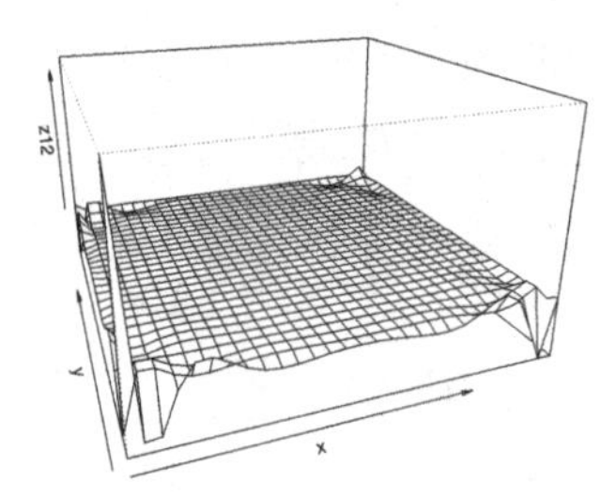

Fig. 2.24 Polynomial $L_{11,11}$.

The functions $V_{m,n}$ and $V_{\mu,\nu}(x, y)$ are orthogonal for all (m, n) and (μ, ν) such that $m + n$ differs from $\mu + \nu$ and

$$\int_D U_{m,n}(x, y)V_{\mu,\nu}(x, y)\, dx\, dy = 0$$

for all $(m, n) \neq (\mu, \nu)$. Moreover

$$\int_D U_{m,n}(x, y)V_{m,n}(x, y)\, dx\, dy = \frac{\pi}{m + n + 1}\frac{(m + n)!}{m!n!}. \tag{2.19}$$

From these definitions

$$\int_D U_{m,n}(x, y)(1 - 2ax - 2by + a^2 + b^2)^{-1}\, dx\, dy = \frac{\pi a^m b^n}{m + n + 1}\frac{(m + n)!}{m!n!},$$

equivalently

$$\int_D \frac{(1 - x^2 - y^2)^{m+n}}{1 - 2ax - 2by + a^2 + b^2} = \frac{\pi}{m + n + 1}.$$

Denoting $y^2 = (1-x^2)z$, this implies

$$\int_{-1}^{1} \frac{(1-x^2)^{p-\frac{1}{2}}}{(1-2rx+r^2)^p} = \frac{\pi}{p!}\frac{1}{2}\frac{3}{2}\cdots(p-\frac{1}{2}).$$

Every function $f(x,y)$ of $L^2(D,\mu)$ has the expansions

$$f = \sum_{m=0}^{\infty}\sum_{n=0}^{\infty} A_{m,n}U_{m,n} = \sum_{m=0}^{\infty}\sum_{n=0}^{\infty} B_{m,n}V_{m,n}$$

with coefficients $A_{m,n}$ and $B_{m,n}$ determined by the equalities

$$\int_D V_{m,n}(x,y)f(x,y)\,dx\,dy = A_{m,n}\frac{\pi}{m+n+1}\frac{(m+n)!}{m!n!},$$
$$\int_D U_{m,n}(x,y)f(x,y)\,dx\,dy = B_{m,n}\frac{\pi}{m+n+1}\frac{(m+n)!}{m!n!}.$$

As a consequence, every function $U_{m,n}$ has an expansion in the basis $(V_{\mu,\nu})_{\mu\geq 0,\nu\geq 0}$ and every $V_{m,n}$ has an expansion in the basis $(U_{\mu,\nu})_{\mu\geq 0,\nu\geq 0}$. Let $V_{m,n} = \sum_{m=0}^{\infty}\sum_{n=0}^{\infty} A_{m,n}U_{m,n}$, then

$$\int_D U_{m,n}V_{m,n}\,dx\,dy = \sum_{m=0}^{\infty} 1_{\{m+n=\mu+\nu\}}A_{\mu,\nu}\int_D U_{m,n}U_{\mu,\nu}\,dx\,dy$$

this sums has a finite number of terms and (2.19) implies a constraint between the coefficients $A_{\mu,\nu}$ of the sum. The norms of the function $U_{m,n}$ and $V_{m,n}$ are easily calculated from (2.19) and they are related to their scalar product

$$\int_D V_{m,n}^2(x,y)\,dx\,dy = A_{m,n}\int_D V_{m,n}(x,y)U_{m,n}(x,y)\,dx\,dy,$$
$$\int_D U_{m,n}^2(x,y)\,dx\,dy = B_{m,n}\int_D V_{m,n}(x,y)U_{m,n}(x,y)\,dx\,dy$$

Proposition 2.21. *The functions $U_{m,n}$ follow second order partial differential equations of wave motions*

$$\frac{\Delta U_{m,n}}{m+n} = \{2(m+n)-1\}\Big(\frac{U'_{m-1,n;x}}{m} + \frac{U'_{m,n-1;x}}{n}\Big)$$
$$-(m+n-1)\Big\{\frac{y^2-1}{m(m-1)}U''_{m-2,n;x,x} + \frac{x^2-1}{n(n-1)}U''_{m,n-2;yy}\Big\},$$

$$U''_{m,n;x,x} - U''_{m,n;x,x} = (m+n)(m+n-1)\Big\{\frac{x^2}{m(m-1)}U''_{m-2,n;x,x}$$
$$-\frac{y^2}{n(n-1)}U''_{m,n-2;y,y}\Big\}.$$

Proof. The second partial derivatives of

$$T = \frac{d^m}{dx^m}\frac{d^n}{dy^n}\{(x^2+y^2-1)^{m+n}\}$$

are calculated as for integers m and n such that $m+n \geq 5$

$$\begin{aligned}
T''_{x,x} &= \frac{d^{m+2}}{dx^{m+2}}\frac{d^n}{dy^n}(x^2+y^2-1)^{m+n} \\
&= 2(m+n)\frac{d^{m+1}}{dx^{m+1}}\frac{d^n}{dy^n}\{x(x^2+y^2-1)^{m+n-1}\} \\
&= 2(m+n)\frac{d^m}{dx^m}\frac{d^n}{dy^n}\{(x^2+y^2-1)^{m+n-1} \\
&\quad + 2(m+n-1)x^2(x^2+y^2-1)^{m+n-2}\} \\
&= 2(m+n)\frac{d^m}{dx^m}\frac{d^n}{dy^n}[\{2(m+n)-1\}(x^2+y^2-1)^{m+n-1} \\
&\quad - 2(m+n-1)(y^2-1)(x^2+y^2-1)^{m+n-2}]
\end{aligned}$$

this implies

$$\begin{aligned}
\frac{d^2}{dx^2}U_{m,n}(x,y) &= \frac{m+n}{m}\{2(m+n)-1\}\frac{d}{dx}U_{m-1,n}(x,y) \\
&\quad + \frac{(m+n)(m+n-1)}{m(m-1)}(1-y^2)\frac{d^2}{dx^2}U_{m-2,n}(x,y), \\
\frac{d^2}{dxdy}U_{m,n}(x,y) &= \frac{m+n}{n}\{2(m+n)-1\}\frac{d}{dy}U_{m,n-1}(x,y) \\
&\quad + \frac{(m+n)(m+n-1)}{n(n-1)}(1-x^2)\frac{d^2}{dxdy}U_{m,n-2}(x,y), \\
\frac{d^2}{dxdy}U_{m,n}(x,y) &= \frac{(m+n)(m+n-1)}{mn}xy\frac{d^2}{dxdy}, U_{m-1,n-1}(x,y).
\end{aligned}$$

□

By differentiation with respect to x, y, a, b the equality (2.18), Didon (1868) proved that the functions $V_{m,n}$ are solutions of the partial differential equations

$$\begin{aligned}
mV_{m,n} &= x\frac{d}{dx}V_{m,n} - \frac{d}{dx}V_{m-1,n}, \\
\frac{d}{dy}V_{m-1,n} &= \frac{d}{dx}V_{m,n-1}, \\
(m+n+1)V_{m-1,n} &= \frac{d}{dx}V_{m,n} - x\frac{d}{dx}V_{m-1,n} - y\frac{d}{dx}V_{m,n-1},
\end{aligned}$$

and, deriving the last equation

$$
\begin{aligned}
0 &= (1-x^2)\frac{d^2}{dx^2}V_{m,n} - xy\frac{d^2}{dxdy}V_{m,n} - (n+3)x\frac{d}{dx}V_{m,n} \\
&\quad + my\frac{d}{dy}V_{m,n} + (m+n)(m+n+2)V_{m,n}, \\
0 &= (1-y^2)\frac{d^2}{dy^2}V_{m,n} - xy\frac{d^2}{dxdy}V_{m,n} - (m+3)y\frac{d}{dy}V_{m,n} \\
&\quad + nx\frac{d}{dx}V_{m,n} + (m+n)(m+n+2)V_{m,n}, \\
0 &= (1-x^2)\frac{d^2}{dx^2}V_{m,n} + (1-y^2)\frac{d^2}{dy^2}V_{m,n} - 2xy\frac{d^2}{dxdy}V_{m,n} \\
&\quad - 3x\frac{d}{dx}V_{m,n} - 3x\frac{d}{dx}V_{m,n} + (m+n)(m+n+2)V_{m,n}.
\end{aligned}
$$

Theorem 2.2. *The term of degree* $m+n$ *of the polynomial function solution of the second order equations has the form* $a_{m,n}x^m y^n$.

2.9 Exercises

2.9.1. Let X be an exponential variable with parameter 1, calculate $\int_0^\infty L_n(x)e^{-x}\,dx$.

2.9.2. Solve the equation $x^2y'' - xy'_x - m(m+2)y = 0$.

2.9.3. Write a second order differential equation for

$$u(x) = A(x^2-1)^m + B(x^2-1)^{-m}.$$

2.9.4. Determine second order differential equations for the derivatives of Legendre's polynomials.

2.9.5. Write the second order differential equation for Legendre polynomials at $x = \cos\theta$.

2.9.6. Chebyshev's polynomials at $x = \cos\theta$ are

$$T_n(\cos\theta) = \cos(n\theta),\ n \geq 1$$

with $T_0(x) = 1$ and $T_1(x) = x$. They are orthogonal in $L^2(]-1,1[,\mu)$ with the measure $d\mu(x) = \{\pi(1-x^2)\}^{-\frac{1}{2}}\,dx$. Write a recurrence formula for the polynomials $T_n(x)$, for x in $]-1,1[$, their generating function and a second order differential equation for T_n.

2.9.7. Solve the differential equation $(1-x^2)y''_x - xy'_x + n^2 y_x = f_x$ and the differential equation of Chebyshev's polynomials.

2.9.8. Calculate the expectation of $g(X-w)$ for a normal variable X and a function g of $L^2(\mathbb{R}, \mu_{\mathcal{N}})$.

2.9.9. Calculate the Fourier transform of Hermite's polynomials.

2.9.10. Calculate the polynomials Q_n of Equation (2.16).

2.9.11. Let X_{mn} be the coefficient of $u^m v^n$ in the expansion of $\{(1-vx)(1-(x-1)u\}^{-1}$, write a differential equation for the m-th derivative of X_{mn}.

2.9.12. Let

$$\varphi_{mn}(x) = \frac{(x-a)^m(x+a)^n}{(x+z)^{n+1}}, \qquad \psi_{mn}(x) = \frac{(x-a)^m(x+a)^n}{(x-z)^{m+1}}$$

write an equation relating the m-th derivative of φ_{mn} and the n-th derivative of ψ_{mn} at z.

Chapter 3

Differential and integral calculus

We present conditions for the differentiability and the existence of a Taylor expansion in complex spaces, the conditions for the existence of an extremum for an implicit function and the Euler-Lagrange conditions for the minimization of the integral of a function $f(t, x_t, x_t')$ in $\mathbb{R}_+$ as x_t and its derivative belong to metric spaces. We generalize them to a function including more derivatives. The applications concern the optimality of areas or volumes, the calculus of parametric integrals as solutions of differential equations, and the elliptic functions.

3.1 Differentiability of functions

A real function f of $C_1(\mathbb{C})$ has a complex derivative $f^{(1)}$ such that for all z and δ in $\mathbb{C}$

$$f(z+\delta) = f(z) + \delta_x f^{(1)}(z) + i\delta_y f^{(1)}(z) + o(\|\delta\|).$$

Cauchy's necessary and sufficient conditions for the differentiability of a complex function $f(x+iy) = f_1(x+iy)+if_2(x+iy)$ on $\mathbb{C}$ are the existence of the partial derivatives of its real and imaginary parts and

$$\frac{\partial f_1}{\partial x} = \frac{\partial f_2}{\partial y}, \qquad \frac{\partial f_2}{\partial x} = -\frac{\partial f_1}{\partial y} \tag{3.1}$$

then the derivative of f at $z = (x + iy)$ is

$$f^{(1)}(x+iy) = \frac{\partial f_1}{\partial x}(z) + i\frac{\partial f_2}{\partial x} = \frac{\partial f_2}{\partial y} - i\frac{\partial f_1}{\partial y}.$$

For the second order derivatives, these equalities imply

$$\frac{\partial^2 f_1}{\partial x^2} = \frac{\partial^2 f_2}{\partial x \partial y} = -\frac{\partial^2 f_1}{\partial y^2}, \tag{3.2}$$
$$\frac{\partial^2 f_2}{\partial y^2} = \frac{\partial^2 f_1}{\partial x \partial y} = -\frac{\partial^2 f_2}{\partial x^2}$$

and the second order derivative of f at z is

$$f^{(2)}(z) = \frac{\partial^2 f_1}{\partial x^2}(z) + i\frac{\partial^2 f_2}{\partial x^2}(z) = \frac{\partial^2 f_2}{\partial x \partial y}(z) - i\frac{\partial^2 f_1}{\partial x \partial y}(z)$$
$$= -\frac{\partial^2 f_1}{\partial y^2}(z) i \frac{\partial^2 f_2}{\partial y^2}(z).$$

In $\mathbb{R}^3$, -1 has two square roots denoted by i and j. The map

$$X = (x, y, z) \mapsto t = x + iy + jz$$

defines a bijection between $\mathbb{R}^3$ and a complex space denoted $\mathbb{C}_2$. The complex conjugate of $t = x + iy + jz$ is $x - iy - jz$ and the space $\mathbb{C}_2$ is a vector space endowed with the Euclidean norm defined as the scalar product of complex conjugates

$$\|t\| = \{(x + iy + jz)(x - iy - jz)\}^{\frac{1}{2}} = \{x^2 + y^2 + z^2\}^{\frac{1}{2}}.$$

Let $\rho = \|x + iy + jz\|_2$, then

$$x + iy + jz = \rho e^{i\theta} e^{j\varphi}, \ \theta \in [0, 2\pi], \ \varphi \in [0, 2\pi],$$
$$\theta = \arctan \frac{x}{y}, \ y \neq 0,$$
$$\varphi = \arctan \frac{x}{z}, \ z \neq 0,$$

and $\theta = 0$ if $y = 0$, $\varphi = 0$ if $z = 0$. The product $u = e^{i\theta} e^{j\varphi}$ belongs to $\mathbb{C}^{\otimes 2}$ and its expansion using trigonometric functions is $u = \cos\varphi \cos\theta + ij \sin\varphi \sin\theta + i \cos\varphi \sin\theta + j \sin\varphi \cos\theta$, where $\cos\varphi(\cos\theta + i\sin\theta)$ is the projection of $e^{i\theta} e^{j\varphi}$ in the horizontal plane and $\sin\varphi(\cos\theta + i\sin\theta)$ is its projection in a vertical section of the sphere.

Cauchy's conditions were extended to the derivability of a complex function of $C(\mathbb{C}_2)$, (Pons, 2012).

Proposition 3.1. *A function $f(x + iy + jz) = P(x, y, z) + iQ(x, y, z) + jR(x, y, z)$ defined from $\mathbb{C}_2$ to $\mathbb{C}_2$ is continuously differentiable at $t = x + iy + jz$ if and only if the real functions P, Q and R belong to $C(\mathbb{R}^3)$ and*

$$\frac{\partial P(x, y, z)}{\partial x} = \frac{\partial Q(x, y, z)}{\partial y} = \frac{\partial R(x, y, z)}{\partial z},$$
$$\frac{\partial Q(x, y, z)}{\partial x} = -\frac{\partial P(x, y, z)}{\partial y},$$
$$\frac{\partial R(x, y, z)}{\partial x} = -\frac{\partial P(x, y, z)}{\partial z},$$
$$\frac{\partial R(x, y, z)}{\partial y} = -\frac{\partial Q(x, y, z)}{\partial z}.$$

Then, its derivative at t is

$$f^{(1)}(t) = \frac{\partial P(x,y,z)}{\partial x} + i\frac{\partial Q(x,y,z)}{\partial x} + j\frac{\partial R(x,y,z)}{\partial x}.$$

The norm of $f^{(1)}(t)$ in $L_2(\mathbb{C}_2)$ is

$$\|f^{(1)}(t)\|_2 = \Big\{\frac{\partial P(x,y,z)}{\partial x}\Big\}^2 + \Big\{\frac{\partial Q(x,y,z)}{\partial x}\Big\}^2 + \Big\{\frac{\partial R(x,y,z)}{\partial x}\Big\}^2$$

and from Proposition 3.1

$$\begin{aligned}\|f^{(1)}(t)\|_2 &= \Big\{\frac{\partial Q(x,y,z)}{\partial y}\Big\}^2 + \Big\{\frac{\partial P(x,y,z)}{\partial y}\Big\}^2 + \Big\{\frac{\partial R(x,y,z)}{\partial y}\Big\}^2 \\ &= \Big\{\frac{\partial R(x,y,z)}{\partial z}\Big\}^2 + \Big\{\frac{\partial Q(x,y,z)}{\partial z}\Big\}^2 + \Big\{\frac{\partial P(x,y,z)}{\partial z}\Big\}^2.\end{aligned}$$

Under the conditions of Proposition 3.1 and as δ tends to zero in $\mathbb{C}_2$, a differentiable function $f = P + iQ + jR$ defined on $\mathbb{C}_2$ and with values in $\mathbb{C}_2$ has a derivative $f^{(1)}$ such that

$$f(z+\delta) = f(z) + \delta f^{(1)}(z) + o(\|\delta\|).$$

Expansions of a $\mathbb{C}_2$-valued complex function are similar to the Taylor expansions of differentiable functions on $\mathbb{R}^3$ in an orthogonal basis, via the representation of the function as $f(x+iy+jz) = P(x,y,z) + iQ(x,y,z) + jR(x,y,z)$ with real functions P, Q and R of $C_n(\mathbb{R}^3)$, satisfying the equalities of Proposition 3.1 for all derivatives upto the nth derivative. As $\|\delta\|$ tends to zero

$$f(t+\delta) = f(t) + \sum_{k=1}^{n} \frac{\delta^k}{k!} f^{(k)}(t) + o(\|\delta\|^k), \tag{3.3}$$

with the derivatives $f^{(k)} = P_x^{(k)} + iQ_x^{(k)} + jR_x^{(k)}$. The isometry between $\mathbb{R}^3$ and $\mathbb{C}_2$ extends to higher dimensions. Let p be an integer larger or equal to 3 and let $p-1$ roots $(i_1, \ldots, i_{p-1})$ of -1, they define a complex space $\mathbb{C}_{p-1}$ by the bijection $(x_1, \ldots, x_p) \mapsto x_1 + \sum_{k=2}^{p-1} i_{k-1}x_k$. Functions of $C_n(\mathbb{C}_{p-1})$ have expansions like (3.3) under Cauchy conditions of dimension p and order n.

3.2 Maximum and minimum of functions

The derivatives of an implicit function $u(x,y) = 0$ of $C_2(\mathbb{R}^{p+1})$ can be written as the derivative with respect to x, considering y as a function of x. Its first derivatives are

$$\begin{aligned}du(x,y_x) &= (u'_x + y'_x u'_y)\,dx, \\ d^2u(x,y_x) &= \{u''_{xx}(x,y) + 2y'_x u''_{xy}(x,y) + y'^2_x u''_{yy}(x,y) + 2y''_{xx}u'_y\}\,dx^2. \qquad (3.4)\end{aligned}$$

The function y_x has an unique maximum or minimum in a subset I of $\mathbb{R}^p$ if its first derivative y'_x is zero at a value x^* of I and its second derivative y''_x has a constant sign in I. If the second derivative is zero at this point, the sign of the first non null even derivative determines the existence of a maximum or a minimum.

In $C_2(\mathbb{R}^2)$, y'_x is zero if $u'_x = 0$ and $u'_y \neq 0$, the second derivative of y with respect to x is the derivative of $y'_x = -u'^{-1}_y u'_x$

$$y''_{xx} = -\frac{u''_{xx}}{u'_y} + u'_x \frac{u''_{yy}}{u'^2_y} = -\frac{u''_{xx}}{u'_y} - y'_x \frac{u''_{yy}}{u'_y}.$$

Sufficient conditions for an unique extremum of y in I at (x^*, y^*) are

$$\begin{aligned} y''_{xx}(x^*) &< 0, \quad \text{for a maximum,} \\ y''_{xx}(x^*) &> 0, \quad \text{for a minimum,} \\ u'_y u''_{xx} - u'_x u''_{yy} & \quad \text{has a constant sign for all } x, y \text{ in } \mathcal{V}*, \end{aligned}$$

where $\mathcal{V}*$ is a neighborhood of (x^*, y^*) in I.

If y has local minimum and maximum in I, y''_{xx} takes the value zero between the extrema. The conditions $y'_x = 0$ and $y''_{xx} = 0$ are satisfied if $u'_x = 0$ and $u''_{xx} = 0$, with $u'_y \neq 0$.

The bivariate function $u(x, y)$ has a maximum or a minimum at (x^*, y^*) where its partial derivatives u'_x and u'_y are zero, and $d^2u(x^*, y^*)$ has a constant sign in a neighborhood of (x^*, y^*). The first condition implies

$$d^2u(x^*, y^*) = \{u''_{xx}(x^*, y^*) + 2y'_x u''_{xy}(x^*, y^*) + y'^2_x u''_{yy}(x^*, y^*)\}\, dx^2.$$

Considering the function $\frac{d^2}{dx^2}u(x, y_x)$ as a polynom in y'_x, sufficient conditions for an extremum at (x^*, y^*) of the function $u(x, y)$ as a multivariate function are

$$\begin{aligned} u''_{xx}(x^*, y^*) &< 0, \quad \text{for a maximum,} \\ u''_{xx}(x^*, y^*) &> 0, \quad \text{for a minimum,} \\ u''^2_{xy}(x, y) &< u''_{xx}(x, y) u''_{yy}(x, y), \quad \text{for all } x, y \text{ in } \mathcal{V}*. \end{aligned}$$

A function $u(x, y, z)$ of $C_2(\mathbb{R}^3)$, with $y = y_x$ and $z = z_x$, has an extremum at (x^*, y^*, z^*) if u'_x, u'_y and u'_z are zero at this point, its second derivative is written as

$$\begin{aligned} \frac{d^2}{dx^2}u(x, y_x, z_x) &= u''_{xx}(x, y, z) + y'^2_x u''_{yy}(x, y, z) + z'^2_x u''_{zz}(x, y, z) \\ &\quad + 2y'_x u''_{xy}(x, y, z) + 2z'_x u''_{xz}(x, y, z) + 2y'_x z'_x u''_{yz}(x, y, z) \\ &\quad + 2y''_{xx} u'_y + 2z''_{xx} u'_z. \end{aligned}$$

At (x^*, y^*, z^*), it is polynomial in $\zeta_x = y'_x$ and $\xi_x = z'_x$ and this bivariate polynomial has no real roots for ζ and ξ at this point. Considering it as a polynomial with respect to ξ, a sufficient condition for an unique extremum is $\{u''_{xz} + \zeta u''_{yz}\}^2 < u''_{zz}\{u''_{xx} + \zeta^2 u''_{yy} + 2\zeta u''_{xy}\}$. Expanding this expression as a polynomial in ζ yields

$$u''^2_{xz} - u''_{zz}u''_{xx} + 2\zeta\{u''_{xz}u''^2_{yz} - u''_{zz}u''_{xy}\} + \zeta_x^2\{u''^2_{yz} - u''_{zz}u''_{yy}\} < 0$$

for every ζ and this is true under the condition

$$\{u''_{xz}u''^2_{yz} - u''_{zz}u''_{xy}\}^2 < \{u''^2_{xz} - u''_{zz}u''_{xx}\}\{u''^2_{yz} - u''_{zz}u''_{yy}\},$$

with $u''_{xx}(x^*, y^*) < 0$ for a maximum at (x^*, y^*) and $u''_{xx}(x^*, y^*) > 0$ for a minimum at (x^*, y^*).

These conditions are sufficient but not necessary. If the polynomial defined by the second order derivative with respect to x of a function u of $C_2(\mathbb{R}^2)$ has two roots, local extrema may occur and there exists an unique minimum at (x^*, y^*) if $u''_{yy}(x^*, y^*) > 0$ and a unique maximum at (x^*, y^*) if $u''_{yy}(x^*, y^*) < 0$.

Let $f(x_1, \ldots, x_n)$ be a real function of $C_1(\mathbb{R}^n)$ with p minima or maxima, its extrema are solutions of p equations $g_k(x) = 0$, $k = 1, \ldots, p$ in $\mathbb{R}^n$. Lagrange proved that they are extrema of the function

$$F(x, \lambda_1, \ldots, \lambda_p) = f(x) + \sum_{k=1}^{p} \lambda_k g_k(x),$$

where $\lambda_1, \ldots, \lambda_p$ are real parameters. This is the Lagrangian method for the search of the solutions of maxima or minima under constraints.

The aera inside a closed curve $y_x = f(x)$ in a subset $(I, f(I))$ of $\mathbb{R}^2$ is $A = \int_I y_x\,dx$, the volume defined by a bivariate curve or by an implicit equation $f(x, y, z_{x,y}, x', y')$, with (x, y) in the subset $\Omega = (I, f(I))$ of $\mathbb{R}^3$ is the integral $u = \int_\Omega z_{x,y}\,dx\,dy$. An ellipse defined by $a^{-2}x^2 + b^{-2}y^2 = 1$ has the coordinates x such that $|x| \leq a$ and $y = a^{-1}b(a^2 - x^2)^{\frac{1}{2}}$. In polar coordinates, $x = a\cos\theta$ and $y = b\sin\theta$, its perimeter is $P = (a + b)\pi$ and its area is

$$A = 2b\int_{-a}^{a}(1 - a^{-2}x^2)^{\frac{1}{2}}\,dx = -ab\int_0^{2\pi}\sin^2\theta\,d\theta = ab\pi,$$

consequently

$$\int_{-a}^{a}(1 - a^{-2}x^2)^{\frac{1}{2}}\,dx = \frac{a\pi}{2}.$$

Let $x = a^2y^2$ be the equation of a parabol along the x-axis, the area it describes as x varies in $[0, X]$ is $A_X = 2a^2\int_0^X y^2\,dy = \frac{2}{3}a^2X^3$.

An ellipsoïd $a^{-2}x^2 + b^{-2}y^2 + c^{-2}z^2 = 1$ has the volume

$$V = 4c \int_{-b}^{b} \int_{-a}^{a} (1 - a^{-2}x^2 - b^{-2}y^2)^{\frac{1}{2}} \, dx \, dy$$

$$= abc \int_{0}^{2\pi} \int_{0}^{2\pi} \sin^2\varphi \cos\varphi \, d\theta \, d\varphi = \frac{4abc\pi}{3}$$

and its spatial area is the derivative of its volume, $\frac{4}{3}\pi(ab + ac + bc)$.

The question of the optimality of the dimensions and the form of solids is always an actual question. It is solved by differentiating integrals. When $x = x_t$, the aera and the volume are written as $A = \int_{t_0}^{t_1} y_t x'_t \, dt$ and $V = \int_{t_0}^{t_1} \int_{t_0}^{t_1} z_{x_t, y_s} x'_t y'_s \, dt \, ds$, they must also be calculated from the implicit equations defining the curves. By the differentiation of an implicit equation $g(x, y) = 0$, the area become

$$A = \int_{t_0}^{t_1} y_t g_x'^{-1}(x_t, y_t) g'_y(x_t, y_t) \, dy_t$$

with x_t defined as a function of y_t from the implicit equation.

Example 4. Let us consider the search of a geometric figure in $\mathbb{R}^2$ with a fixed perimeter and maximum surface. For a triangle with basis x and height h, the length of the two other equal sides is $y = (h^2 + \frac{1}{4}x^2)^{\frac{1}{2}}$, its perimeter is $a = x + 2y$ and its surface is $s = \frac{1}{2}hx$. We have to determine $h = (y^2 - \frac{1}{4}x^2)^{\frac{1}{2}}$ or $y = \frac{1}{2}(a - x)$ such that $2s = x(y^2 - \frac{1}{4}x^2)^{\frac{1}{2}}$ is minimum. It is proportional to

$$u(x) = x\{(a - x)^2 - x^2\}^{\frac{1}{2}} = x(a^2 - 2ax)^{\frac{1}{2}},$$

such that $u'(x) = x(a^2 - 2ax) - 2ax^2 = 2ax(a - 3x) = 0$ and its second derivative is $u''(x) = 2a^2 > 0$ so that $x = y = \frac{1}{3}a$, $h = \frac{\sqrt{3}}{2}x$ and $s = \frac{\sqrt{3}}{4}x^2$. The same question for a rectangle leads to a square having the surface $\frac{1}{4}a^2$.

Example 5. In the space, the volume of a pyramid having the height h and a square basis is $V = \frac{1}{3}hx^2$, the faces of the pyramid are identical triangles with one side of length x and two sides of length $y = (h^2 + \frac{1}{2}x^2)^{\frac{1}{2}}$ hence $x < \sqrt{2}y$, their height in a plane is $h_T = (h^2 + \frac{1}{4}x^2)^{\frac{1}{2}}$ with $x < 2h_T$, and their perimeter is $a = x + 2y$. The outer surface of the pyramid is $S = 2xh_T$ and its volume

$$V = \frac{1}{3}x^2 h = \frac{1}{6}x(S^2 - x^4)^{\frac{1}{2}}$$

is minimum at fixed S if $S^2 = 3x^4$ which implies $x = \sqrt{2}h$ and $h = (\frac{2}{3})^{\frac{1}{2}} h_T$.

With equilateral pyramids, the ratio $h_T^{-1}h$ is approximately equal to .86 and $h_T \approx 1.16h$, the ratio $h_T^{-1}h$ is smaller than the optimum value. For a pyramid with the slope one, $h_T = \sqrt{2}h$ and $x = 2h$, the ratio is larger than the optimum value.

Guldin's theorems (Sturm, 1861) states that the surface generated by a curve turning around an axis in its plane equals the product of the length of the curve and the perimeter drawn by its center of gravity, the volume generated by a planar surface turning around an axis in its plane equals the product of the surface of the planar set and the perimeter drawn by its center of gravity. Their optimization is equivalent to the optimization of the parameters of the curve and, respectively, the surface, and their distance to the axis.

3.3 Euler-Lagrange conditions

The Euler-Lagrange conditions concern the minimization of an integral function and the derivatives of integrals with respect to parameters. The maximum or minimum of a function $u(x,y)$ on $\mathbb{R}^2$ is reached at (x^*, y^*) where the first derivative of u is zero and the sign of its second order derivative (3.4) at this point determines whether u has a maximum or a minimum at (x^*, y^*). In particular, the primitive of a function having a Taylor expansion has often been calculated as the series of the integrated terms of the expansion. The same properties enable to solve differential equations using a Taylor expansion of the solution.

Let x_t be a function in a metric space $\mathbb{X}$ of $C_2(\mathbb{R}_+)$ with a first derivative x'_t in $\mathbb{X}'$ and let $f(t, x_t, x'_t)$ be a functional of x_t and its derivative. Conditions for a function x^*_t to minimizes the integral over an interval $[t_0, t_1]$ of a function f of $C_2(\mathbb{R}_+ \times \mathbb{X} \times \mathbb{X}')$ are conditions for the first two derivatives of f. The first condition for the function f that ensures the existence of a function x^*_t that minimizes the integral

$$I = \int_{t_0}^{t_1} f(t, x_t, x'_t)\, dt$$

with respect to x is a null first derivative of I with respect to x at x^*. For every real function η_t of the tangent space to the space $\mathbb{X}$ at x^*, a second order local approximation of x_t is written as

$$x_{t,\theta} = x^*_t + \theta\eta_t + \frac{1}{2}\theta^2\eta'_t + o(\theta^2)$$

uniformly in t, as θ converges to zero. The first derivative of $x_{t,\theta}$ with respect to θ is $x'_{t,\theta} = \eta_t + \theta\eta'_t + o(\theta)$ in a neighborhood of x^*.

Theorem 3.1. *The first necessary condition for the existence of a function x_t^* that minimizes the finite integral $I = \int_{t_0}^{t_1} f(t, x_t, x_t')\,dt$ with respect to x is*

$$\frac{\partial f}{\partial x}(t, x_{t,\theta}^*, x_{t,\theta}^{*\prime}) - \frac{d}{dt}\frac{\partial f}{\partial x'}(t, x_{t,\theta}^*, x_{t,\theta}^{*\prime}) = 0. \tag{3.5}$$

Proof. The first derivative of the function $I(\theta) = \int_{t_0}^{t_1} f(t, x_{t,\theta}, x_{t,\theta}')\,dt$ is

$$\begin{aligned} I'(\theta) &= \int_{t_0}^{t_1} \frac{\partial}{\partial\theta} f(t, x_{t,\theta}, x_{t,\theta}')\,dt \\ &= \int_{t_0}^{t_1} \Big\{\eta_t \frac{\partial}{\partial x} f(t, x_{t,\theta}, x_{t,\theta}') + \eta_t' \frac{\partial}{\partial x'} f(t, x_{t,\theta}, x_{t,\theta}')\Big\}\,dt + o(1). \end{aligned}$$

Integrating by parts the second term of the integral, for every function η with values zero at t_0 and t_1

$$\int_{t_0}^{t_1} \eta_t' \frac{\partial}{\partial x'} f(t, x_{t,\theta}, x_{t,\theta}')\,dt = -\int_{t_0}^{t_1} \eta_t \frac{d}{dt}\frac{\partial}{\partial x'} f(t, x_{t,\theta}, x_{t,\theta}')\,dt.$$

It follows that

$$I'(\theta) = \int_{t_0}^{t_1} \eta_t \Big\{\frac{\partial}{\partial x} f(t, x_{t,\theta}, x_{t,\theta}') - \frac{d}{dt}\frac{\partial}{\partial x'} f(t, x_{t,\theta}, x_{t,\theta}')\Big\}\,dt + o(1)$$

and the first necessary condition follows from the equality $I'(0) = 0$ for every function η of the tangent space to $\mathbb{X}$ at x^*, with boundary values zero. $\square$

The second condition for a minimum of I at x^* is $I'' \geq 0$, it is expressed in different forms in the following propositions.

Theorem 3.2. *The inequality $I'' \geq 0$ at x^* is satisfied under one of the following conditions*

$$\frac{\partial^2}{\partial x^2} f(t, x_{t,\theta}^*, x_{t,\theta}^{*\prime}) \geq 0,$$

$$\Big\{\frac{\partial^2}{\partial x \partial x'} f(t, x_{t,\theta}^*, x_{t,\theta}^{*\prime})\Big\}^2 - \Big\{\frac{\partial^2}{\partial x^2} f(t, x_{t,\theta}^*, x_{t,\theta}^{*\prime})\Big\}\Big\{\frac{\partial^2}{\partial x'^2} f(t, x_{t,\theta}^*, x_{t,\theta}^{*\prime})\Big\} \leq 0.$$

Proof. The second derivative of the integral $I(\theta)$ with respect to θ is

$$\begin{aligned} I''(\theta) &= \int_{t_0}^{t_1} \frac{\partial^2}{\partial\theta^2} f(t, x_{t,\theta}, x_{t,\theta}')\,dt \\ &= \int_{t_0}^{t_1} \Big\{\eta_t^2 \frac{\partial^2}{\partial x^2} f(t, x_{t,\theta}, x_{t,\theta}') + 2\eta_t\eta_t' \frac{\partial^2}{\partial x \partial x'} f(t, x_{t,\theta}, x_{t,\theta}') \\ &\qquad + \eta'^2 \frac{\partial^2}{\partial x'^2} f(t, x_{t,\theta}, x_{t,\theta}')\Big\}\,dt + o(1). \end{aligned}$$

The discriminant of the polynom in $y_t = \eta_t^{-1}\eta_t'$ in the integrand must be negative for every η_t and this is true under the condition. $\square$

Proposition 3.2. *Necessary and sufficient conditions for $I'' \geq 0$ at x^* are*

$$\frac{\partial^2}{\partial x^2} f(t, x^*_{t,\theta}, x^{*\prime}_{t,\theta}) - \frac{d}{dt} \frac{\partial^2}{\partial x' \partial x} f(t, x^*_{t,\theta}, x^{*\prime}_{t,\theta}) \geq 0,$$

$$\frac{\partial^2}{\partial x'^2} f(t, x^*_{t,\theta}, x^{*\prime}_{t,\theta}) \geq 0.$$

Proof. Integrating by parts the integrand of $I''(\theta)$ in the previous proof, for every function η with values zero at t_0 and t_1

$$2 \int_{t_0}^{t_1} \eta_t \eta'_t \frac{\partial^2}{\partial x \partial x'} f(t, x_{t,\theta}, x'_{t,\theta})\, dt = -\int_{t_0}^{t_1} \eta_t^2 \frac{d}{dt} \frac{\partial^2}{\partial x' \partial x} f(t, x_{t,\theta}, x'_{t,\theta})\, dt,$$

it follows that

$$I''(\theta) = \int_{t_0}^{t_1} \Big[\eta_t^2 \Big\{ \frac{\partial^2}{\partial x^2} f(t, x_{t,\theta}, x'_{t,\theta}) - \frac{d}{dt} \frac{\partial^2}{\partial x' \partial x} f(t, x_{t,\theta}, x'_{t,\theta}) \Big\} + \eta_t'^2 \frac{\partial^2}{\partial x'^2} f(t, x_{t,\theta}, x'_{t,\theta}) \Big]\, dt + o(1)$$

and the inequality $I''(0) \geq 0$ for every function η_t implies the conditions. □

The minimization of higher order functionals requires conditions with higher order derivatives with respect to t. Let f be a functional of x_t and its first k derivative and let

$$I = \int_{t_0}^{t_1} f(t, x_t, x_t^{(1)}, \ldots, x_t^{(k)})\, dt.$$

The derivatives of I with respect to x includes partial derivatives of f with respect to $x_t, x_t^{(1)}, \ldots, x_t^{(k)}$. A kth order local approximation of x_t in a neighborhood of x_t^* is written as

$$x_{t,\theta} = x_t^* + \theta \eta_t + \sum_{m=1}^{k} \frac{1}{m!} \theta^m \eta_t^{(m)} + o(\theta^k)$$

uniformly in t, for θ in a neighborhood of zero and its jth derivative is

$$x_{t,\theta}^{(m+1)} = \sum_{m=0}^{k-j} \eta_t^{(m)} + o(1)$$

uniformly in t, for θ close to zero.

Proposition 3.3. *The first necessary condition for the existence of a function x_t^* that minimizes $I = \int_{t_0}^{t_1} f(t, x_t, x'_t)\, dt$ with respect to x is*

$$\frac{\partial f}{\partial x}(t, x_t^*, \ldots, x_t^{*(m)}) + \sum_{m=1}^{k} (-1)^m \frac{d^m}{dt^m} \frac{\partial f}{\partial x^{(m)}}(t, x_t^*, \ldots, x_t^{*(m)}) = 0.$$

Proof. The first derivative of the integral $I(\theta) = \int_{t_0}^{t_1} f(t, x_{t,\theta}, \ldots, x_{t,\theta}^{(k)})\,dt$ with respect to θ is

$$\begin{aligned} I'(\theta) &= \int_{t_0}^{t_1} \sum_{m=0}^{k} \Big\{ \eta_t^{(m)} \frac{\partial}{\partial x^{(m)}} f(t, x_{t,\theta}, \ldots, x_{t,\theta}^{(k)}) \Big\}\,dt + o(1) \\ &= \int_{t_0}^{t_1} \eta_t \Big\{ \frac{\partial}{\partial x} + \sum_{m=1}^{k} (-1)^m \frac{d^m}{dt^m} \frac{\partial}{\partial x^{(m)}} \Big\} f(t, x_{t,\theta}, \ldots, x_{t,\theta}^{(k)})\,dt + o(1), \end{aligned}$$

using m integrations by parts for the derivative of f with respect to $x^{(m)}$, with functions η having derivatives zero at t_0 and t_1. Then the equality $I'(0) = 0$ for every function η yields the condition. □

Proposition 3.4. *Necessary and sufficient conditions for $I'' \geq 0$ at x^* are*

$$\begin{aligned} &\frac{\partial^2}{\partial x'^2} f(t, x_{t,\theta}^*, x_{t,\theta}^{*\prime}) \geq 0, \\ &\frac{\partial^2}{\partial x^{(m)2}} f(t, x_t^*, \ldots, x_t^{*(k)}) \\ &\qquad + \sum_{j=1}^{m-1} (-1)^{m-j} \frac{d^{m-j}}{dt^{m-j}} \frac{\partial^2}{\partial x^{(j)} \partial x^{(m)}} f(t, x_t^*, \ldots, x_t^{*(k)}) \geq 0, \end{aligned}$$

for every $m = 1, \ldots, k$.

Proof. The second derivative of $I(\theta)$ is

$$\begin{aligned} I''(\theta) = \int_{t_0}^{t_1} \sum_{m=0}^{k} \Big\{ &\eta^{(m)2} \frac{\partial^2}{\partial x^{(m)2}} f(t, x_{t,\theta}, \ldots, x_{t,\theta}^{(k)}) \\ &+ 2 \sum_{j=1}^{m-1} \eta^{(m)} \eta^{(j)} \frac{\partial^2}{\partial x^{(j)} \partial x^{(m)}} f(t, x_{t,\theta}, \ldots, x_{t,\theta}^{(k)}) \Big\}\,dt + o(1). \end{aligned}$$

Integrating by parts, for every function η having derivatives with values zero at t_0 and t_1 and for all integer m and $j < m$

$$\begin{aligned} &\int_{t_0}^{t_1} \eta_t^{(m)} \eta_t^{(j)} \frac{\partial^2}{\partial x^{(j)} \partial x^{(m)}} f(t, x_{t,\theta}, \ldots, x_{t,\theta}^{(k)})\,dt \\ &\quad = -\int_{t_0}^{t_1} \eta_t^{(m-1)} \eta_t^{(j)} \frac{d}{dt} \frac{\partial^2}{\partial x^{(j)} \partial x^{(m)}} f(t, x_{t,\theta}, x'_{t,\theta})\,dt \\ &\quad = \int_{t_0}^{t_1} (-1)^{m-j} \eta_t^{(j)2} \frac{d^{m-j}}{dt^{m-j}} \frac{\partial^2}{\partial x^{(j)} \partial x^{(m)}} f(t, x_{t,\theta}, \ldots, x_{t,\theta}^{(k)})\,dt, \end{aligned}$$

and the inequality $I'(0) \geq 0$ for every function η_t implies the conditions. □

3.4 Integral calculus

Let f be a function of $C_\infty(\mathbb{R})$ and let r be a real such that the expansion $f(x+r)$ is a series according to r^k, $k \geq 0$, converges. Cauchy's integration formula

$$f(x) = \frac{1}{2\pi} \int_0^{2\pi} f(x + re^{iy})\,dy \tag{3.6}$$

relies and expansion of $f(x+re^{iy})$ at x, its primitive reduces to the first term with $\int_0^{2\pi} e^{iy}\,dy = 0$.

Liouville's formula is a generalization of the inversion formula for Laplace or Fourier's transforms with respect to an arbitrary function φ.

Theorem 3.3. *Let f be a continuous function in $\mathbb{R}$ and let φ be a real function having a primitive $\phi(x) = \int_0^x \varphi(s)\,ds$, such that $A = \int_0^\infty s^{-1}\phi(s)\,ds$ is finite and non null. For every x*

$$f(x) = A^{-1} \lim_{z\to\infty} \int_{-\infty}^{\infty} z^{-1} \int_0^z \varphi(\zeta y - \zeta x) f(y)\,d\zeta\,dy.$$

Proof. Using the change of variable $y = x + z^{-1}s$ in the integral over $(0,z)$, the right-hand member of the equality is

$$\begin{aligned} u &= A^{-1} \int_{-\infty}^{\infty} (zy - zx)^{-1} \phi(zy - zx) f(y)\,dy, \\ &= A^{-1} \int_{-\infty}^{\infty} s^{-1}\phi(s) f(x + z^{-1}s)\,ds. \end{aligned}$$

As z tends to infinity, $f(x + z^{-1}s)$ tends to $f(x)$ therefore $u = f(x)$. □

In the following applications, the calculus of integrals of parametric functions of C_1 is performed using the properties of the derivative of the integrals.

Example 6. Let

$$X = \int_0^\infty e^{-x^2} \cos(2ax)\,dx, \quad Y = \int_0^\infty e^{-x^2} \sin(2ax)\,dx$$

with a real a. We have $\sqrt{\pi} = \int_{-\infty}^\infty e^{-x^2}\,dx = 2\int_0^\infty e^{-x^2}\,dx$, so $|X|$ and $|Y|$ are bounded by $\frac{\sqrt{\pi}}{2}$

$$\begin{aligned} X &= \frac{1}{2} \int_0^\infty e^{-x^2} (e^{2aix} + e^{-2aix})\,dx \\ &= \frac{e^{-a^2}}{2} \int_{-\infty}^\infty e^{-(x+ai)^2}\,dx = e^{-a^2} \frac{\sqrt{\pi}}{2}. \end{aligned}$$

Deriving the expression of X with respect to the parameter a, we get

$$\int_0^\infty xe^{-x^2}\sin(2ax)\,dx = ae^{-a^2}\frac{\sqrt{\pi}}{2},$$

$$\int_0^\infty x^2e^{-x^2}\cos(2ax)\,dx = (\frac{1}{2}-a^2)e^{-a^2}\frac{\sqrt{\pi}}{2},$$

$$\int_0^\infty x^3e^{-x^2}\sin(2ax)\,dx = a(\frac{3}{2}-a^2)e^{-a^2}\frac{\sqrt{\pi}}{2},$$

$$\int_0^\infty x^4e^{-x^2}\cos(2ax)\,dx = \frac{1}{2}(\frac{3}{2}-6a^2+a^4)e^{-a^2}\frac{\sqrt{\pi}}{2},$$

$$\int_0^\infty x^5e^{-x^2}\sin(2ax)\,dx = \frac{a}{2}(\frac{15}{2}-8a^2+a^4)e^{-a^2}\frac{\sqrt{\pi}}{2},$$

and so on. Expanding the sine function

$$\begin{aligned} Y &= \sum_{k=0}^\infty (-1)^k \int_0^\infty \frac{(2ax)^{2k+1}}{(2k+1)!}e^{-x^2}\,dx \\ &= a\sum_{k=0}^\infty (-1)^k \frac{(2a)^{2k}}{(2k+1)!}\int_0^\infty y^k e^{-y}\,dy \\ &= \sum_{k=0}^\infty (-1)^k \frac{(2a)^{2k}(k+1)!}{(2k+1)!}. \end{aligned}$$

Example 7. The integral

$$I_A(a) = \int_0^A \frac{\cos(ax)}{1+x^2}\,dx \tag{3.7}$$

has the second derivative

$$I_A''(a) = \int_0^A \cos(ax)\Big(\frac{1}{1+x^2}-1\Big)\,dx = I_A(a) - \frac{\sin(aA)}{a},$$

with initial value as A tends to infinity

$$I(0) = \int_0^\infty \frac{dx}{1+x^2} = \frac{\pi}{2},$$

moreover $I(a) < I(0)$ hence it is bounded for every a. Since the equation $I_A''(a) = I_A(a)$ has the exponential solutions $c_1e^{-a} + c_2e^a$, we shall consider a solution of (3.7) in the form $I(a) = \{I(0) + u(a)\}e^a$ with a function u such that $u'' - 2u' = -a^{-1}\sin(aA)e^a$. Let $u'(a) = v(a)e^{2a}$, where v has the derivative $v'(a) = a^{-1}\sin(aA)e^{-a}$ and

$$\begin{aligned} v(a) &= -\int_0^a x^{-1}\sin(xA)e^{-x}\,dx \\ &= A^{-1}\Big\{\cos(aA) - 1 - \int_0^a x^{-2}(x+1)\cos(xA)e^{-x}\,dx\Big\}. \end{aligned}$$

The limit of u as A tends to infinity is zero and $I_A(a)$ converges to

$$I(a) = \frac{\pi}{2} e^{-a}.$$

Deriving $I(a)$, it follows that

$$\begin{aligned} I(a) = -I'(a) &= \int_0^\infty \frac{x \sin(ax)}{1+x^2}\,dx, \\ = I''(a) &= \int_0^\infty \frac{x^2 \cos(ax)}{1+x^2}\,dx, \\ = (-1)^k I^{(k)}(a) &= \int_0^\infty \frac{x^k \cos(ax)}{1+x^2}\,dx. \end{aligned} \tag{3.8}$$

The function $I(a)$ is the derivative of

$$J(a) = \int_0^\infty \frac{\sin(ax)}{x(1+x^2)}\,dx = -I(a).$$

The function Gamma. The function Gamma $\Gamma_p = (p-1)!$ has the generating function

$$G(z) = 1 + ze^z$$

and $\Gamma_{p+1} = p\Gamma_p$, for every integer $p \geq 1$. Its definition extends to a real $z > 1$ as

$$\Gamma_z = \int_0^\infty x^{z-1} e^{-x}\,dx. \tag{3.9}$$

By the change of variable $x = p^{-1}y^p$

$$\Gamma_z = p^{1-z} \int_0^\infty y^{pz-1} e^{-\frac{y^p}{p}}\,dy$$

is defined for all $z \geq p^{-1}$ and p integer larger than one, it is therefore defined for every real $z > 0$. With $p = 2$ and the normal density $f_{\mathcal{N}}$

$$\Gamma_z = 2^{1-z}\sqrt{2\pi} \int_0^\infty y^{2z-1} f_{\mathcal{N}}(y)\,dy.$$

In the same way, by the change of variable $x = p^{-1}y^p$

$$\Gamma_z = p^{1-z} \int_0^\infty y^{pz-1} e^{-\frac{y^p}{p}}\,dy \tag{3.10}$$

is defined for all $z \geq p^{-1}$ and p integer larger than one, it is therefore defined for every real $z > 0$.

For $z > 1$, its first derivatives are

$$\Gamma'_z = \int_0^\infty \log(x) x^{z-1} e^{-x}\, dx,$$

$$\Gamma_z^{(k)} = \int_0^\infty \log^k(x) x^{z-1} e^{-x}\, dx,\ k > 0,$$

the function Γ is monotonically increasing on $[1, \infty]$ and it diverges at infinity. From Hölder's inequality $(\int fg)^2 \leq \int f . \int fg^2$ for all functions f and g of L^2, the function Gamma is log-convex on $\mathbb{R}_+$. The values of Γ'_z as $z > 0$ are $\Gamma'_z = E_{\mathcal{E}}(X^{z-1} \log X)$ where X is an exponential variable with parameter 1 and $\Gamma'_1 = E_{\mathcal{E}} \log X$.

Integrating Γ by parts on $[1, \infty[$, we obtain the same recurrence formula as the integer function

$$\Gamma_z = (z-1)\Gamma_{z-1}.$$

At zero, the function Gamma tends to $+\infty$ as z tends to 0_+. With a variable $z < 0$, the calculus of the integral (3.9) is infinite like Γ_0 and the integration by parts which provided the recurrence formula is not definite. The definition of the function Gamma does extend to negative values. The extension of the recurrence formula to $\mathbb{R}_-$ is however commonly used

$$\Gamma_{-z} = -\frac{\Gamma_{1-z}}{z} = \frac{\Gamma_{2-z}}{z(1-z)} = (-1)^k \frac{\Gamma_{k-z}}{z(1-z)\cdots(k-1-z)}$$

for every z in $]0, k[$, it yields

$$\Gamma_{-\frac{1}{z}} = -z\Gamma_{\frac{z-1}{z}}.$$

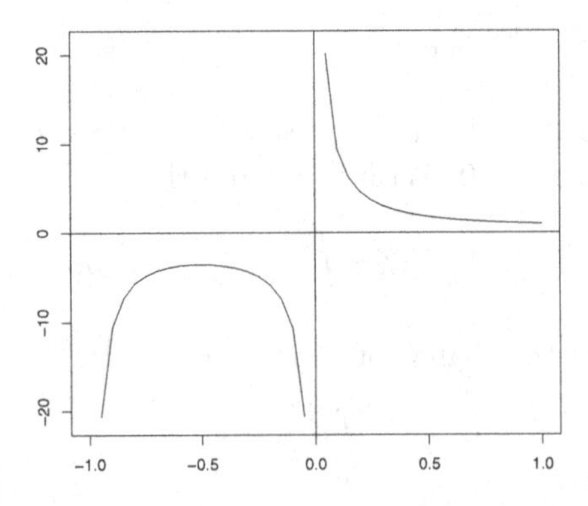

Fig. 3.1 Graph of the function Gamma on $]-1, 1]$.

The negative part of Gamma defined by this recurrence is singular at every negative integer k where it has vertical asymptotes, its right limits are $(-1)^k$ infinity and its left-limits are $(-1)^{k+1}$ infinity. There exists infinitely many curves of the same form, located alternatively above and respectively under the x-axis.

From the recurrence formula of Γ and the approximation of Γ_z by $\int_0^1 x^{z-1}e^{-x}\,dx + \lim_{A\to\infty}\Gamma(z;1,A)$ where

$$\begin{aligned}\Gamma(z;1,A) &= \int_1^A x^{z-1}e^{-x}\,dx\\ &= -e^{-1} - A^{z-1}e^{-A} + (z-1)\Gamma(z-1;1,A),\\ \lim_{A\to\infty}\Gamma(z;1,A) &= -e^{-1} + (z-1)\lim_{A\to\infty}\Gamma(z-1;1,A),\end{aligned}$$

we obtain

$$\int_0^1 x^{z-1}e^{-x}\,dx - \frac{1}{e} = (z-1)\int_0^1 x^{z-2}e^{-x}\,dx.$$

By a change of scale in the integrand, the function Gamma becomes

$$\int_0^\infty x^{z-1}e^{-ax}\,dx = a^{-z}\Gamma_z$$

and for every integer $k \geq 1$

$$\int_0^\infty (x+y)^{k-1}e^{-x}\,dx = \sum_{n=1}^{k}\binom{k-1}{n-1}y^{k-n}\Gamma_n = \Gamma_k\sum_{j=0}^{k-1}\frac{y^j}{j!}.$$

3.5 Partial derivatives of elliptic functions

The spherical coordinates (r,θ) in $\mathbb{R}_+^* \times \mathbb{R}$ are

$$x = r\cos\theta, \quad y = r\sin\theta, \tag{3.11}$$

where $r = (x^2+y^2)^{\frac{1}{2}}$ is the modulus of (x,y) and

$$\theta = \arccos\frac{x}{r} = \arcsin\frac{y}{r} = \arctan\frac{y}{x}.$$

According to the curves, r may depend on x and/or y, it may also depend on a variable t, with x_t and y_t. The partial derivatives of r are

$$\begin{aligned}r'_x &= \frac{\partial r}{\partial x} = \frac{1}{r}(x + y'_x y)\\ r'_y &= \frac{\partial r}{\partial y} = \frac{1}{r}(x'_y x + y)\end{aligned} \tag{3.12}$$

and by differentiation of (3.11), for all non null x and y

$$\theta'_x = \frac{x^2 + xy'_x y}{r^2 y} - \frac{1}{y} = \frac{xr'_x}{yr} - \frac{1}{y},$$
$$\theta'_y = \frac{1}{x} - \frac{y^2 + xx'_y y}{r^2 x} = \frac{1}{x} - \frac{yr'_y}{rx}. \tag{3.13}$$

Deriving the equality $\theta = \arctan x^{-1}y$ with the derivative $\tan'\theta = x^{-2}r^2$, we get equivalent expressions

$$\theta'_x = \frac{xy'_x - y}{r^2},$$
$$\theta'_y = \frac{x - yx'_y}{r^2}. \tag{3.14}$$

The second derivatives are calculated from (3.13) or (3.14)

$$\theta''_{xx} = \frac{ry''_x - 2(x + y'_x y)}{r^3}, \quad \theta''_{yy} = \frac{2(x'_y x + y) - rx''_{yy}}{r^3}.$$

In the equation of a circle, considering x and y as independent variables depending on r and θ, $rr'_x = x$ and $rr'_y = y$, and $r^3 r''_{xx} =$ and $rr'_y = y$ therefore $xr'_x + yr'_y = r$ and

$$r'^2_x + rr''_{xx} = 1,$$
$$r'^2_y + rr''_{yy} = 1.$$

As Fourier (1822) proved, the expressions of the first two derivatives of r entails the following properties.

Lemma 3.1. *The first two derivatives of the radius of the circle C centered at zero, with respect to the coordinates satisfy*

$$r'^2_x + r'^2_y = 1,$$
$$r''_{xx} + r''_{yy} = \frac{1}{r}.$$

The partial derivatives of a function $u(x, y)$ on C, depending only on r are

$$u'_x = u'_r r'_x = \frac{x}{r} u'_r, \quad u'_y = u'_r r'_y = \frac{y}{r} u'_r,$$
$$u''_{xx} = u''_{rr} r'^2_x + u'_r r''_{xx} = u''_{rr} \frac{x^2}{r^2} + u'_r \frac{y^2}{r^3}, \tag{3.15}$$
$$u''_{yy} = u''_{rr} r'^2_y + u'_r r''_{yy} = u''_{rr} \frac{y^2}{r^2} + u'_r \frac{x^2}{r^3}.$$

With Lemma (3.1), this entails properties for u similar to those of the partial derivatives of r.

Lemma 3.2. *The partial derivatives of a function $u(x, y)$ on C, depending only on r satisfy*

$$u'^2_x + u'^2_y = u'^2_r,$$

$$u''_{xx} + u''_{yy} = u''_{rr} + \frac{u'_r}{r}.$$

In the polar coordinates (3.11), the first derivatives of θ with respect to x and y depend on the derivatives of r with respect to x and y. On a circle and from (3.13), they are written as

$$\theta'_x = -\frac{1}{y}\Big(1 - \frac{x^2}{r^2}\Big) = -\frac{y}{r^2},$$

$$\theta'_y = \frac{1}{x}\Big(1 - \frac{y^2}{r^2}\Big) = \frac{x}{r^2}, \tag{3.16}$$

therefore

$$y\theta'_x + x\theta'_y = \frac{x^2 - y^2}{r^2},$$

$$x\theta'_y - y\theta'_x = 1,$$

$$x\theta'_x + y\theta'_y = 0, \tag{3.17}$$

$$\theta'^2_x + \theta'^2_y = \frac{1}{r^2}.$$

The partial derivatives of a function $u(x, y)$ depending on r and θ are

$$u'_x = u'_r r'_x + u'_\theta \theta'_x,$$

$$u'_y = u'_r r'_y + u'_\theta \theta'_y.$$

Lemma 3.3. *The partial derivatives of a function $u(x, y)$ on C satisfy*

$$xu'_y - yu'_x = u'_\theta,$$

$$yu'_x + xu'_y = 2\frac{xy}{r}u'_r + \frac{x^2 - y^2}{r^2}u'_\theta.$$

Proof. From (3.16) and the expression of r'_x and r'_y, $xu'_y - yu'_x = u'_\theta(x\theta'_y - y\theta'_x)$, and the first result is due to (3.17), the second one uses the same arguments. □

The second derivatives of θ C are calculated as

$$\theta''_{xx} = 2\frac{xy}{r^4}, \quad \theta''_{yy} = -2\frac{xy}{r^4}, \quad \theta''_{xy} = \frac{y^2 - x^2}{r^4}.$$

Lemma 3.4. *The second order partial derivatives of θ satisfy the following differential equations*

$$0 = \theta''_{xx} + \theta''_{yy},$$

$$0 = \theta''_{xy} - \theta'^2_x + \theta'^2_y.$$

Proposition 3.5. *The second order partial derivatives of $u(x,y)$ satisfy the equality*

$$u''_{xx} + u''_{yy} = u''_{rr} + \frac{u'_r}{r} + u''_{\theta\theta}.$$

Proof. The second derivatives of u are

$$u''_{xx} = u''_{rr}r'^2_x + u'_r r''_{xx} + 2r'_x\theta'_x u''_{r\theta} + u''_{\theta\theta}\theta'^2_x + u'_\theta\theta''_{xx},$$
$$u''_{yy} = u''_{rr}r'^2_y + u'_r r''_{yy} + 2r'_y\theta'_y u''_{r\theta} + u''_{\theta\theta}\theta'^2_y + u'_\theta\theta''_{yy}$$

and the properties of the derivatives of θ in (3.17) and in Lemmas (3.1) and (3.4) yield the result. □

In the equation of an ellipse $b^2x^2 + a^2y^2 = c$ with fixed parameters a, b and c, the squared l^2-norm of (x,y) is

$$r^2 = x^2(1 - a^{-2}b^2) + a^{-2}c = y^2(1 - a^2b^{-2}) + b^{-2}c,$$

denoted $r^2 = k_1x^2 + c_1 = k_2y^2 + c_2$, with non null constants k_1 and k_2. Its partial derivatives differ from those of the circle by the constants

$$r'_x = k_1\frac{x}{r}, \quad r'_y = k_2\frac{y}{r},$$
$$r''_{xx} = \frac{k_1}{r}\left(1 - \frac{xr'_x}{r}\right) = \frac{k_1}{r}\left(1 - k_1\frac{x^2}{r^2}\right),$$
$$r''_{yy} = \frac{k_2}{r}\left(1 - \frac{yr'_y}{r}\right) = \frac{k_2}{r}\left(1 - k_2\frac{y^2}{r^2}\right).$$

With a radius function of x and y, (3.13) must be used and from the partial derivatives of r^2

$$y'_x = \frac{r}{y}r'_x - \frac{x}{y} = \frac{r^2}{x}\theta'_x + \frac{y}{x},$$
$$x'_y = \frac{r}{x}r'_y - \frac{y}{x} = \frac{x}{y} - \frac{r^2}{y}\theta'_x, \tag{3.18}$$

the partial derivatives of r depend on those of θ since x and y depend on r and θ

$$r'_x = \frac{r}{x}(y\theta'_x + 1),$$
$$r'_y = \frac{r}{y}(1 - x\theta'_y).$$

The second derivatives are deduced by elementary calculus.

It is not always possible to write the coordinates of an elliptic curve defined by trigonometric functions as a function $f(x,y) = 0$ that do not depend on the angle. The polar equations of the spiral curves are defined by a varying radius r_θ and by trigonometric coordinates $x = r_\theta \cos\theta$ and $y = r_\theta \sin\theta$. For instance, the polar equation with $r_\theta = k\theta^{-\alpha}$ and positive constants k and α defines spirals, in that case $r^{-1}r'_\theta = -\alpha\theta^{-1}$. For a regular spiral, r_θ is proportional to θ. The exponential spiral of Figure (3.2) has the same equation with $r_\theta = \exp(a^{-1}\theta)$ hence r'_θ is proportional to r

$$\begin{aligned} ax'_\theta &= x + ay, \\ ay'_\theta &= y - ax, \\ ax'_\theta - a^2 y'_\theta &= x(1 + a^2). \end{aligned}$$

Its cartesian derivatives satisfy

$$axx'_y + ay = r^2, \quad ayy'_x + ax = r^2,$$

they are the partial derivatives of the circle

$$x^2 + y^2 = \frac{2}{a} r^2.$$

It turns to the right if the integer part of a is even and to the left if it is odd.

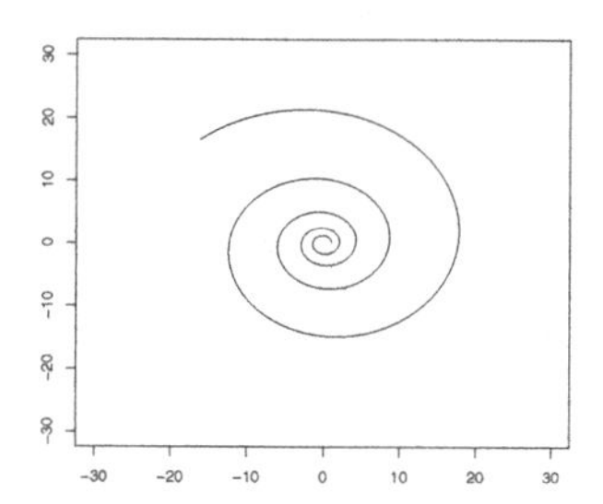

Fig. 3.2 Graph of a spiral.

The equation of the curve such as spiral curves determines the derivative r'_θ and the expressions of x'_θ and y'_θ. More generally, the radius is parametrized by the angle and by the cartesian coordinates, the derivatives r'_x, r'_y and r'_θ are then dependent.

For an ellipse, the derivatives of x and y with respect to the angle satisfy $b^2 xx'_\theta + a^2 yy'_\theta = rr'_\theta$ with

$$\begin{aligned} x'_\theta &= \frac{a^2 rr'_\theta}{x(a^2 - b^2)}, \\ y'_\theta &= \frac{b^2 rr'_\theta}{y(b^2 - a^2)}. \end{aligned}$$

The relationships between r'_θ and r'_x or r'_y do not depend on the parameters $r'_\theta = r'_x x'_\theta = r'_y y'_\theta$, they imply

$$r'_\theta = \frac{(a^2 - b^2)xx'_\theta}{ra^2} = \frac{(b^2 - a^2)yy'_\theta}{rb^2}.$$

In $\mathbb{R}^3$, the Cartesian coordinates of a point (x, y, z) are related to the polar coordinates (r, θ, φ) as

$$x = r\cos\varphi\cos\theta = \rho\cos\theta, \quad y = r\cos\varphi\sin\theta = \rho\sin\theta, z = r\sin\varphi, \quad (3.19)$$

with the l^2-norms $r = (x^2 + y^2 + z^2)^{\frac{1}{2}}$ and $\rho = (x^2 + y^2)^{\frac{1}{2}} = r\cos\varphi$, and the angles

$$\begin{aligned}
\varphi &= \arcsin\frac{z}{r} = \arctan\frac{z}{\rho},\\
\theta &= \arcsin\frac{y}{\rho} = \arccos\frac{x}{\rho} = \arctan\frac{y}{x}.
\end{aligned}$$

The partial derivatives of r and ρ are

$$\begin{aligned}
r'_x &= \frac{1}{r}(x + y'_x y + z'_x z),\\
r'_y &= \frac{1}{r}(x'_y x + y + z'_y z),\\
r'_z &= \frac{1}{r}(x'_z x + y'_z y + z),\\
\rho'_x &= \frac{1}{r}(x + y'_x y),\\
\rho'_y &= \frac{1}{r}(x'_y x + y).
\end{aligned}$$

By differentiation of (3.19), the partial derivative of x and y are

$$\begin{aligned}
xr'_x - ry\theta'_x - rz\varphi'_x\cos\theta &= r,\\
yr'_y + xr\theta'_y - rz\varphi'_y\sin\theta &= r,\\
xr'_y - ry\theta'_y - rz\varphi'_y\cos\theta &= rx'_y,\\
xr'_z - ry\theta'_z - rz\varphi'_z\cos\theta &= rx'_z,\\
yr'_x + xr\theta'_x - rz\varphi'_x\sin\theta &= ry'_x,\\
yr'_z + xr\theta'_z - rz\varphi'_z\sin\theta &= ry'_z,
\end{aligned}$$

and the partial derivative of z are

$$\begin{aligned}
zr'_z + r^2\varphi'_z\cos\varphi &= r,\\
zr'_x + r^2\varphi'_x\cos\varphi &= rz'_x,\\
zr'_y + r^2\varphi'_y\cos\varphi &= rz'_y,
\end{aligned}$$

the partial derivatives of r, θ and φ are deduced from these equations. Moreover

$$\begin{aligned} \rho'_x &= \frac{\rho(\rho'_x + \varphi'_x)}{r} + \frac{zr'_x}{r^2}, \\ \rho'_y &= \frac{\rho(\rho'_y + \varphi'_y)}{r} + \frac{zr'_y}{r^2}, \\ \rho'_z &= \frac{\rho\rho'_z}{r} + \frac{z}{r}. \end{aligned} \tag{3.20}$$

The partial derivatives of θ satisfy equations like (3.13) where r is replaced by ρ

$$\begin{aligned} \theta'_x &= \frac{xy'_x - y}{\rho^2}, \\ \theta'_y &= \frac{x - yx'_y}{\rho^2}, \\ \theta'_z &= \frac{xy'_z - yx'_z}{\rho^2}. \end{aligned} \tag{3.21}$$

In the same way, $z + \rho\rho'_z = rr'_z$ and

$$\begin{aligned} \varphi'_x &= \frac{\rho z'_x - z\rho'_x}{r^2}, \\ \varphi'_y &= \frac{\rho z'_y - z\rho'_y}{r^2}, \\ \varphi'_z &= \frac{\rho - z\rho'_z}{r^2}. \end{aligned} \tag{3.22}$$

Combining these derivatives with the expressions of the derivatives of r with respect to the coordinates in terms of the partial derivatives of the coordinates y'_x, y'_z, etc. provides other relations between the partial derivatives of θ and φ with respect to the coordinates. If the angles are independent, the first partial derivatives of θ and φ with respect to x and y satisfy the relations

$$\begin{aligned} x'_\theta &= r'_\theta \cos\theta \cos\varphi - r \sin\theta \cos\varphi = x\frac{r'_\theta}{r_\theta} - y, \\ y'_\theta &= r'_\theta \sin\theta \cos\varphi + r \cos\theta \cos\varphi = y\frac{r'_\theta}{r_\theta} + x, \\ x'_\varphi &= -r \sin\varphi \cos\theta = -z \cos\theta, \\ y'_\varphi &= -r \sin\varphi \sin\theta = -z \sin\theta, \\ z'_\varphi &= r \cos\varphi. \end{aligned}$$

The equation of a sphere $x^2+y^2+z^2=r^2$, has the partial derivatives $rr'_x=x$, $rr'_y=y$ and $rr'_z=z$ therefore

$$\begin{aligned} r'^2_x + rr''_{xx} &= 1, \\ r'^2_y + rr''_{yy} &= 1, \\ r'^2_z + rr''_{zz} &= 1. \end{aligned}$$

Fourier's lemma for the elliptic functions in $\mathbb{R}^2$ is generalized in $\mathbb{R}^3$ with a modified constant.

Lemma 3.5. *The partial derivatives of the radius of a sphere satisfy*

$$\begin{aligned} r'^2_x + r'^2_y + r'^2_z &= \frac{x^2+y^2+z^2}{r^2} = 1, \\ r''_{xx} + r''_{yy} + r''_{zz} &= \frac{2}{r}. \end{aligned}$$

The partial derivatives of a function $u(x,y,z)$ depending only on the radius of a sphere $\mathcal{S}$ such that $x^2+y^2+z^2=r^2$ are given by (3.15) and

$$\begin{aligned} u'_r r'_z &= \frac{z}{r} u'_r, \\ u''_{zz} &= u''_{rr} r'^2_z + u'_r r''_{zz}. \end{aligned}$$

From Lemmas (3.1) and (3.5), the partial derivatives of u have the following properties.

Lemma 3.6. *The partial derivatives of a function $u(x,y,z)$ on the sphere $\mathcal{S}$, depending only on r satisfy*

$$\begin{aligned} u'^2_x + u'^2_y + u'^2_z &= u'^2_r, \\ u''_{xx} + u''_{yy} + u''_{zz} &= u''_{rr} + 2\frac{u'_r}{r}. \end{aligned}$$

On the sphere $\mathcal{S}$ the coordinates are independent, θ depends only on x and y, and Lemma 3.4 is unchanged. From (3.21)-(3.22), the partial derivatives of θ and φ reduce to

$$\begin{aligned} \theta'_x &= -\frac{y}{\rho^2}, \quad \theta'_y = \frac{x}{\rho^2}, \\ \varphi'_x &= -\frac{z\rho'_x}{r^2} = -\frac{xz}{\rho r^2}, \\ \varphi'_y &= -\frac{z\rho'_y}{r^2} = -\frac{yz}{\rho r^2}, \quad \varphi'_z = \frac{\rho}{r^2}. \end{aligned}$$

A function u depending on the radius and the angles of a sphere has the first order partial derivatives

$$\begin{aligned}
u'_x &= u'_r r'_x + u'_\theta \theta'_x + u'_\varphi \varphi'_x = \frac{x}{r}u'_r - \frac{y}{\rho^2}u'_\theta - \frac{xz}{\rho r^2}u'_\varphi,\\
u'_y &= u'_r r'_y + u'_\theta \theta'_y + u'_\varphi \varphi'_y = \frac{y}{r}u'_r + \frac{x}{\rho^2}u'_\theta - \frac{yz}{\rho r^2}u'_\varphi,\\
u'_z &= u'_r r'_z + u'_\varphi \varphi'_z = \frac{z}{r}u'_r + \frac{\rho}{r^2}u'_\varphi.
\end{aligned}$$

Its second order partial derivatives are

$$\begin{aligned}
u''_{xx} &= u''_{rr} r'^2_x + u'_r r''_{xx} + 2r'_x\theta'_x u''_{r\theta} + u''_{\theta\theta}\theta'^2_x + u'_\theta\theta''_{xx}\\
&\quad + 2r'_x\varphi'_x u''_{r\varphi} + u''_{\varphi\varphi}\varphi'^2_x + u'_\varphi\varphi''_{xx},\\
u''_{yy} &= u''_{rr} r'^2_y + u'_r r''_{yy} + 2r'_y\theta'_y u''_{r\theta} + u''_{\theta\theta}\theta'^2_y + u'_\theta\theta''_{yy}\\
&\quad + 2r'_y\varphi'_y u''_{r\varphi} + u''_{\varphi\varphi}\varphi'^2_y + u'_\varphi\varphi''_{yy},\\
u''_{zz} &= u''_{rr} r'^2_z + u'_r r''_{zz} + 2r'_z\varphi'_z u''_{r\varphi} + u''_{\varphi\varphi}\varphi'^2_z + u'_\varphi\varphi''_{zz},
\end{aligned}$$

they depend on the partial derivatives already studied in $\mathbb{R}^2$ and on the partial derivatives with respect to z, in particular

$$\begin{aligned}
\varphi''_{xx} &= \frac{(x^2 - y^2)z}{\rho r^4} + \frac{x^2 z}{\rho^3 r^2},\\
\varphi''_{yy} &= \frac{(y^2 - x^2)z}{\rho r^4} + \frac{y^2 z}{\rho^3 r^2},\\
\varphi''_{zz} &= -2\frac{z\rho}{r^3}.
\end{aligned}$$

Using the same arguments as in the circle, on $\mathcal{S}$

$$\begin{aligned}
r'_x\varphi'_x + r'_y\varphi'_y + r'_z\varphi'_z &= 0,\\
\varphi'^2_x + \varphi'^2_y + \varphi'^2_z &= \frac{1}{r^2},\\
\varphi''_{xx} + \varphi''_{yy} + \varphi''_{zz} &= \frac{z}{\rho r^4}(r^2 + z^2).
\end{aligned}$$

Proposition 3.6. *The partial derivatives of $u(x, y, z)$ on $\mathcal{S}$ satisfy $yu'_x - xu'_y = u'_\theta$ and*

$$(u'_x - u'_r r'_x)^2 + (u'_y - u'_r r'_y)^2 + (u'_z - u'_r r'_z)^2 = \frac{u'^2_\theta}{\rho^2} + u'^2_\varphi,$$

$$\Delta u = u''_{rr} + \frac{2u'_r}{r} + \frac{u''_{\theta\theta} + u''_{\varphi\varphi}}{r^2} + u'_\varphi\frac{z(r^2 + z^2)}{\rho r^4}.$$

The equation of an ellipsoïd is $A^2x^2 + B^2y^2 + C^2z^2 = c$ and the sum $r^2 = x^2 + y^2 + z^2$ is also denoted

$$r^2 = x^2\Big(1 - \frac{A^2}{B^2}\Big) + z^2\Big(1 - \frac{C^2}{B^2}\Big) + \frac{c}{B^2}$$

or $r^2 = k_1x^2 + k_4^2 + c_1 = k_2y^2 + k_5z^2 + c_2 = k_3x^2 + k_6y^2 + c_3$ with independent variables x and z and, respectively, y and z, or x and y. The partial derivatives of a function u defined on the ellipsoïd satisfy

$$u_x'^2 + u_y'^2 + u_z'^2 = u_r'^2(r_x'^2 + r_y'^2 + r_z'^2)$$
$$u''_{xx} + u''_{yy} + u''_{zz} = u''_{rr}(r_x'^2 + r_y'^2 + r_z'^2) + u'_r(r''_{xx} + r''_{yy} + r''_{zz})$$

with

$$r_x'^2 + r_y'^2 + r_z'^2 = \frac{k_1^2x^2 + k_2^2y^2 + k_3^2z^2}{r^2},$$
$$r''_{xx} + r''_{yy} + r''_{zz} = \frac{k_1}{r}\Big(1 - \frac{k_1x^2}{r^2}\Big) + \frac{k_2}{r}\Big(1 - \frac{k_2y^2}{r^2}\Big) + \frac{k_3}{r}\Big(1 - \frac{k_3z^2}{r^2}\Big).$$

Considering the coordinates of a point of $\mathbb{R}^3$ as functions of a common variable t, let $x_t = \rho_t\cos\theta_t$, $y_t = \rho_t\sin\theta_t$ and let $z_t = r_t\sin\varphi_t$, with $\rho_t^2 = x_t^2 + y_t^2$ and $r_t^2 = \rho_t^2 + z_t^2$, $\theta_t = \arctan(x_t^{-1}y_t)$ and $\varphi_t = \arctan(\rho_t^{-1}z_t)$.

Proposition 3.7. *The coordinates x_t, y_t and z_t satisfy the differential equations*

$$\begin{aligned} x'_ty_t - x_ty'_t &= -\theta'_t\rho_t^2, \\ x_t'^2 + y_t'^2 &= \rho_t'^2 + \rho_t^2\theta_t'^2, \\ x_t'^2 + y_t'^2 + z_t'^2 &= r_t'^2 + r_t^2\varphi_t'^2 + \rho_t^2\theta_t'^2. \end{aligned} \tag{3.23}$$

Moreover

$$x'_tz_t - x_tz'_t = -\theta'_tz_ty_t - \varphi'_t(z_t^2\cos\theta_t + \rho_tx_t),$$
$$y'_tz_t - y_tz'_t = -\theta'_tz_tx_t - \varphi'_t(z_t^2\sin\theta_t - \rho_ty_t).$$

3.6 Applications

Waves oscillating between two curves $f(t)$ and $-f(t)$ symmetric with respect to the x-axis have the coordinates

$$x_t = f(t)\sin(\omega t), \quad y_t = f(t)\cos(\omega t).$$

Their graph is oriented along a line with slope a by adding the equation of the straight line with slope a to x_t and y_t , they become

$$x_t = f(t)\sin(\omega t) + t, \quad y_t = f(t)\cos(\omega t) + at.$$

The wave equation of Figure (3.3) is defined by the polar coordinates

$$x = \theta - \frac{a}{r}\sin\left(\frac{r}{a}\theta\right), \quad y = a - \frac{a}{r}\cos\left(\frac{a}{r}\theta\right).$$

Let $\alpha_r = a^{-1}r$, the coordinate vector $X = (x, y)$ satisfies the equation

$$(x-\theta)^2 + (y-a)^2 = \alpha_r^2 \sin^2(\alpha_r\theta) + \alpha_r^{-2}\cos^2(\alpha_r^{-1}\theta)$$

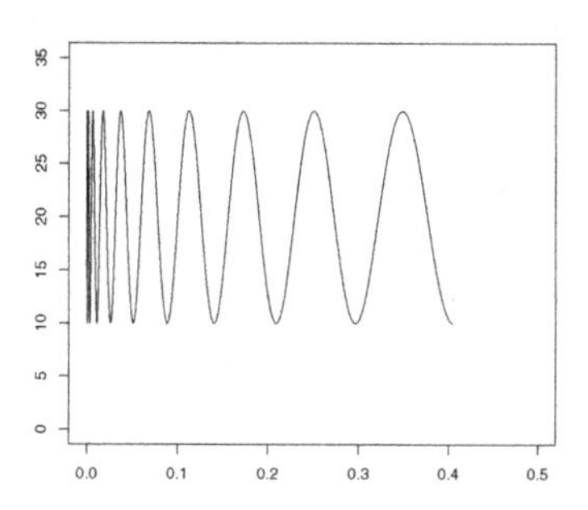

Fig. 3.3 Wave motion.

this is the equation of a cycloïd having increasing periods constant size and moving along the abscisse with the angle. The second derivatives of $x = \theta - \alpha_r^{-1}\sin(\alpha_r\theta)$ and $y = a - \alpha_r^{-1}\cos(\alpha_r^{-1}\theta)$ with respect to θ satisfy the differential equations

$$x''_{\theta\theta} + \alpha_r^2(x-\theta) = 0,$$
$$y''_{\theta\theta} + \alpha_r^{-2}(y-a) = 0.$$

The function $u(x,y) = \alpha_r^{-2}x + \alpha_r^2 y$ is therefore solution of the differential equations

$$\Delta_{\theta,\theta}u(x,y) + x + y = \theta + a$$

and

$$\Delta_{\theta,r}u(x,y) + \frac{\theta}{r}\Big\{\frac{x-\theta}{\alpha_r} + \alpha_r(y-a)\Big\} = 0.$$

The second derivatives of x and y with respect to r are

$$x''_{rr} - \frac{1}{r^2}(2-\theta^2\alpha_r^2)(x-\theta) = -2\frac{\alpha_r^2}{r^2}\theta\cos(\alpha_r\theta),$$
$$y''_{rr} + \frac{\alpha_r}{r^2}\Big(1+\frac{\theta^2}{\alpha_r}\Big)(y-a) = -\frac{\theta}{r^2}\sin(\alpha_r^{-1}\theta)$$

and for the function u

$$r^2\Delta_{r,r}u(x,y) - (\theta^2 - \frac{2}{\alpha_r^2})(x-\theta) + \frac{1}{\alpha_r}\Big(1+\frac{\theta^2}{\alpha_r}\Big)(y-a)$$
$$= -2\theta\cos(\alpha_r\theta) - \frac{\theta}{\alpha_r^2}\sin(\alpha_r^{-1}\theta).$$

From (5.13), the Laplacien of u satisfies

$$x^2y^2\Delta_{x,y}u = r^2(\theta+a-x-y) - 2\theta xy\Big\{\frac{x-\theta}{\alpha_r} + \alpha_r(y-a)\Big\}$$
$$+\frac{1}{r^2}\Big\{(\theta^2 - \frac{2}{\alpha_r^2})(x-\theta) - \frac{1}{\alpha_r}\Big(1+\frac{\theta^2}{\alpha_r}\Big)(y-a)$$
$$- 2\theta\cos(\alpha_r\theta) - \frac{\theta}{\alpha_r^2}\sin(\alpha_r^{-1}\theta)\Big\}.$$

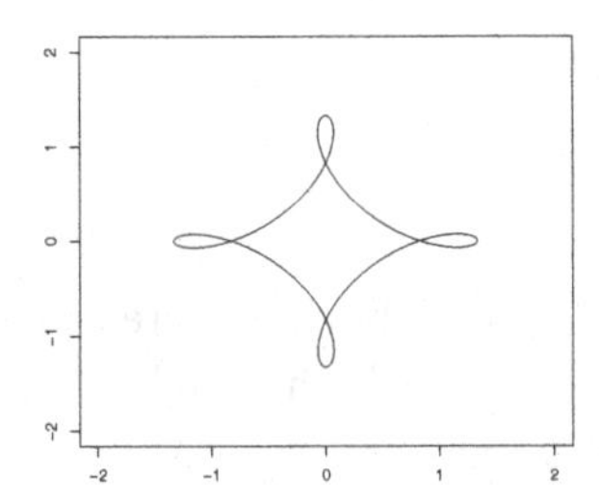

Fig. 3.4 Epicycloïd.

The equations of the epicycloïd of Figure (3.4) is defined by two embedded circles with angles varying with the radius

$$x = \frac{3}{4}r\cos\frac{\theta}{r} + \frac{r}{4}r\cos\frac{3\theta}{r}, \quad y = \frac{3}{4}r\sin\frac{\theta}{r} + \frac{r}{4}r\sin\frac{3\theta}{r}.$$

Its derivatives are also the equations of a regular spiral. The graph of a curved spring of Figure (3.6) has similar equations

$$x = \cos\theta + \sin(a\theta), \quad y = \cos\theta + \cos(a\theta)$$

with a scaling constant a for the period.

The cycloïds are defined by combining the equations of a curve and moving circles. The Rhodonea curve of Figure (3.8) has the coordinates $x_t = r\cos t$, $y_t = r\sin t$ with a periodic modulus $r = a\sin(kt)$, it has $2k$ folia if k is even and k folia if k is odd. This is also the curve described by

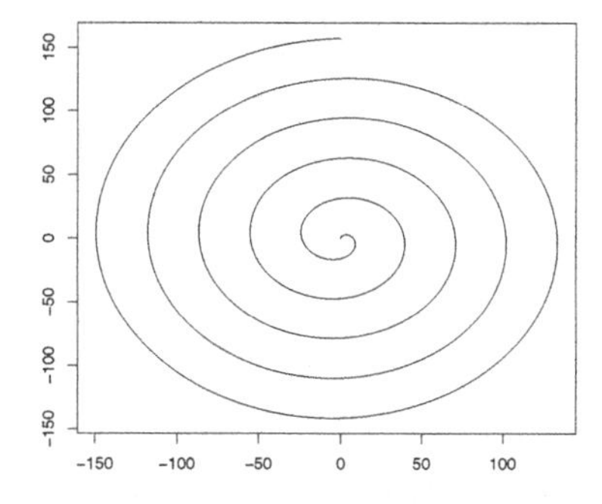

Fig. 3.5 Graph of a regular spiral.

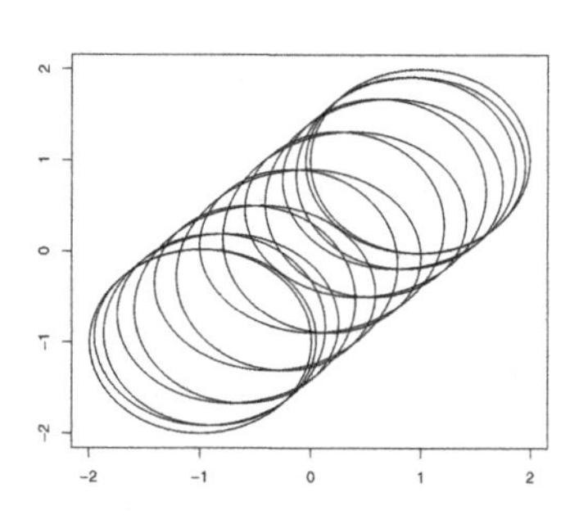

Fig. 3.6 Graph of a spring.

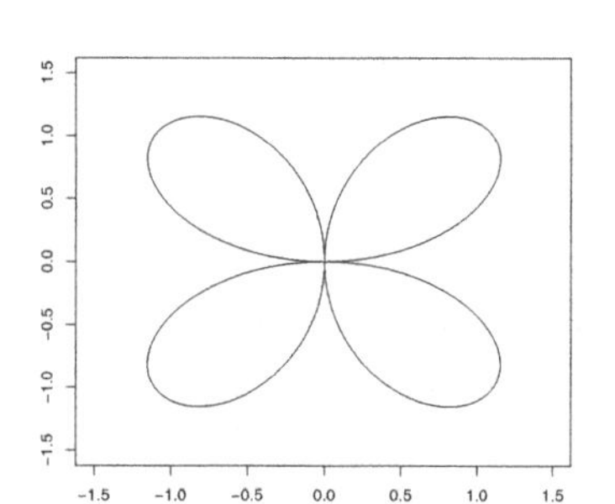

Fig. 3.7 Graph of a four-leaf-clover with $k = 3$.

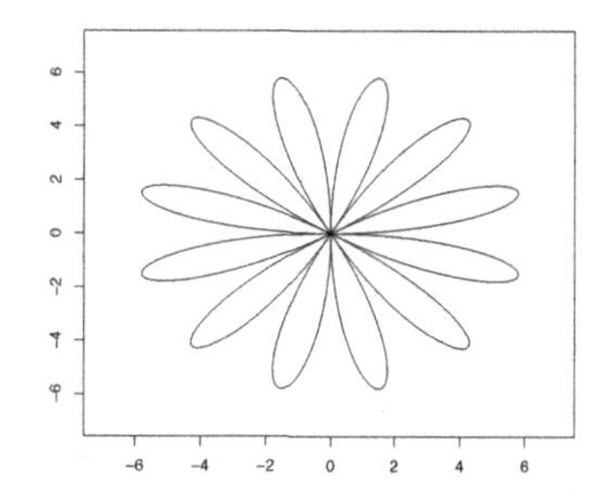

Fig. 3.8 Graph of a Rhodonea curve with $k = 6$.

a pendulum in a plane orthogonal to its axis, when its axis is in rotation, as k is large.

The cone of (3.9) is defined by

$$x_t = 5\sin(2t)\cos^4 t, \quad y_t = 5\cos(2t)\sin^4 t,$$

equivalently its coordinates satisfy the equation

$$\Big(\frac{x_t}{a_t}\Big)^2 + \Big(\frac{y_t}{b_t}\Big)^2 = k,$$

with trigonometric functions of the same angle $a_t = \cos^4 t$ and, respectively, $b_t = \sin^4 t$. This is an elliptic equation with normalizing functions a_t and b_t. The equations of Figure (3.8) defined also an elliptic function with a variable radius.

Another class of epicycloïd is defined by the equations

$$\begin{aligned} x_t &= k\cos t + (k+1)\cos(kt), \\ y_t &= -k\sin t + (k+1)\sin(kt). \end{aligned}$$

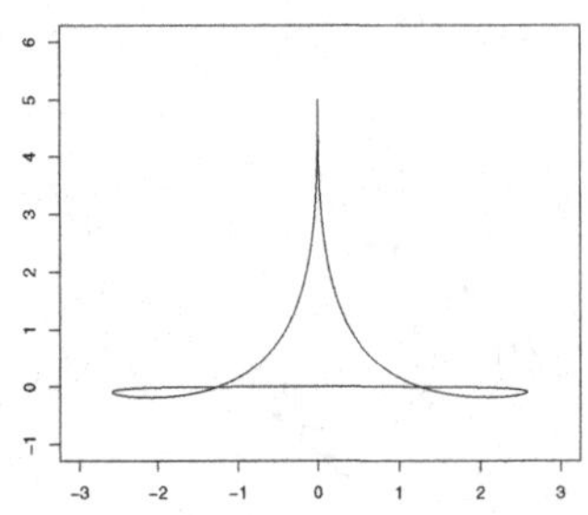

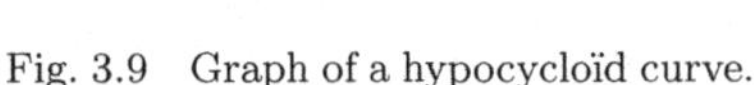

Fig. 3.9 Graph of a hypocycloïd curve.

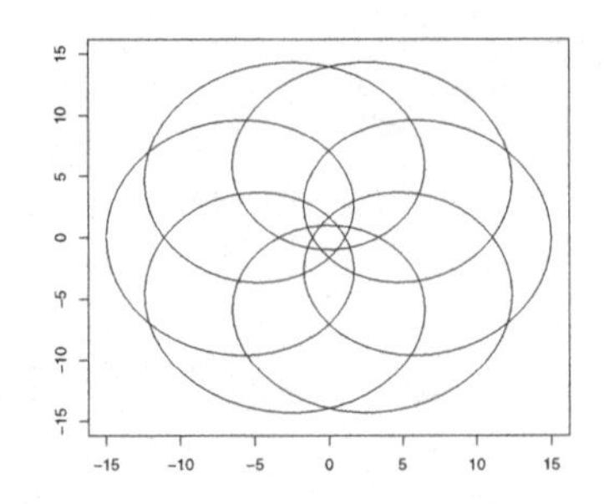

Fig. 3.10 Graph of a hypocycloïd.

This is the equation of Figure (3.10) with the parameter $k = 7$. They satisfy

$$\{x_t - (k+1)\cos(kt)\}^2 = \{y_t - (k+1)\sin(kt)\}^2 = k^2,$$

this is the equation of a circle with radius k with center $((k+1)\cos(kt), (k+1)\sin(kt))$ moving around an inner circle with radius $k+1$ and center at zero.

Poisson (1811, 1838) described the movement of a simple pendulum by its angle θ with the vertical axis, its length a and an inertial constant k

$$\theta''_x + \frac{g}{a}\sin\theta = 0.$$

Multiplying this equation by $2\theta'_x$ and integrating it yields

$$\theta'^2_x - 2\frac{g}{a}\cos\theta + C = 0.$$

The initial conditions with an angle α imply

$$\theta'_x = \left(2\frac{g}{a}\right)^{\frac{1}{2}}(\cos\theta - \cos\alpha)^{\frac{1}{2}},$$

The primitive of this equation determines the duration of a complete oscillation

$$T = \left(\frac{g}{a}\right)^{\frac{1}{2}} \int_0^{1-\cos\alpha} \{(1-\cos\alpha)x - x^2\}^{-\frac{1}{2}} \left(1 - \frac{x}{2}\right)^{-\frac{1}{2}} dx.$$

The angle θ of the gravity center of a composite pendulum satisfies Sturm's differential equation (1861)

$$\theta''_x = \Omega^2 + \frac{2ag}{a^2+k^2}(\cos\theta - \sin\theta)$$

where Ω is the initial angular velocity. It is similar to a simple pendulum with length $l = a^{-1}(a^2 + k^2)$.

If the pendulum does not move vertically, its trajectory draws an ellipse and the time for a revolution is the same as for a pendulum moving vertically. A pendulum drawing a circle on the surface of a sphere with a constant velocity v has the differential equations

$$xx''_t + yy''_t + \frac{d(x^2 + y^2)}{dt^2} = 0$$

in the plane containing the circle, and $v = k^{-1}gr^2$.

A system of solids $(m_i)_{i=1,\ldots,k}$ moving together satisfies the differential equations

$$\sum_{i=1}^{k} m_i x''_{i,t} = \sum_{i=1}^{k} F_{i,x}, \quad \sum_{i=1}^{k} m_i y''_{i,t} = \sum_{i=1}^{k} F_{i,y}, \quad \sum_{i=1}^{k} m_i z''_{i,t} = \sum_{i=1}^{k} F_{i,z},$$

where m_i is submitted to the strength $F_i = (F_{i,x}, F_{i,y}, F_{i,z})$.

The energy in a system of two free spheres is expressed as a mixture of the energies of its components, with squared mixture coefficients, when their exchanges are negligibale. The coefficients are functions of the distance between the spheres. With two components, it is written in the form $E(\alpha) = \alpha^2 e_1 + (1 - \alpha)^2 e_2$, where α belongs to $]0, 1[$, e_1 and e_2 are the energies of each sphere. The energy is minimum as

$$\alpha_{min} = \frac{e_2}{e_1 + e_2}$$

and the minimum of the energy is

$$E_{min} = \frac{e_1 e_2}{e_1 + e_2}$$

it depends only on the sum and the product of the energies. In a system with n components, the same equations lead to a global energy

$$E_n(\alpha) = \sum_{k=1}^{n} \alpha_k^2 e_k,$$

with coefficients α_k summing to one and with the discrete energies e_k, $k = 1, \ldots, n$. It is minimal as the mixture coefficients satisfy $\alpha_k e_k = \alpha_n e_n$ for every $k = 1, \ldots, n - 1$, which implies

$$\alpha_k = \frac{1}{e_k} \Big(\sum_{j=1}^{n} \frac{1}{e_j} \Big)^{-1}, \; k = 1, \ldots, n.$$

The minimum of the energy attained at these values is

$$E_{n,min} = \sum_{k=1}^{n} \frac{1}{e_k} \Big(\sum_{j=1}^{n} \frac{1}{e_j}\Big)^{-2} = \Big(\sum_{k=1}^{n} \frac{1}{e_k}\Big)^{-1},$$

it is the geometric mean of the energies.

The energy of interactions between particles increases the total energy of a system. With constant interactions, it becomes

$$E(\alpha) = \sum_{k=1}^{n} \alpha_k^2 e_k + \Big(\sum_{j\neq k, j=1}^{n} \alpha_k \alpha_j \Big) I,$$

under the constraint $\sum_{k=1}^{n} \alpha_k = 1$. The energy is minimum with the mixture proportions

$$\alpha_{k,min} = \frac{1}{e_k - I} \Big(\sum_{j=1}^{n} \frac{1}{e_j - I}\Big)^{-1},$$

and it can be reached only if all e_j are strictly larger than I. In the model of two interacting particles, the smallest energy is

$$E_{min} = \frac{e_1 e_2}{e_1 + e_2 - 2I} + \frac{I^3}{(e_1 + e_2 - 2I)^2},$$

it is always larger than in the system of free particles and the existence of a minimal energy requires that the interactions are bounded.

If the interactions depend on the distances between the particles, $E(\alpha) = \sum_{k=1}^{n} \alpha_k^2 e_k + (\sum_{j\neq k,j=1}^{n} \alpha_k \alpha_j) I_{jk}$ and the mixture proportions of its minimum satisfy the equations

$$\alpha_k(e_k - 2I_{nk}) = \alpha_n e_n - I_{nk} - \sum_{j=1}^{n-1} \alpha_j I_{jk},$$

where $I_{kk} = 0$, for $k = 1, \ldots, n$.

3.7 Exercises

3.7.1. Determine the extrema of an ellipse centered at (x_0, y_0) from its equation.

3.7.2. Minimum or maximum of y such that $y^2 + 2x^2 y + 6x - 3 = 0$.

3.7.3. Minimum or maximum of y such that $u(x, y) = y^3 + 2x^3 + axy = 0$.

3.7.4. Minimum or maximum of $u(x, y) = y^2 + x^2 - 2axy$.

3.7.5. Integration of $f(x) = (a + bx + cx^2)^{-\frac{1}{2}}$.

3.7.6. Integration of $f(x) = (\cos x)^{-1}$.

3.7.7. Integration of $f(x) = x \log(x)(1 - x^2)^{-\frac{1}{2}}$.

3.7.8. Calculate $u_n = \int_0^{\frac{\pi}{2}} \sin^n x \, dx$.

3.7.9. Calculate $y_{m,n} = \int_0^{\frac{\pi}{2}} \sin^m x \cos^n x \, dx$, with integers $n > m > 0$.

3.7.10. Calculate $u_{m,n} = \int_0^{\frac{\pi}{2}} \sin^m x \cos^{-n} x \, dx$, with integers m and $n > 0$.

3.7.11. Prove that $y_{m,n} = y_{n,m}$ and $u_{m,n} = u_{n,m}$.

3.7.12. Calculate $A = \int_0^\infty e^{-ax} \cos(bx) \, dx$ and $B = \int_0^\infty e^{-ax} \sin(bx) \, dx$.

3.7.13. Calculate $I_a(y) = I_a(y_0) + \int_{y_0}^{y} x^{-1} \cos(ax) \, dx$, with $a > 0$.

3.7.14. Calculate $\int_0^{2\pi} (1 - e^{i\theta})^{-1} d\theta$.

3.7.15. Calculate $I(a) = \int_0^\infty e^{-ax^2} \, dx$, $a > 0$.

3.7.16. Calculate $I_k = \int_0^1 e^{x-1} x^k (1 - x)^{k-1} \, dx$, for every integer $k \leq 1$.

3.7.17. Calculate $I_k = \int_1^\infty (\log x)^{k-1} x^{-2} \, dx$, for every integer $k \leq 1$.

3.7.18. Calculate $\int_0^\infty (u^{x-1} + u^{x+1}) e^{-u} \, du$.

3.7.19. Find the length of the axis of an ellipsoïd having the smallest spatial area, with a fixed volume.

3.7.20. Calculate $\int_0^1 x^m \log^{n-1} \, dx$, for integers $m > 0$ and $n > 0$.

3.7.21. Primitive of $x^{-n} e^{ax}$, for $n \geq 1$ and $a > 0$.

3.7.22. Find the limit of $\sum_{k\geq 0}(-1)^{2k+1}\frac{x^{2k+1}}{(2k+1)(2k+1)!}$ as x tends to infinity.

3.7.23. Prove $\int_0^{\frac{\pi}{2}} \sin^{2k} x\, dx = \int_0^{\frac{\pi}{2}} \cos^{2k} x\, dx$.

3.7.24. Prove

$$n(n-1)\int_0^{\infty} \sin^{n-2}(x)e^{-x}\, dx = \{n+(-1)^n\}\int_0^{\infty} \sin^{n}(x)e^{-x}\, dx$$

and generalize to a relationship between $\int_0^{\infty} \sin^{n-2k}(x)e^{-x}\, dx$ and $\int_0^{\infty} \sin^{n}(x)e^{-x}\, dx$.

3.7.25. Prove $(n-1)\int_0^{\frac{\pi}{2}} \sin^{n-2}(x)\cos^m x\, dx = \int_0^{\frac{\pi}{2}} \sin^{n}(x)\cos^m x\, dx$.

3.7.26. Calculate $I = \int_0^{\infty}(1+x^2)^{-2}\, dx$.

3.7.27. Calculate $\Gamma_a\Gamma_{1-a}$.

3.7.28. Find the function f of $L^2([0,1])$ such that f' belongs to $L^2([0,1])$ and the integral $\int_0^1(f^2-\lambda^2 f'^2)\, dx$ is minimal and find its minimum.

3.7.29. Find conditions that ensure $\int_0^{\infty}(f^2 - af'^2 + bf''^2)\, dx > 0$ for every f such that $f^{(k)}$ belongs to $L^2(\mathbb{R}_+)$ for $k = 0, 1, 2$.

Chapter 4

Linear differential equations

This chapter presents methods for the resolution of linear differential equations and special functions defined by second order linear differential equations. The solutions of the homogeneous equations with constant coefficients are defined by boundary conditions, otherwise the solutions are generally not unique. The other methods use the expansion in an orthonormal basis for inhomogeneous differential equations and reparametrizations. We present equivalences between linear and nonlinear second order equations and between classes of first order nonlinear equations such as Bernoulli's and Riccati's equations.

4.1 First order differential equations in $\mathbb{R}_+$

Let $(E, \|.\|)$ be a metric space of real functions and let f be a continuous function in an open convex set of $\mathbb{R} \times E$. A first order differential equation in U may be written as

$$u'_x + f(x, u_x) = 0.$$

Under integrability conditions of u'_x and $f(x, u_x)$, the function u_x is solution of the implicit equation $F(x, u_x) = C$ with the primitive function $F(x, u_x) = u_x + \int_{x_0}^{x} f(u_s, s)\, ds$ of the differential equation and an arbitrary constant C. The initial value u_0 of u_x at an arbitrary x_0 defines the constant and the derivative of F is $\frac{d}{dx}F(x, u_x) = u'_x + f(u_x, x)$. In the simplest case, u is a primitive of a derivative u'_x defined by an equation $f(x, u'_x) = 0$ and conditions for the existence of a primitive are required. Reversely, an implicit equation $F(x, u_x) = 0$ has the first order derivative

$$F'_x(x, u_x) + u'_x F'_u(x, u_x) = 0$$

with $F'_u(x, u_x)$ non identically null, this is a first order differential equation for u. Asuming the existence of two distinct solutions u_1 and u_2 such that $F(x, u_{1x}) = 0 = F(x, u_{2x})$ for every x of a real interval I implies $F'_x(x, u_{1x}) = F'_x(x, u_{2x})$ and $F'_u(x, u_{1x}) = F'_u(x, u_{2x})$ for every x of I, therefore u_{1x} and u_{2x} are identical upto a constant determined by boundary conditions. Consider a nonlinear first order differential equation

$$f(x, u_x, u'_x) = 0,$$

its primitive is a function $F(x, u_x, u'_x)$ such that the solution u_x of the differential equation is solution of the implicit equation $F(x, u_x, u'_x) = C$, with a constant C. By differentiation

$$\frac{d}{dx} F(x, u_x, u'_x) = F'_1(x, u_x, u'_x) + u'_x F'_2(x, u_x, u'_x) + u''_x F'_3(x, u_x, u'_x),$$

with $F'_k(x_1, x_2, x_3) = \frac{d}{dx_k} F(x)$ for $x = (x_1, x_2, x_3)$ and $k = 1, 2, 3$. In the first order differential equations $F'_3(x, u_x, u'_x) = 0$ and

$$u'_x = \frac{1}{F'_2(x, u_x, u'_x)} \left\{ \frac{d}{dx} F(x, u_x, u'_x) - F'_1(x, u_x, u'_x) \right\}$$

for every x such that $F'_2(x, u_x, u'_x)$ is different from zero. The differential equation becomes $u'_x + f(x, u_x, u'_x) = 0$ with

$$f(x, u_x, u'_x) = -\frac{1}{F'_2(x, u_x, u'_x)} \left\{ \frac{d}{dx} F(x, u_x, u'_x) - F'_1(x, u_x, u'_x) \right\}.$$

A first order differential equation has an explicit solution if it is written in a separable form

$$u'_x f_2(u_x) = g(x), \tag{4.1}$$

where the function g is continuous and has a primitive G in an interval $[x_0, x_1]$ and the function f_2 is continuous and has a primitive F_2 in the interval $[u(x_0), u(x_1)]$. Its solution u_x satisfies

$$F_2(u_x) = F_2(u_0) + G(x) - G(x_0).$$

Under the additional condition $f_2 \neq 0$ in $[u(x_0), u(x_1)]$, its primitive is monotone and has an inverse function F_2^{-1}, then

$$u_x = F_2^{-1}\{F_2(u_0) + G(x) - G(x_0)\}$$

is the solution of (4.1).

An autonomous first order differential equation for a funtion u has the form $f(u_x, u'_x) = 0$, with a function f depending only on u and its derivative. For example, the function $u_x = \sin x + c$ with an arbitrary constant c is solution of the autonomous first order differential equation $u'_x = (1-u_x^2)^{\frac{1}{2}}$. This is the unique non singular solution of this nonlinear equation. A first order differential equation $u'_x = f(u_x, x)$ with boundary value at u_0 at x_0, has a solution

$$u_x = u_0 + \int_0^x f(u_s, s)\, ds.$$

Example 8. A necessary condition for a real function u_x to be solution of a differential equation $a_x u'_x + b_x = 0$ is

$$a'_x = \frac{\partial b_x}{\partial u_x}$$

under the Euler-Lagrange conditions, then there exists a function F on $\mathbb{R}^2$ suh that $F(x, u_x)$ is constant and a'_x is the second order derivative of $F(x, u_x)$ with respect to x and u.

Let γ be a continuous function in a real interval (x_0, x_1), the first order linear differential equation $L_{\gamma,x} u_x = f_x$ in $\mathbb{R}_+$ defined by an operator $L_{\gamma,x}$ such that

$$L_{\gamma,x} u_x = \gamma_x u_x + \frac{du_x}{dx} \tag{4.2}$$

in an interval (x_0, x_1), with the initial condition $u(0) = u_0$, has a solution

$$u_x = \exp\Big(-\int_0^x \gamma_s\, ds\Big)\Big\{u_0 + \int_0^x f_t \exp\Big(\int_t^x \gamma_s\, ds\Big)\, dt\Big\},$$

for every x in (x_0, x_1). Let $G(x) = e^{-\int_{x_0}^x \gamma_s\, ds}$ be the solution of the differential equation $L_{\gamma,x} u_x = 0$.

Theorem 4.1. *The unique solution u_x of the first order linear differential equation $L_{\gamma,x} u_x = f_x$ is the sum of the unique solution $V_x = u_0 G(x)$ of the homogeneous equation $u'_x + \gamma_x u_x = 0$ with the initial condition $u(x_0) = u_0$ at an arbitrary x_0 and of a particular solution of the inhomogeneous equation*

$$u_x = u_0 G(x) + \int_{x_0}^x f_t G(x-t)\, dt.$$

The last term is a convolution of the solution of the homogeneous equation and of the second member of (4.2) under the initial condition. Theorem 4.1 proves the existence of solutions. The solution of a differential equation with operator $a_x L_{\gamma,x}$ is defined on each sub-interval where a_x is non null and continuous.

The solutions of a first order linear differential equation $L_{\gamma,x}u_x = f_x$ in $\mathbb{R}_+$ are uniquely defined by the initial conditions: assuming that u and v are solutions, their difference $\varphi = u - v$ satisfies

$$\frac{d\varphi_x}{dx} = -\gamma_x \varphi_x$$

and the unique solutions of this equation is $\varphi_x = \varphi_0 \exp\Big(-\int_{x_0}^x \gamma_s \, ds\Big)$ where $\varphi_0 = 0$ by the initial condition $u_0 = v_0$ at x_0.

Proposition 4.1. *Let γ be in $C^k(I)$, then the solution of the first order linear differential equation $L_{\gamma,x}u_x = 0$ belongs to $C^{k+1}(I)$. If moreover f belongs to $C^k(I)$, then the solution of the equation $L_{\gamma,x}u_x = f_x$ belongs to $C^{k+1}(I)$.*

Proposition 4.1 is deduced from Theorem 4.1.

Example 9. Let a_x and b_x be continuous functions in an interval I, with primitives $A_x = \int_{x_0}^x a_s \, ds$ and, respectively $B_x = \int_{x_0}^x b_s \, ds$. The differential equation

$$u'_x + b_x u_x = (a_x + b_x)e^{A_x}$$

with initial value $u(x_0) = u_0$ has the solutions $u_x = e^{A_x} + e^{-B_x})$ with $u_0 = 2$, $u_x = e^{A_x}$ with $u_0 = 1$, and the solution given by Theorem 4.1, the initial value determines the unique solution.

Example 10. Let a_x and b_x be continuous functions of $C_1(I)$, the differential equation

$$b_x u'_x + b'_x u_x = b_x a'_x + b'_x a_x$$

has the solutions $u_x = a_x + c_1 b_x^{-1}$ and $u_x = a_x + c_2$ with constants $c_1 > 0$ and c_2. Let m be an integer and let c be a strictly positive real constant, the differential equation

$$y'_x + c\frac{y_x}{x} = bx^{m-1}$$

has the solutions $y_x = ax^m + kx^{-c}$, with the constants $b = a(m + c)$ and k depends on the initial value y_0 at x_0.

A real function f on $\mathbb{R}$ is periodic with period T if $f(x + T) = f(x)$ for every real x. An inhomogeneous first order linear differential equation (4.2) with a periodic right-member f has a speudo-periodic solution i.e. its graph has oscillations along the solution of the homogeneous differential equation but the period of the oscillations is not always preserved. At the equilibrium of the differential equation, $y'_x = 0$ and $y = f$. The solution given by Theorem (4.1) satisfies the property

$$\begin{aligned} u(x) &= u_0 G(x) + \int_{x_0}^{x} f(t) G(x-t)\, dt \\ &= u_0 G(x) + \int_{x_0}^{x} f(t+T) G(x-t)\, dt \\ &= u_0 G(x) + \int_{x_0+T}^{x+T} f(s) G(x-s+T)\, ds. \end{aligned}$$

This property extends as a direct consequence of Theorem 4.1.

Proposition 4.2. *The solution of the first order linear differential equation $L_{\gamma,x} u_x = 0$ with a periodic function γ is periodic. The solution of the equation $L_{\gamma,x} u_x = f_x$ with periodic functions γ and f is periodic.*

A homogeneous first order differential equation has the form

$$M(x,y)\, dx + N(x,y)\, dy = 0 \tag{4.3}$$

with homogeneous functions M and N of x and y with the same degree, i.e. the ratio of $M(tx,ty)$ and $N(tx,ty)$ is free of t, as in the equation $(ax+by)\, dx + (\alpha x + \beta y)\, dy = 0$. The reparametrization $y = x u_x$ allows us to write (4.3) as

$$M(1,u)\, dx + N(1,u)(u\, dx + x\, du) = 0$$

for all non null x and y. By separation of the variables x and u, the differential equation becomes

$$\begin{aligned} 0 &= \{M(1,u) + uN(1,u)\}\, dx + xN(1,u)\, du, \\ \frac{dx}{x} &= -\frac{N(1,u)}{M(1,u) + uN(1,u)}\, du := \varphi(u)\, du. \end{aligned}$$

A solution u_x of this differential equation with initial condition $u_0 = u(x_0)$ is such that

$$\log \frac{x}{x_0} = \int_{u_0}^{u_x} \varphi(s)\, ds := \phi(u_x) - \phi(u_0). \tag{4.4}$$

If the function φ has a constant sign, its primitive ϕ is invertible and the expressions of u_x and $y_x = x u_x$ are obtained in a close form as

$$u_x = \phi^{-1}\Big\{\log \frac{x}{x_0} + \phi(u_0)\Big\}.$$

Otherwise, the primitive ϕ is invertible in sub-intervals where φ has a constant sign and u_x is defined in these sub-intervals.

A differential equation (4.3) with the functions $M(x,y) = a + mx + ny$ and $N(x,y) = b + px + qy$ is equivalent to an equation with homogeneous functions of x and y, $P(x',y') = mx' + ny'$ and $Q(x',y') = px' + qy'$, by the change of variables $x = x' + \alpha$ and $y' = y + \beta$ with constants α and β such that $a + m\alpha + n\beta = 0$ and $b + p\alpha + q\beta = 0$ (Poisson).

Proposition 4.3. *The first order differential equation*

$$M(x, y_x) + N(x, y_x)y'_x = f(x, y_x)$$

with homogeneous functions M, N and f of x and y has a solution $y_x = xu_x$ such that

$$x = x_0 \exp\{\phi(u_x, f) - \phi(u_0, f)\},$$

with $\phi(u_x, f)$ a primitive of the function

$$\phi(u, f) = \frac{N(1,u)}{f(1,u) - M(1,u) - uN(1,u)}.$$

Proof. The solution of the differential equation with homogeneous second member f is deduced in the same form as (4.4) for (4.3) and depending of $f(1,u)$ through the function $\phi(u_x, f)$ in the form $x = x_0 \exp\{\phi(u_x, f) - \phi(u_0, f)\}$

$$\frac{dx}{x} = -\frac{N(1,u)}{M(1,u) + uN(1,u) - f(1,u)}\, du := \varphi(u,f)\, du$$

$$x = x_0 \exp\Big\{\int_{u_0}^{u_x} \varphi(s, f(1,s))\, ds\Big\} := x_0 \exp\{\phi(u_x, f) - \phi(u_0, f)\}. \quad \square$$

Let y_x be a real function of $C_1(\mathbb{R})$ satisfying the equation of Proposition 4.3 with polynomials M, N and f of degree p in x and y, then y_x is a polynomial function of degree $m \leq p$. With $p = 1$, the equation has the form $a + bx + (c + dx)y'_x = k_1 + k_2 x$ for every x and all non constant terms of y'_x are necessarily equal to zero. The same argument applies to $p > 1$.

If M, N and f are non homogeneous continuous functions, the reparametrization $y_x = xu_x$ is generally not sufficient to separate the variables x and y in the differential equation and there is no function such as $\varphi(u)$ above. Necessary conditions for the existence of a solution $f(x,y) = C$ for (4.3), with a constant C, is

$$M = f'_x, \quad N = f'_y$$

or the existence of a continuous function $A(x,y)$ such that

$$AM = f'_x, \quad AN = f'_y.$$

Useful tricks to solve explicitly nonlinear and inhomogeneous first order differential equations are reparametrizations and the separation of the variables of the equation. The equation $xy'_x - (1+y)^2 = 0$ has the solution y such that $\log x^{-1}x_0 + c = (1+y)^{-1}$ with a constant $c = (1+y_0)^{-1}$ defined by the value y_0 at an initial value x_0. The equation

$$kxy' - by + a\alpha x^\alpha y = 0,$$

with initial value y_0 at x_0 and a real constants a, b and α, has the solution

$$y = y_0\left(\frac{x}{x_0}\right)^{\frac{b}{k}} e^{\frac{a}{k}(x_0^\alpha - x^\alpha)}.$$

The equation

$$y'_x + p(x)y = f(x)y^\alpha$$

is linearized as

$$(1-\alpha)^{-1}u'_x + p(x)u = f(x)$$

using the reparametrization by $u = y^{1-\alpha}$ and its solution is obtained from Theorem (4.1).

The solution of a first order nonlinear differential equation is not necessarily exponential.

Example 11. Let $x_t = e^{A_t}$ with derivatives $x'_t = a_t x_t$ and $x''_t = (a'_t + a_t^2)x_t$. The first derivative of $xy'_x = x_t(x'_t)^{-1}y'_t$ with respect to t is

$$\begin{aligned}\frac{d}{dt}(xy'_x) &= \left\{1 - \frac{x_t x''_t}{x'^2_t}\right\}y'_t + \frac{x_t}{x'_t}y''_t \\ &= \left\{1 - \frac{(a'_t + a_t^2)}{a_t^2}\right\}y'_t + \frac{1}{a_t}y''_t.\end{aligned}$$

The first order differential equation for y'_t

$$y''_t + \left\{a_t - \frac{(a'_t + a_t^2)}{a_t}\right\}y'_t = f_t$$

has solutions y_x such that $xy'_x = \int_{t_0}^t a_s f_s\, ds + k_1$, where the right-hand member is denoted $G(t) = H(x_t)$ with x such that $\log x = A_t$. It follows that

$$y_x = \int_{x_0}^x s^{-1}H(s)\, ds + k_1 \log x + k_2$$

with arbitrary constants k_1 and k_2.

4.2 Existence and unicity of solutions

More general first order differential equations on $\mathbb{R}$ have the form

$$\frac{dx_t}{dt} = f(t, x_t), \qquad (4.5)$$
$$x(t_0) = x_0$$

in an open real interval I, where x is $C^1(E)$, where $E \subset \mathbb{R}$. The homogeneous equation (4.2) is obtained with $f(t, x_t) = -\gamma_t x_t$ and we have $f(t, x_t) = -\gamma_t x_t + h_t$ for the inhomogeneous equation $L_{\gamma,t} x_t = h_t$.

Let f be a continuous function on $I \times E$, satisfying the local Lipschitz property

$$\sup_{t \in I} |f(t, x) - f(t, y)| \leq k|x - y| \qquad (4.6)$$

for every real x in E and every y in a neighborhood of x. The property (4.6) is satisfied is f is differentiable with respect to x.

Theorem 4.2. *Let I be a bounded interval and let f be locally a Lipschitz function, then there exists an unique solution of equation (4.5).*

Proof. Let $(I_i)_{i \leq n}$ be a partition of I with a path $\delta > 0$, an approximated solution x_ε of a solution x is defined linearly on $(I_i)_{i \leq n}$ with the slope $f(t_i, x_{t_i})$ in I_{i+1}, from (4.5) let

$$\begin{aligned} x_{\delta,t} &= x_0 + (t - t_0) f(t_0, x_0), \text{ on } I_1, \\ x_{\delta,t_1} &= x_0 + \delta f(t_0, x_0), \\ x_{\delta,t} &= x_{\delta,t_1} + (t - t_1) f(t_1, x_{\delta,t_1}), \text{ on } I_2, \end{aligned}$$

and so on, so the linear approximation $x_{\delta,t}$ of x_t satisfies

$$|x_{\delta,t} - x_{\delta,t_i}| \leq \delta |f(t_i, x_{\delta,t_i})|$$

for all t and t_i in I_{i+1}. By continuity of $f(t, x)$ with respect to t and by the local Lipschitz property, for every $\varepsilon > 0$ there exists a partition with a sufficiently small path $\delta > 0$ to have

$$\begin{aligned} \sup_{t \in I_{i+1}} |f(t_i, x_{\delta,t_i}) - f(t, x_{\delta,t})| &\leq \sup_{t \in I_{i+1}} |f(t_i, x_{\delta,t_i}) - f(t_i, x_{\delta,t})| \\ &\quad + \sup_{t \in I_{i+1}} |f(t_i, x_{\delta,t}) - f(t, x_{\delta,t})| \\ &\leq \delta k + \varepsilon \sup_{x \in E} |f(t, x)| \end{aligned}$$

which may be written as $\sup_{t\in I_{i+1}} |f(t_i, x_{\delta,t_i}) - f(t, x_{\delta,t})| \leq M\varepsilon$, equivalently

$$\sup_{t\in I_{i+1}} |x'_{\delta,t} - f(t, x_{\delta,t})| \leq M\varepsilon.$$

Letting ε tend to zero, it follows that the linear approximation converges to a solution. To prove the unicity of the solution, let u and v be solutions of the equation with initial values u_0 and respectively v_0, then for every t in I

$$|u'_t - v'_t| = |f(t, u_t) - f(t, v_t)| \leq k|u_t - v_t|,$$

the unique solution of the equation is

$$|u_t - v_t| \leq |u_0 - v_0|e^{k(t-t_0)},$$

and the bound is zero if $u_0 = v_0$. □

Theorem 4.3. *Let $I = \mathbb{R}_+$ and let f be a bounded function on $I \times E$ satisfy (4.6), then there exists an unique solution of equation (4.5).*

Proof. The proof of the existence of a solution of Theorem 4.2 extends to every real bounded interval bounded, with a partition $(I_i)_{i\leq n}$ on this bounded interval. As f is bounded, the proof holds on $I \times E$. The proof of the unicity of the solution is unchanged. □

Proposition 4.4. *Let f have k-th order derivatives with respect to (t, x) in $I \times E$, then the solution of the first order differential equation (4.5) has a $(k + 1)$-th order derivative with respect to t in I.*

Proof. Let x_t be solution of (4.5), such that $x_t - x_0 = \int_{t_0}^t f(s, x_s)\, ds$ then

$$x''_t = f'_t(t, x_t) + x'_t f'_x(t, x_t) = f'_t(t, x_t) + f_x(t, x_t) f'_x(t, x_t)$$

where f'_t and f'_x belong to $C^{k-1}(I)$ hence x''_t has $(k - 1)$ order derivatives on I. □

Theorem 4.2 implies the existence and the unicity of the solution of (4.2) on every open real interval

$$\frac{du_x}{dx} = A_x u_x$$

with initial conditions. The local Lipschitz property of the linear operator is then written

$$\sup_{x\in I} |A_x(u - v)| \leq \sup_{x\in I} |A_x| \, \|u - v\|$$

for all real valued functions u and v on I. The existence and the unicity extend to the solution of the inhomogeneous first order differential equation.

Second order differential equations in an open real interval I of $\mathbb{R}$ have the form

$$x''_t = f(t, x_t, x'_t), \qquad (4.7)$$
$$x_0 = x_{t_0}, \quad x'_0 = \frac{dx}{dt}(t_0),$$

for x in $C^2_b(E_1)$ and x' in $C_b(E_2)$, where E_1 and E_2 are real sets, and f is a continuous function on $I \times E_1 \times E_2$. Let f have the local Lipschitz property

$$\sup_{t\in I} |f(t, x, x') - f(t, y, y')| \leq k_1|x - y| + k_2|x' - y'|$$

for all x and y in E_1, x' and y' in E_2.

Theorem 4.4. *Under the conditions, there exists an unique solution of equation (4.7).*

Proof. An approximated solution x_δ of a solution x is defined from (4.5) on a partition $(I_i)_{i=1,\dots,n}$ of I with a path $\delta > 0$, as

$$x_{\delta,t} = x_0 + (t - t_0)x'_0 + \frac{1}{2}(t - t_0)^2 f(t_0, x_0, x'_0)$$

on I_1, which determines x_{t_1}, and recursively on I_{i+1}

$$x_{\delta,t} = x_{t_i} + (t - t_i)x'_{t_i} + \frac{1}{2}(t - t_i)^2 f(t_0, x_{t_i}, x'_{t_i}).$$

For every $\varepsilon > 0$, there exists a partition with a sufficiently small path $\delta > 0$ so that for all t and t_i in I_{i+1}, the quadratic approximation $x_{\delta,t}$ of x_t satisfies

$$|x_{\delta,t} - x_{\delta,t_i}| \leq \delta|x'_{t_i}|,$$
$$|x'_{\delta,t} - x'_{\delta,t_i}| \leq \delta \sup_{t\in I} |f(t, x_t, x'_t)|.$$

By continuity of $f(t, x, x')$ with respect to t and by the local Lipschitz property, for every $\varepsilon > 0$ there exists a partition with a sufficiently small path $\delta > 0$ to have

$$\begin{aligned}
\sup_{t\in I_{i+1}} & |f(t_i, x_{\delta,t_i}, x'_{\delta,t_i}) - f(t, x_{\delta,t}, x'_{\delta,t})| \\
&\leq \sup_{t\in I_{i+1}} |f(t_i, x_{\delta,t_i}, x'_{\delta,t_i}) - f(t_i, x_{\delta,t}, x'_{\delta,t})| \\
&\quad + \sup_{t\in I_{i+1}} |f(t_i, x_{\delta,t}, x'_{\delta,t}) - f(t, x_{\delta,t}, x'_{\delta,t})| \\
&\leq \delta k + \varepsilon \sup_{x\in E} |f(t, x)|,
\end{aligned}$$

equivalently there exists a constant M such that

$$\sup_{t\in I} |x''_{\delta,t} - f(t, x_{\delta,t}, x'_{\delta,t})| \leq M\varepsilon.$$

Letting ε tend to zero, it follows that the linear approximation converges to a solution.

To prove the unicity of the solution, let u and v be solutions of the equation with unitial values u_0 and respectively v_0, then for every t in I

$$|u''_t - v''_t| = |f(t, u_t, u'_t) - f(t, v_t, v'_t)| \leq k_1|u_t - v_t| + k_2|u'_t - v'_t|,$$

where

$$\sup_{t\in I} |u_t - v_t| \leq c \sup_{t\in I} |u'_t - v'_t|$$

by first order expansions of u_t and v_t. Then there exists a constant k such that

$$\sup_{t\in I} |u''_t - v''_t| \leq k \sup_{t\in I} |u'_t - v'_t|$$

and the unique solution of the equation satisfies

$$|u'_t - v'_t| \leq |u'_1 - v'_1| e^{k(t-t_0)},$$

where the bound is zero if $u'_1 = v'_1$. □

Proposition 4.5. *Let $f(t, x, y)$ have k-th order derivatives with respect to (t, x, y) in $I \times E$, then the solution of the second order differential equation (4.7) has a $(k+2)$-th order derivative in I.*

Proof. Let x_t be solution of (4.7), such that $x'_t - x'_0 = \int_{t_0}^t f(s, x_s, x's)\, ds$ then

$$x_t^{(3)} = f'_t(t, x_t, x'_t) + x'_t f'_x(t, x_t, x'_t) + f_x(t, x_t, x'_t) f'_{x'}(t, x_t, x'_t)$$

where f'_t, f'_x and $f'_{x'}$ belong to $C^{k-1}(I \times E)$ hence $x_t^{(3)}$ has a $(k-1)$ order derivative on I. □

4.3 Behavior of the solutions under small perturbations

A small linear perturbation of the initial condition or of the linear operator induces small linear perturbation of the solution of a first order differential equation, this generalizes to other perturbations of first order differential equations.

Proposition 4.6. *Let u_x and, respectively v_x, be solutions of the differential equation $L_{\gamma,x}u_x = f_x$ on an interval I where the function γ is bounded, under the initial conditions $u(x_0) = u_0$ and, respectively $v(x_0) = v_0$. Then*

$$\sup_{x\in I} |u_x - v_x| \le |u_0 - v_0| \exp(|x - x_0| \sup_{x\in I} |\gamma_x|)$$

if moreover the function γ is strictly positive, there exists a strictly positive constant $c < 1$ such that $\sup_{x\in I} |u_x - v_x| \le c|u_0 - v_0|$.

Proof. By Theorem 4.1, for the homogeneous equation $L_{\gamma,x}u_x = 0$, we have

$$\sup_{x\in I} |u_x - v_x| \le |u_0 - v_0| \sup_{x\in I} \exp\Big(-\int_{x_0}^{x} \gamma_s \, ds\Big),$$

where there exists a x_m in $]x_0, x[$ such that $\int_{x_0}^{x} \gamma_s \, ds = (x - x_0)\gamma_{x_m}$ and the exponential has the bound

$$\exp\Big(\sup_{x\in I} \int_{x_0}^{x} (-\gamma_s) \, ds\Big) \le \exp(|x - x_0| \sup_{x\in I} |\gamma_x|).$$

The solution of the inhomogeneous equation is the sum of the solution of the homogeneous equation and of an integral of the function f which does not depend on the initial conditions so the bound is the same as for the homogeneous equation. □

Proposition 4.7. *Let u_x be the solution of the differential equation $L_{\gamma,x}u_x = 0$ and let $v_x = \int_{x_0}^{x} f_t \exp(\int_t^x \gamma_s \, ds)$, with the initial condition $u(x_0) = u_0$. As η tends to zero, the approximated equation*

$$(\gamma_x + \eta)u_x + \frac{du_x}{dx} = f_x$$

has a solution approximated by

$$u_{\eta,x} = u_x\{1 - \eta(x - x_0)\} + \eta \int_{x_0}^{x} (s - x_0)v_s \, ds + o(\eta).$$

Proof. The homogeneous equation has the solution

$$\begin{aligned} u_{\eta,x} &= u_0 \exp\Big(-\int_{x_0}^{x} (\gamma_s + \eta) \, ds\Big) \\ &= u_0 \exp\{-\eta(x - x_0)\} \exp\Big(-\int_{x_0}^{x} \gamma_s \, ds\Big) \end{aligned}$$

and first order expansion of the exponential as η tends to zero gives

$$u_{\eta,x} = u_x\{1 - \eta(x - x_0) + o(\eta)\}.$$

A function $y_{\eta,x} = u_{\eta,x}v_{\eta,x}$ is solution of the inhomogeneous equation, with

$$\begin{aligned} v_{\eta,x} &= u_0 + \int_{x_0}^{x} \exp\Big(\int_{x_0}^{t} (\gamma_s + \eta)\, ds\Big) f_t\, dt \\ &= u_0 + \int_{x_0}^{x} \exp\{\eta(t - x_0)\} \exp\Big(\int_{x_0}^{t} \gamma_s\, ds\Big) f_t\, dt \\ &= u_0 + \int_{x_0}^{x} \{1 + \eta(t - x_0) + o(\eta)\} \exp\Big(\int_{x_0}^{t} \gamma_s\, ds\Big) f_t\, dt \\ &= v_t + \eta \int_{x_0}^{x} (t - x_0) \exp\Big(\int_{x_0}^{t} \gamma_s\, ds\Big) f_t\, dt + o(\eta). \end{aligned}$$

□

Let f be a continuous bounded function on $I \times \mathbb{R}$ having the local Lipschitz property (4.6). We consider now a functional perturbation of the differential equation (4.5) defined by the existence of a continuous bounded function g on $I \times \mathbb{R}$ having the local Lipschitz property (4.6) and such that

$$\lim_{\eta \to 0} \sup_{t \in I} \sup_{x \in E} |f_\eta(t, x) - f(t, x) - \eta g(t, x)| = 0. \tag{4.8}$$

Theorem 4.5. *Let x be solution of (4.5) and let f_η satisfy,(4.6) and (4.8), the solution of the approximated equation*

$$\frac{dx_{\eta,t}}{dt} = f_\eta(t, x_{\eta,t}) \tag{4.9}$$

on an open real interval I and with the initial condition $x_{\eta,t_0} = x_{\eta,0}$, has the approximation $x_{\eta,t} = x(t) + \eta \int_{t_0}^{t} g(s, x_{\eta,s})\, ds + o(\eta)$, as η tends to zero.

Proof. The existence and the unicity of a solution $x_{\eta,t}$ are proved like for Theorem 4.2, with the function f_η. To prove the approximation of $x_{\eta,t}$, we shall prove an approximation of the approximated solution linear by parts on a partition $(I_i)_{i \le n}$ of I with a small path $\delta > 0$. On $I_1 =]t_0, t_1]$, let

$$\begin{aligned} x_{\delta,\eta,t} &= x_{\eta,0} + (t - t_0) f_\eta(t_0, x_{\eta,0}) \\ &= x_{\eta,0} + (t - t_0) f(t_0, x_{\eta,0}) + \eta(t - t_0) g(t_0, x_{\eta,0}) \\ &= x_{\delta,t} + \eta(t - t_0) g(t_0, x_{\eta,0}), \end{aligned}$$

which defines $x_{\delta,\eta,t_1} = x_{\delta,t_1} + \eta\delta g(t_0, x_{\eta,0})$. On $I_2 =]t_1, t_2]$, let

$$\begin{aligned} x_{\delta,\eta,t} &= x_{\delta,\eta,t_1} + (t - t_1) f_\eta(t_1, x_{\delta,\eta,t_1}) \\ &= x_{\delta,t_1} + \eta\delta g(t_0, x_{\eta,0}) + (t - t_1) f(t_1, x_{\delta,\eta,t_1}) + \eta(t - t_1) g(t_1, x_{\delta,\eta,t_1}) \\ &= x_{\delta,t} + (t - t_1)\{f(t_1, x_{\delta,\eta,t_1}) - f(t_1, x_{\eta,t_1})\} \\ &\quad + \eta\{\delta g(t_0, x_{\eta,0}) + (t - t_1) g(t_1, x_{\eta,t_1})\} \\ &\quad + \eta(t - t_1)\{g(t_1, x_{\delta,\eta,t_1}) - g(t_1, x_{\eta,t_1})\}, \end{aligned}$$

and so on. The linear approximation $x_{\delta,\eta,t}$ of $x_{\eta,t}$ satisfies

$$|x_{\delta,\eta,t} - x_{\delta,\eta,t_i}| \leq \delta |f_\eta(t_i, x_{\delta,\eta,t_i})|$$

for all t and t_i in I_{i+1}. Using the same arguments as in the proof of Theorem 4.2, $x_{\delta,\eta,t}$ converges to a solution $x_{\eta,t}$ as ε tends to zero, it follows that $f(t_i, x_{\delta,\eta,t_i}) - f(t_i, x_{\eta,t_i})$ and $g(t_i, x_{\delta,\eta,t_i}) - g(t_i, x_{\eta,t_i})$ converge to zero as ε tends to zero, then

$$x_{\eta,t} = x_t + \eta\{\delta \sum_{k\leq n} 1_{\{t_k \leq t\}} g(t_k, x_{\eta,t_k}) + (t - t_{N(t)}) g(t_{N(t)}, x_{\eta,t_{N(t)}})\} + o(\eta)$$

where $N(t) = \sum_{k\leq n} 1_{\{t_k \leq t\}}$. For η tending to zero, this is equivalent to the approximation

$$x_{\eta,t} = x_t + \int_{t_0}^{t} g(s, x_{\eta,s})\, ds + o(\eta).$$

□

4.4 Second order linear differential equations in $\mathbb{R}_+$

The general form of the linear second order differential equation is defined by an operator

$$L_x = a(x)\frac{d^2}{dx^2} + b(x)\frac{d}{dx} + c(x)Id, \tag{4.10}$$

with continuous functions $a(x)$, $b(x)$ and $c(x)$, and by conditions for the initial values of the solution and its first derivative

$$\begin{aligned} Lu_x &= f_x,\ x \in I, \\ Du : u(x_0) &= u_0,\ u'(x_1) = u_1', \end{aligned} \tag{4.11}$$

with initial values at x_0 and x_1 in the domain I of $\mathbb{R}$ where the differential equation is defined. With absolutely continuous functions a_x, b_x, c_x in the interval (x_0, x_1), if a function u_x is solution of the differential equation (4.10) at $x = t$ in (x_0, x_1), then it is also solution of (4.10) in a neighborhood of t.

In the intervals where the function $a(x)$ is different from zero, the differential operator is written as

$$L_x = \frac{d^2}{dx^2} + p(x)\frac{d}{dx} + q(x)Id. \tag{4.12}$$

In the special case with $q_x = p'_x$, the differential operator (4.11) is the derivative of the operator $L_{p,x}$ of the first order differential equation defined with $\gamma_x = p_x$.

A condition for the existence of solutions to the equation (4.11) is the existence of a function $F(x, u_x)$ such that

$$\frac{\partial^2 F(x, u_x)}{\partial x^2} = f(x),$$

this condition is also written as

$$F''_{xx} + 2u'_x F''_{xu} + u''_{xx} F''_{uu} = a(x)u''_{xx} + b(x)u'_x + c(x)u_x$$

for every x of I. This is equivalent to

$$\begin{aligned} F''_{xx} &= c(x)u_x, \\ 2F''_{xu} &= b(x), \\ F''_{uu} &= a(x). \end{aligned}$$

Let $u_{1,x}$ be solution of the second order differential equation $L_x u_{1,x} = 0$ with (4.10), a function $u_x = u_{1,x}v_x$ is a solution of the same equation if the function v is solution of the first order differential equation $a_x u_{1,x} v''_x + (b_x u_{1,x} + 2a_x u'_{1,x})v'_x = 0$ hence

$$v'_x = u_{1,x}^{-2} \exp\Big\{-\int_{x_1}^{x} \frac{b_s}{a_s}\, ds\Big\}$$

if $[x_1, x]$ in included in $D = \{x : a_x \neq 0\}$ and u is deduced by integration of $v'x$.

Equation (4.11) has a unique solution on $I \cap \{x \geq x_1 \wedge x_0\}$: Let u and v be solutions such that a_0 and a_1 are not zero, the difference $\varphi = u' - v'$ satisfies

$$a_x \varphi'_x = -b_x \varphi_x$$

and the unique solutions of this equation is $\varphi_x = \varphi_1 \exp\Big(-\int_{x_1}^{x} b_s a_s^{-1}\, ds\Big)$ for every x such that a is not zero on $[x_1, x]$, the initial condition at x_1 implies $\varphi_x = 0$ on $[x_1, x]$ and the initial condition at x_0 implies $u = v$ on intervals $[x_1 \wedge x_0, x]$ where a is not zero. On subsets where a is zero, (4.11) is a first order differential equation and its solution is unique under the initial condition at x_0.

Example 12. The differential equation

$$4x^2(x-2)u''_x + 2x(x-2)u'_x + \Big(\frac{x}{2} - 3\Big)u_x = 0,$$

denoted $a_x u''_x + b_x u'_x + c_x u_x = 0$, has a solution $u_x = x^{\frac{1}{2}} + k$ with an arbitrary constant k. Looking for a solution $y_x = x^{\frac{1}{2}} v_x$, the function v is solution of the first order differential equation $xv''_x + v'_x = 0$, therefore $v'_x = x_0 v'_0 x^{-1}$ and $v_x = c_1 \log x + c_2$ with constants depending on the initial conditions. The solution is $y_x = y_0 x^{\frac{1}{2}} + y'_0 x^{\frac{1}{2}} \log(x x_0^{-1})$.

Let u_x be solution of two second order differential equations (4.12) with different coefficients p_{kx} and q_{kx}, for $k = 1, 2$. By difference of the equations, u_x is also solution of the first order differential equation $(p_{1x} - p_{2x})u'_x + (q_{1x} - q_{2x})u_x = 0$ and it is their unique solution determined by Theorem 4.1.

Proposition 4.8 (Sturm 1868). *On every bounded interval $[x_0, x]$ where p and q are bounded, if the homogeneous differential equation (4.11)-(4.12) has a solution $u(x) = y(x)z(x)$ where y and z are not constant, then*

$$y(x) = y_0 + \exp\Big\{-\frac{1}{2}\int_{x_0}^x p(s)\,ds\Big\}$$

and z is a bounded solution of the implicit equation

$$z(x) = z_0 + z_0'(x - x_0) - \int_{x_0}^x \int_{x_0}^t R(s)z(s)\,ds\,dt$$

with $R = y^{-1}(y'' + py' + qy)$.

Proof. Let $u(x) = y(x)z(x)$ be a solution, the differential equation is writte as

$$yz''_x + (2y' + py)z'_x + (y''_x + py' + qy)z = 0$$

and it is equivalent to the system of differential equations

$$2y' + py = 0,$$

$$yz''_x + (y''_x + py'_x + qy)z = 0.$$

The first equation has the unique solution $y(x) = y_0 + \exp\{-\frac{1}{2}\int_{x_0}^x p(s)\,ds\}$, let $R = y^{-1}(y''_x + py' + qy)$ then the second differential equation has the solution

$$z(x) = z_0 + z_0'(x - x_0) - \int_{x_0}^x \int_{x_0}^t R(s)z(s)\,ds\,dt.$$

Replacing iteratively the function $z(s)$ in the integral by the same expression leads to an infinite series which converges under the condition, as proved by Sturm (1868). ☐

Proposition 4.9. *In every bounded interval $[x_0, x]$ where u, p, p'_x and q are bounded, the differential equation (4.11)-(4.12), with a differentiable function p, has a solution*

$$u(x) = y(x)\Big[v_0 + v_0'\int_{x_0}^x e^{-2\int_{x_0}^t p_s y'_s y_s^{-1}\,ds}\,dt$$

$$+ \int_{x_0}^x e^{-2\int_{x_0}^t p_s y'_s y_s^{-1}\,ds}\Big\{\int_{x_0}^t y^{-1}(\xi)f(\xi)e^{2\int_{x_0}^\xi p_z y'_z y_z^{-1}\,dz}\,d\xi\Big\}\,dt\Big]$$

where y is solution of the homogeneous differential equation.

Proof. Let y be solution of the homogeneous differential equation and let $u(x) = y(x)v(x)$ be a solution of the inhomogeneous equation then v is solution of the differential equation

$$yv''_x + 2py'_x v'_x = f.$$

The equation $yh'_x + 2py'_x h = 0$ has the solution

$$h(x) = h_0 \exp(-2 \int_{x_0}^{x} p_s y'_s y_s^{-1} \, ds),$$

denoting $v' = hw$ where the function w is such that

$$y(x)w'(x) = f(x) \exp(2 \int_{x_0}^{x} p_s y'_s y_s^{-1} \, ds)$$

we obtain

$$w(x) = w_0 + \int_{x_0}^{x} y^{-1}(t) f(t) e^{2 \int_{x_0}^{t} p_s y'_s y_s^{-1} \, ds} \, dt,$$

$$v(x) = v_0 + h_0 \int_{x_0}^{x} e^{-2 \int_{x_0}^{t} p_s y'_s y_s^{-1} \, ds} w(t) \, dt,$$

with initial values $v'_0 = h_0 w_0$ and v_0. Then $u(x) = y(x)v(x)$ with initial values $u_0 = y_0 v_0$ and $u'_0 = y'_0 v_0 + y_0 v'_0$. □

By integration of a linear second order homogeneous differential equation, it is equivalent to an implicit equation.

Theorem 4.6. *On every interval (x_0, x_1) where the function u'_x is different from zero, the solutions of the homogeneous differential equation (4.12) with initial values $u_0 = u(x_0)$ and $u'_1 = u'(x_1)$ satisfy the implicit equation*

$$u_x = u_0 + u'_1 \int_{x_0}^{x} \exp\Big\{ \int_{y}^{x_1} \Big(p_s + q_s \frac{u_s}{u'_s} \Big) \, ds \Big\} \, dy.$$

This result does not proves the existence of a solution. In every interval where u'_x is different from zero, the equation $u''_x + p_x u'_x + q_x u_x = 0$ is equivalent to

$$\frac{u''_x}{u'_x} + p_x + q_x \frac{u_x}{u'_x} = 0$$

and Theorem 4.6 is obtained by two consecutive integrations of this equation. The initial condition $u'_1 = 0$ implies u_x is a constant.

Let $\psi(x) = \int_0^x \left(p_s + q_s \frac{u_s}{u'_s}\right) ds$, the derivative of $u_x - u_0$ is

$$\frac{u'_x}{u_x - u_0} = \frac{\exp\{\psi(x_1) - \psi(x)\}}{\int_{x_0}^x \exp\{\psi(x_1) - \psi(y)\}\, dy}$$
$$= \frac{1}{\int_{x_0}^x \exp\{\psi(x) - \psi(y)\}\, dy}.$$

It is lower than $(x - x_0)^{-1}$ if ψ is an increasing function. This entails an upper bound for u_x.

Corollary 4.1. *Let u, p and q be functions on $[x_0, x_1]$ such that $p_x + q_x u_x u'_x \geq 0$ on $[x_0, x_1]$, there exist a constant $k > 0$ such that on this interval*

$$u_x \leq u_0 + k(x - x_0).$$

Example 13. The differential equation

$$xu''_x - xu'_x - u_x = 0$$

for $x \geq x_0 > 0$ is defined by the functions $p_x = -1$ and $q_x = -x^{-1}$, it has the solution $u_x = xe^x + c$.

Proposition 4.10. *The solutions of the homogeneous differential equation (4.11)-(4.12) with a non null constant p and a continuous function q satisfy the implicit equation*

$$u_x = u_0 - \frac{1}{p} u'_0 e^{-p(x-x_0)} + \frac{1}{p} \int_{x_0}^x q_s u_s \{e^{-p(x-s)} - 1\}\, ds.$$

Proof. A solution u such that $u'_x = v_x e^{-px}$ has the derivative

$$u''_x = \frac{v'_x}{v} u'_x - pu'_x = -pu'_x - q_x u_x$$

therefore $v'_x e^{-px} + q_x u_x = 0$ and $v_x = v_0 - \int_{x_0}^x q_s u_s e^{ps}\, ds$, with $u'_0 = v_0$. the result is obtained by integration of the equation

$$u'_x = v_x e^{-px} = u'_0 e^{-px} - e^{-px} \int_{x_0}^x q_s u_s e^{ps}\, ds.$$

□

Corollary 4.2. *Let u and q be positive functions on $[x_0, x_1]$ and let $p \geq 0$, if the functions q and $q_s e^{ps}$ are integrable on $[x_0, x_1]$, the function u is bounded on this interval as*

$$u_x \leq u_0 + (u_1 - u_0) e^{-p(x_1 - x)}.$$

Proof. The derivative of the logarithm of u_x is

$$-\frac{u'_x}{u_x - u_0} = \frac{pe^{-px}\int_{x_0}^{x} q_s u_s e^{ps}\,ds}{e^{-px}\int_{x_0}^{x} q_s u_s e^{ps}\,ds - \int_{x_0}^{x} q_s u_s\,ds}$$

and it is lower than p. The result follows from a integration on $[x, x_1]$. $\square$

With constant functions $p(x)$ and $q(x)$, the equation becomes $u''_x + pu'_x + qu_x = 0$, with initial conditions. The exponential function is a solution since u''_x, u'_x and u_x are proportional. Let $u_x = u_0 e^{zx}$, with z a root of the characteristic equation

$$z^2 + pz + q = 0$$

and $u'_0 = zu_0$. According to the sign of the discriminant $\Delta = p^2 - 4q$, there exist two conjugate real or complex roots z_1 and z_2 of the characteristic equation or a single double root and u_x is a linear combination of the related exponential functions $u_x = u_0\{ce^{z_1 x} + (1-c)e^{z_2 x}\}$, with the constant $c = -(2u'_0 + u_0\sqrt{\Delta} + u_0 p)(u_0\sqrt{\Delta})^{-1}$, if z_1 and z_2 are unequal. If $p^2 = 4q$ the characteristic equation is $(2z+p)^2 = 0$, there is a single root $z_1 = -\frac{1}{2}p$ and, according to d'Alembert, the solutions are linear combinations of $e^{z_1 x}$ and $xe^{z_1 x}$. Combining the exponentials, the solutions are written as follows.

Proposition 4.11. *The homogeneous equation $u''_x + pu'_x + qu_x = 0$ with the initial values $u(x_0) = u_0$, $u(x_1) = u_1$ and $u'(x_1) = u'_1$, has the unique solutions*

$$\begin{aligned}
u_x &= u_0 e^{-\alpha(x-x_0)}[c_1 \sin\{\beta(x-x_0) + c_2 \cos\{\beta(x-x_0)\}], \quad \textit{if } \Delta < 0,\\
u_x &= u_0 e^{-\alpha(x-x_0)}[c_1 \sinh\{\beta(x-x_0)\} + c_2 \cosh\{\beta(x-x_0)\}], \quad \textit{if } \Delta > 0,\\
u_x &= u_0 e^{-\alpha(x-x_0)} + (u'_1 + \alpha u_1)(x-x_0)e^{\alpha(x-x_1)}, \quad \textit{if } \Delta = 0,
\end{aligned}$$

with $\Delta = p^2 - 4q$, $\beta = \frac{1}{2}\sqrt{|\Delta|}$, $\alpha = \frac{p}{2}$, and with constant c_1 and c_2 determined by u'_1.

The trigonometric function u_x may be extended by periodicity of $\mathbb{R}$.

Example 14. The functions $u_{1x} = e^{\alpha x}\cos(\theta x)$ and $u_{2x} = e^{\alpha x}\sin(\theta x)$ are solutions of the differential equation

$$u''_x - 2\alpha u'_x + (\alpha^2 + \theta^2)u_x = 0.$$

Their initial values are $u_{10} = 1$, $u'_{10} = \alpha$, $u_{20} = 0$ and $u'_{20} = -\theta$. Every linear combination $c_1 u_{1x} + c_2 u_{2x}$ with real or complex coefficients is also solution of this equation, such as $e^{(\alpha+i\theta)x}$, $e^{\alpha x}\cosh(\theta x)$ and $e^{\alpha x}\sinh(\theta x)$.

Example 15. The function $u_{1x} = xe^{-\alpha x}\cos(\theta x)$ and $u_{2x} = xe^{-\alpha x}\sin(\theta x)$ are solutions of the differential equation

$$u''_x - 2(\alpha + x)u'_x + 2\alpha x u_x + (\alpha^2 + \theta^2 + 2)u_x = 0.$$

If $\alpha = 0$, $u_{1x} = x\cos(\theta x)$ and $u_{2x} = x\sin(\theta x)$ are solutions of the differential equation $u''_x - 2xu'_x + (\theta^2 + 2)u_x = 0$.

Example 16. By the change of variable $y = x^{-1}u$, the differential equation

$$xy''_x + 2y'_x + axy = 0$$

with $a > 0$, becomes $u''_x + au = 0$ the solutions of which are linear combinations of sine and cosine functions, hence

$$y_x = x^{-1}\sum_{k=0}^{+\infty}\{b_k\sin(\sqrt{a}x + \omega_{1k}) + c_k\cos(\sqrt{a}x + \omega_{2k})\}$$

is solution of the differential equation, with arbitrary constants.

Example 17. By the change of variable $y = \exp u(x)$, the differential equation $ay''_x + by'_x + cy = 0$ is equivalent to the nonlinear first order differential equation

$$a(v'_x + v_x^2) + bv_x + c = 0$$

for the first derivative of u, $v_x = u'_x$. By integration, it is equivalent to the implicit equation $v_x = av_0\{a + cv_0(x - x_0) + bv_0(V_x - V_0)\}^{-1}$, with V a primitive function of v.

Example 18. By the change of variable $y(x) = u(x)e^{-cx}$ on $\mathbb{R}_+$, the second order differential equation $y''_x - c^2y = 0$ with initial values y_0 and y'_0 at x_0 is equivalent to the first order differential equation for u'

$$u''_x - 2cu'_x = 0$$

with initial values $u_0 = y_0e^{cx_0}$ and $u'_0 = y'_0e^{cx_0} + cy_0$ at x_0. Its solution is

$$u(x) = u_0 + \frac{u'_0}{2c}e^{2c(x-x_0)}$$

then $y(x) = \alpha\sinh(cx) + \beta\cosh(cx)$ with the constants $\alpha = c^{-1}y'_0$ and $\beta = y_0$.

The functions defining the solutions in Proposition 4.11 are denoted

$$y_1(x) = e^{-\alpha(x-x_0)}\sin\{\beta(x - x_0)\},$$
$$y_2(x) = e^{-\alpha(x-x_0)}\cos\{\beta(x - x_0)\}$$

in the trigonometric case, in the same way they are

$$y_1(x) = e^{-\alpha(x-x_0)} \sinh\{\beta(x - x_0)\},$$
$$y_2(x) = e^{-\alpha(x-x_0)} \cosh\{\beta(x - x_0)\}$$

in the hyperbolic case . With a double root they are $y_1(x) = e^{-\alpha(x-x_0)}$ and respectively $y_2(x) = (x - x_0)e^{\alpha(x-x_1)}$.

Proposition 4.12. *The differential equation*

$$u''_x + pu'_x + qu_x = f$$

has the solutions

$$\begin{aligned} u_x &= \Big[c_1 + \frac{1}{\beta}\int_{x_0}^{x} e^{\alpha(s-x_0)} \cos\{\beta(s - x_0)\} f_s\, ds\Big] y_{1x} \\ &\quad + \Big[c_2 + \frac{1}{\beta}\int_{x_0}^{x} e^{\alpha(s-x_0)} \sin\{\beta(s - x_0)\} f_s\, ds\Big] y_{2x}, \quad \textit{if } \Delta < 0, \\ u_x &= \Big[c_1 + \frac{1}{\beta}\int_{x_0}^{x} e^{\alpha(s-x_0)} \cosh\{\beta(s - x_0)\} f_s\, ds\Big] y_{1x} \\ &\quad + \Big[c_2 + \frac{1}{\beta}\int_{x_0}^{x} e^{\alpha(s-x_0)} \sinh\{\beta(s - x_0)\} f_s\, ds\Big] y_{2x}, \quad \textit{if } \Delta > 0, \end{aligned}$$

$$\begin{aligned} u_x &= \Big[c_1 + \int_{x_0}^{x} e^{\alpha(s-x_1)}(s - x_0) f_s\, ds\Big] y_{1x} \\ &\quad + \Big[c_2 + \int_{x_0}^{x} e^{\alpha(s-x_0)} f_s\, ds\Big] y_{2x}, \quad \textit{if } \Delta = 0. \end{aligned}$$

If the right-hand side of the differential equation is a trigonometric function, a particular solution of the inhomogeneous equation may be proportional to a function of the same form (Poisson).

Example 19. On $[0, 2\pi]$, the differential equation

$$y''_x + ay = \sin(\omega x), \; a > 0,$$

under the initial conditions $y(0) = y_0$ and $y'(0) = y'_0$, has the solution

$$y(x) = c_1 \sin(\sqrt{a}x) + y_0 \cos(\sqrt{a}x) - \frac{1}{a - \omega^2} \sin(\omega x),$$

such that $y'_0 + \sqrt{a}c_1 + \frac{\sqrt{a}}{a-\omega^2} = 0$, if $a - \omega^2$ is not zero. If $\omega^2 = a$, a solution of the differential equation is

$$y(x) = c_1 \sin(\omega x) + (y_0 + c_4 x) \cos(\omega x)$$

with $2\omega c_4 + 1 = 0$ and with $y'_0 + \omega(c_1 + y_0) = c_4$.

On $[0,2\pi]^n$, the differential equation $y''_x + ay = \sum_{j=1}^n \sin(\omega x_j)$, with $a > 0$, has the solution

$$y(x) = \sum_{j=1}^n \{c_{1j} \sin(\sqrt{a}x_j) + y_0 \cos(\sqrt{a}x_j) - \frac{1}{a-\omega^2} \sin(\omega x_j)\},$$

if $a - \omega^2 \neq 0$. If $a - \omega^2 = 0$, a solution is

$$y(x) = \sum_{j=1}^n \{c_{1j} \sin(\omega x_j) + (y_0 + c_{4j} x_j) \cos(\omega x_j),$$

under the initial conditions for the constants.

Example 20. On $[0,2\pi]$, the differential equation

$$y''_x + ay = f(x),\ a > 0,$$

has the solution

$$y(x) = \{c_1 \sin(\sqrt{a}x) + cy_0 \cos(\sqrt{a}x)\}v_x := u_x v_x$$

with a function v_x such that $2u'_x v'_x + u_x v''_x = f$. A solution of the homogeneous equation $2u'_x v'_x + u_x v''_x = 0$ satisfies $v'_x = cu_x^{-2}$, let $v'_x = h_x u_x^{-2}$ for the solution of the inhomogeneous equation, then $h'_x = f_x u_x$ so $h_x = h_0 + \int_0^x f_t u_t \, dt$ and the function v is deduced by integration of v'_x. The constants h_0 and c_1 are deduced from $y'(0)$.

The differential equation with $p_x \equiv 0$ and q constant is a special case of Proposition (4.12). With $q_x \equiv 0$ and a function q_x, Theorem 4.6 implies that the solution is an integral of $P(x) = \int_{x_0}^x p_s \, ds$

$$u_x = u_0 + u'_1 \int_{x_0}^x \exp\{P(x_1) - P(y)\} \, dy.$$

In the case $p = 0$ with a non constant function q_x, an exponential function $u(x) = e^{a(x)}$ is solution of $u''_x + q_x u_x = 0$ if the function $v_x = a'_x$ is solution of the first order nonlinear differential equation $v'_x + v_x^2 + q_x = 0$.

Theorem 4.7. *The inhomogeneous equation defined by (4.11) and (4.12), with the initial conditions u_0 and u'_0 at x_0, y_1 and y'_1 at x_1, has a multiplicative solution $y_x = u_x v_x$ such that u_x is the solution of the homogeneous equation if v_x is solution of the inhomogeneous equation*

$$u_x v''_x + (2u'_x + p_x u_x) v'_x = f_x \tag{4.13}$$

with the initial values $v_0 = 1, v'_0 = 0$, $y_1 = u_1 v_1$ and $y'_1 = y_1 u_1^{-1} u'_1 + u_1 v'_1$. For every x in an interval (x_0, x_1) where u_x has no zero value, v_x is solution of the implicit equation

$$v_x = 1 + v'_1(x - x_0) + \int_{x_0}^x \int_t^{x_1} \Big\{\Big(\frac{2u'_s}{u_s} + p_s\Big) dv_s - u_s^{-1} f_s \, ds\Big\} \, dt.$$

Proof. Applying the inhomogeneous equation to $y_x = u_x v_x$ with u_x solution of the homogeneous equation implies

$$\begin{aligned} y_x'' + p_x y_x' + q_x y_x &= u_x'' v_x + 2u_x' v_x' + u_x v_x'' + p_x(u_x' v_x + u_x v_x') + q_x y_x \\ &= 2u_x' v_x' + u_x v_x'' + p_x u_x v_x' \end{aligned}$$

and it equals f_x. The initial conditions at zero are $v_0 = 1$ and $v_0' = 0$. At x_1, the values of u_x and u_x' are determined by the expression of the function u_x and the values of v_x and v_x' by those of u_x and y_x. By integration of the differential equation $v_x'' + 2u_x^{-1}u_x'v_x' + p_x v_x' = u_x^{-1} f_x$

$$\begin{aligned} v_x' &= v_1' + \int_x^{x_1} (2u_s^{-1}u_s' + p_s)\, dv_s - \int_x^{x_1} u_s^{-1} f_s\, ds, \\ v_x &= 1 + v_1'(x - x_0) + \int_{x_0}^{x} \int_t^{x_1} \Big\{ \Big(\frac{2u_s'}{u_s} + p_s \Big)\, dv_s - u_s^{-1} f_s\, ds \Big\}\, dt. \end{aligned}$$

□

The solution v_x of the inhomogeneous differential equation of Theorem 4.7 is the sum of v_{1x} solution of the homogeneous differential equation $u_x v_x'' + (2u_x' + p_x u_x) v_x' = 0$, with the initial values for v_x, and of the particular solution of the inhomogeneous equation with the right term

$$\varphi_x = - \int_{x_0}^{x} \int_t^{x_1} u_s^{-1} f_s\, ds\, dt. \tag{4.14}$$

The function v_{1x} is solution of the implicit equation

$$v_{1x} = 1 + v_1'(x - x_0) + \int_{x_0}^{x} \int_t^{x_1} \Big(\frac{2u_s'}{u_s} + p_s \Big)\, dv_{1s}\, dt$$

and the derivatives of φ_x are $\varphi_x' = - \int_x^{x_1} u_s^{-1} f_s\, ds$ and $\varphi_x'' = u_x^{-1} f_x$ so that $u_x \varphi_x'' = f_x$. The initial values of φ_x are $\varphi_{x_0} = 0$ and $\varphi_{x_1}' = 0$. The homogeneous differential equation of Theorem 4.7 can be explicitly solved as a function of u_x and p_x, the solution of the inhomogeneous equation is therefore entirely defined from the solution of the homogeneous equation.

Theorem 4.8. *On every interval $(x_0, x_1]$ where the functions u_x and u_x' are different from zero, the inhomogeneous equation defined by (4.11) and (4.12), with the initial conditions y_0 and y_0' at x_0, y_1 and y_1' at x_1, has a solution $y_x = u_x(v_x + \varphi_x)$ such that u_x is the solution of the homogeneous equation if φ_x is defined by (4.14) and*

$$v_x = 1 + v_1' u_1^2 \int_{x_0}^{x} u_t^{-2} \exp\Big\{ - \int_t^{x_1} p_s\, ds \Big\}\, dt.$$

Proof. If v_x is not constant, the homogeneous differential equation $u_x v''_x + (2u'_x + p_x u_x)v'_x = 0$ is equivalent to

$$\frac{v''_x}{v'_x} = -2\frac{u'_x}{u_x} - p_x$$

therefore $v'_x = v'_1 u_1^2 u_x^{-2} \exp\left\{-\int_x^{x_1} p_s\,ds\right\}$ and the result follows by integration. □

Theorem 4.7 is valid in particular for inhomogeneous equations with constant p and q. In that case, a closed form expression of the solution is deduced from the functions u_{1x} and u_{2x} of Proposition 4.11 or 4.6. The special case with $p_x \equiv 0$ and a function q_x is considered in the next section. If a particular solution of the inhomogeneous equation is known, the reverse of Theorem 4.7 is true.

Theorem 4.9. *Let v be particular solution of the inhomogeneous equation, a general solution of this equation is $u = v + yv$ such that y is solution of the first order homogeneous differential equation*

$$y''_x + (p + 2v^{-1}v'_x)y'_x = 0$$

for y'_x and

$$y_x = y_0 + y'_0 \int_{x_0}^{x} v^{-2}(s)e^{-P(s)}\,ds,$$

with a primitive $P(x) = \int_{x_0}^{x} p + P(x_0)$ of p.

Conditions for the existence of oscillatory solutions along the x-axis for second order inhomogeneous or nonlinear equations have been studied by many authors (Atkinson, 1955). They rely on the boundedness and integrability conditions at infinity of the control function f and of the coefficients.

If a solution is not oscillatory, one can assume that it is strictly positive for every $x > x_0$ sufficiently large. For a differential equation $u''_x + p_x u'_x + q_x u_x = f_x$ such that $q_x > 0$ and $f_x < 0$ as $x > x_0$, we get $u''_x + p_x u'_x < f_x < 0$ therefore $u'_x < u'_0 e^{P_0 - P_x}$ for every $x > x_0$, with a primitive P_x of p_x. The limiting behaviour of u'_x and u_x is then determined by their values at x_0 and by the behaviour of P_x as x tends to infinity. If P_x goes to infinity as x tends to infinity, $u'_x < 0$ and u is decreasing on $(x_0, \infty[$. From the assumption it must converge to a positive limit. If $q_x < 0$, $u''_x + p_x u'_x > f_x$ and by integration, $u'_x > 0$ therefore u is increasing.

Let

$$L_{a,x}u_x = a^2u_x + 2au'_x + u''_x$$

with the initial conditions Du. The homogeneous equation $L_{a,x}u_x = 0$ with the initial conditions $u(x_0) = u_0$, $u(x_1) = u_1$ and a constant a has the solution

$$u_x = u_0 + e^{-a(x-x_0)}.$$

The differential equation $a^2u_x - 2au'_x + u''_x = 0$ has a similar solution, replacing a by $-a$.

The solutions of the inhomogeneous equation $L_{a,x}u_x = f_x$ depend on the integral of f with respect to e^{ax}. Let $G(x) = e^{-ax}$.

Proposition 4.13. *The unique solution of the equation $L_{a,x}u_x = f_x$ with the boundary conditions u_0 and u_0 at zero is*

$$u_x = u_0 + \theta'_0 xe^{-ax} + e^{-ax}\int_0^x \int_0^t e^{as}f_s\,ds\,dt,$$

with $\theta'_0 = u'_0 + au_0$.

Proof. The solution of this equation with a non zero function f has the form $u_x = \theta_x G_x$ with a function θ of $C_2(\mathbb{R}_+)$. The derivatives of u_x are $u'_x = \theta'_x G_x + \theta_x G'_x = (\theta'_x - a\theta_x)G_x$ and $u''_x = (\theta''_x - 2a\theta'_x + a^2\theta_x)G_x$ and the differential equation reduces to $\theta''_x = e^{ax}f_x$. Its solution is such that

$$\theta'_x = \theta'_0 + \int_{x_0}^x e^{as}f_s\,ds,$$

$$\theta_x = \theta_0 + \theta'_0(x - x_0) + \int_{x_0}^x \int_{x_0}^t e^{as}f_s\,ds\,dt,$$

with $\theta_0 = u_0e^{ax_0}$ and $\theta'_0 = u'_0e^{ax_0} + a\theta_0$ and $u_x = G_x\theta_x$ is a solution of the inhomogeneous equation. □

Let a_x be a function of $C_2(\mathbb{R})$ defining the second order differential equation

$$L_{a_x,x}u_x = a_x^2u_x + 2a_xu'_x + u''_x = f_x$$

with the initial conditions Du at zero and let $A_x = \int_0^x a_s\,ds$. The function $y_x = y_0e^{-A_x}$ is solution of the homogeneous equation defined by $L_{a_x,x}$ if and only a_x is constant, otherwise it is solution of the homogeneous equation

$$L_{a_x,x}u_x = a_x^3u_x + a'_xu'_x - a_xu''_x = 0.$$

With a non constant function a_x, the derivatives of u are $u'_x = -a_xu_x$ and $u''_x = -a'_xu_x + a_x^2u_x$ then $L_{a_x,x}u_x = -a'_xu_x$ and the solution of the inhomogeneous equation satisfy the implicit equation $a_x = \int_x^{x_0} u_s^{-1}f_s\,ds$.

Solutions of the inhomogeneous equation may be obtained by varying the constant in y_x. Let $F_x = \int_0^x f_s\,ds + F_0$.

Proposition 4.14. *The differential equation $L_{a_x,x}u_x = f_x$ with the initial conditions $u(0) = u_0$ and $u'(0) = u'_0$ has a solution $u_x = v_x e^{-A_x}$ in $C_2(\mathbb{R})$ if and only if v_x is the solution of the differential equation $v''_x - a'_x v_x = f_x$ under the initial conditions Dv at zero. Then*

$$u_x = e^{-A_x}\Big\{u_0 + (u'_0 + F_x)x - \int_0^x s\,dF_s + \int_0^x (x-s)v_s\,da_s\Big\}.$$

Proof. Let $G_x = e^{-A_x}$ and $u_x = v_x G_x$, its derivatives are

$$\begin{aligned}
u'_x &= (v'_x - a'_x v_x)G_x,\\
u''_x &= (v''_x - 2a'_x v'_x - a''_x v_x + a^2 v_x)G_x.
\end{aligned}$$

The differential equation $u''_x + 2a'_x u'_x + a^2_x u_x = f_x$ is therefore equivalent to

$$v''_x - a'_x v_x = f_x.$$

By successive integrations, $v'_x = \int_0^x v_s\,da_s + F_x$ and v_x is deduced. □

Theorem 4.4 implies the existence and the unicity of the solution of the linear second order differential equation (4.12)

$$\frac{d^2u_x}{dx^2} = a_x \frac{du_x}{dx} + b_x u_x$$

with initial conditions. The local Lipschitz property of the linear operator is then written

$$\sup_{x\in I}\Big|a_x\Big(\frac{du_x}{dx} - \frac{dv_x}{dx}\Big) + b_x(u_x - v_x)\Big| \le \sup_{x\in I}|a_x|\|u - v\| + \sup_{x\in I}|b_x|\|u' - u'\|$$

for all u and u in E_1, u' and v' in E_2. The existence and the unicity extend to the solution of the inhomogeneous second order differential equation.

A reparametrization of the variable x as x_t implies

$$\begin{aligned}
y'_t &= x'_t y'_x,\\
y'_x &= \frac{y'_t}{x'_t},
\end{aligned} \tag{4.15}$$

$$\begin{aligned}
y''_t &= x''_t y'_x + x'^2_t y''_x,\\
y''_x &= \frac{x'_t y''_t - x''_t y'_t}{x'^3_t}
\end{aligned}$$

and the differential equation $y''_x + p_x y'_x + q_x y_x = 0$ is equivalent to

$$x'_t y''_t + (p_{x_t} x'^2_t - x''_t) y'_t + q_{x_t} x'^3_t y_t = 0$$

where $y_t = y(x_t)$. Denoting $A_t y''_t + B_t y'_t + C_t y_t = 0$ this equation, we have $B_t = 0$ if $x''_t = p_x x'^2_t$, i.e.

$$\log \frac{x'_t}{x'_{t_0}} = \int_{t_0}^{t} p_x x'_s \, ds = \int_{x_0}^{x} p_u \, du := P(x),$$
$$x'_t = x'_0 e^{P(x)},$$
$$x_t = x'_0 \int_{t_0}^{t} e^{P(x_s)} \, ds + x_0$$

the differential equation is then equivalent to $y''_t + q_{x_t} x'^2_t y_t = 0$ and

$$y''_t + x'_t q_{x_t} \{(x_t y_t)'_t - x_t y'_t\} = 0.$$

Let $x_t = k_1 e^{at} + k_2$, $x'_t = a(x_t - k_2)$ and $x''_t = a^2(x_t - k_2)$, the differential equation $y''_x + p_x y'_x + q_x y_x = 0$ becomes

$$y''_t + a\{(x_t - k_2) p_{x_t} - 1\} y'_t + a^2 q_{x_t} (x_t - k_2)^2 y_t = 0.$$

It reduces to $y''_t + a^2 q_{x_t} (x_t - k_2)^2 y_t = 0$ if $p_x = (x - k_2)^{-1}$.

Example 21. The differential equation $y''_t + a^2 y_t = 0$ has the trigonometric solutions $y_t = b\sin(at) + c\cos(at)$, with arbitrary constants b and c. By the reparametrization $x_t = ke^{at}$, with an arbitrary constant k, the differential equation is equivalent to

$$x^2 y''_x + x y'_x + y_x = 0.$$

Example 22. By the reparametrization $x_t = \log t$, the differential equation $y''_t + b y'_t + a^2 y_t = 0$ is equivalent to

$$y''_x + (be^x - 1) y'_x + a^2 e^{2x} y_x = 0$$

its solutions are given by Proposition 4.11 and a second member f_x is modified as f_{x_t}.

Trigonometric changes of variables also provide solutions of second order differential equations.

Example 23. (Bertrand, 1870). With the reparametrization by $x = \sin(t)$, $y'_t = \cos(t) y'_x$, $\cos^2(t) y''_x = y''_t + tan(t) y'_t$ and the differential equation

$$(1 - x^2) y''_x - x y'_x + a^2 y_x = 0$$

also known as Chebyshev's equation, is equivalent to $y''_t + a^2 y_t = 0$. This equation has the solutions $y_x = b\sin(a \arcsin x) + c\cos(a \arcsin x)$.

Expanding the function f_x and the solution y_x of a inhomogeneous linear second order differential equations in an orthonormal basis of functions $(\psi_n)_{n\geq 0}$ is a general method to get explicit solutions. The unicity of the expansion provides equations for the coefficients of the expansions and solving them defines a solution y_x. The functions are expanded in series as $y_x = \sum_{n\geq 0} a_n\psi_{nx}$ and $f_x = \sum_{n\geq 0} b_n\psi_{nx}$ such that for every $x \geq 0$

$$\sum_{n\geq 0} a_n\{\psi''_{nx} + p_x\psi'_{nx} + q_x\psi_{nx}\} = f_x.$$

This equation has a solution if the derivatives ψ'_{nx} and ψ''_{nx} have convergent expansions in the basis.

For functions f_x and y_x in $L_2(\mathbb{R}_+, \mu_\mathcal{E})$, Laguerre's basis provides solutions having a convergent expansions. Their coefficients satisfy the conditions

$$\begin{aligned} f_x &= \frac{1}{x^2}\sum_{n\geq 0} a_n n(n-1)\{L_{nx} - 2L_{n-1,x} + L_{n-2,x}\} \\ &\quad + \frac{p_x}{x}\sum_{n\geq 0} a_n n\{L_{nx} - L_{n-1,x}\} + q_x\sum_{n\geq 0} a_n L_{nx} \\ &= a_0 q_x L_{0x} + \Big\{\frac{p_x}{x} + q_x\Big\}a_1 L_{1x} + \sum_{n\geq 2}\Big[\frac{1}{x^2}\{n(n-1)a_n - 2n(n+1)a_{n+1} \\ &\quad + (n+1)(n+2)a_{n+2}\} + \frac{p_x}{x}\{na_n - (n+1)a_{n+1}\} + q_x a_n\Big]L_{nx}. \end{aligned}$$

By the unicity of the expansion, the coefficients of L_{0x} and L_{1x} must be zero for f_x and y_x, unless q is constant and $p_x \equiv px$. For every $n \geq 2$, a_n is solution of the recurrence equation for every $x \geq 0$

$$\begin{aligned} b_n x^2 &= a_n n(n-1) - 2n(n+1)a_{n+1} + (n+1)(n+2)a_{n+2} \\ &\quad + xp_x\{na_n - (n+1)a_{n+1}\} + x^2 q_x a_n \end{aligned}$$

and the existence of solutions in $L_2(\mathbb{R}, \mu_\mathcal{E})$ depends on the expression of p_x and q_x. Let $p_x = px$ and $q_x \equiv q$, the differential equation has a solution if and only if

$$\begin{aligned} b_n &= p\{na_n - (n+1)a_{n+1}\} + qa_n(n+1), \\ 0 &= a_n n(n-1) - 2n(n+1)a_{n+1} + (n+1)(n+2)a_{n+2}. \end{aligned}$$

The last equality defines a_{n+2} recursively from the values of a_0 and a_1, for every $n \geq 2$.

Functions in $L_2(\mathbb{R}, \mu_{\mathcal{N}})$ have a convergent expansion in Hermite's basis. The differential equation and the recurrence equation (2.8) imply that for every real x

$$\begin{aligned} f_x &= \sum_{k\geq 0} a_k[(1+x^2)H_k(x) - 2xH_{k+1}(x) - H_{k+2}(x) \\ &\qquad + p_x\{xH_k(x) - H_{k+1}(x)\} + q_x H_k(x)] \\ &= \sum_{k\geq 2} H_k(x)\{a_k(1+x^2) - 2xa_{k-1} - a_{k-2} + xp_x a_k - p_x a_{k-1} + a_k q_x\} \\ &\quad + H_1(x)\{a_1(1+x^2) + xp_x a_1 - 2xa_0 - p_x a_0\} \\ &\quad + a_0 H_0(x)\{(1+x^2) + xp_x + q_x\}. \end{aligned}$$

The coefficients in the expansions of f_x and y_x must satisfy the conditions

$$\begin{aligned} b_0 &= a_0\{(1+x^2) + xp_x + q_x\}, \\ b_1 &= a_1(1+x^2) + xp_x a_1 - 2xa_0 - p_x a_0, \\ b_k &= a_k(1+x^2) - 2xa_{k-1} - a_{k-2} + xp_x a_k - p_x a_{k-1} + a_k q_x,\ k \geq 2. \end{aligned}$$

The existence of solutions depends on the expressions of the functions p_x and q_x and of the compatibility of these equations.

The Airy function is defined in $\mathbb{R}_+$ by the second order linear differential equation

$$x^2 u_x'' - xu_x = 0. \tag{4.16}$$

As in Theorem 4.6, the solution u_x of the equation with initial values at zero has the derivative $u_x' = u_0' + \int_0^x s^{-1} u_s\, ds$ for every x such that the integral $\int_0^x s^{-1} u_s\, ds$ is finite, with the convention $\frac{0}{0} = 0$.

By integration of u_x', the solution of the Airy equation satisfies the implicit equation

$$\begin{aligned} u_x &= u_0 + u_0' x + \int_0^x \int_0^y s^{-1} u_s\, ds\, dy \\ &= u_0 + u_0' x + \int_0^x (s^{-1}x - 1) u_s\, ds. \end{aligned}$$

Assuming that $u_x = e^{h_x}$, (4.16) is equivalent to the nonlinear second order differential equation

$$h_x'' + h'^2 = \frac{1}{x}.$$

Looking for a solution in the form $h(x) = \{x + k(x)\}^{-1}$, with a function $k(x)$ such that

$$xk_x' + (x+k)^2 = 0,$$

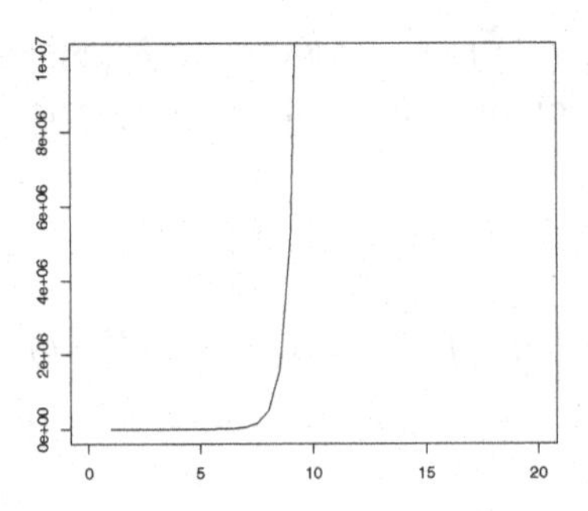

Fig. 4.1 Graph of the Airy function in $\mathbb{R}^+$.

it appears that $k'_x < 0$ and k is a decreasing function in $\mathbb{R}^+$. Replacing k by $k + x$ in this equality yields

$$0 = x(k'_x + 1) + (x + k - 1)^2 + k^2 - (k - 1)^2 + x$$

thus $x + k(x)$ is decreasing and the functions $h(x)$ and e^{h_x} are increasing. Moreover, e^{h_x} increases faster than e^x. These properties prove the existence of a solution and the implicit equation provides an algorithm for the graph Figure (4.1) of this function. In $\mathbb{R}^+$, the Airy function A solution of the differential equation (4.16) is therefore a positive and increasing function.

4.5 Sturm-Liouville second order differential equations

Let p and q be continuous functions, the Sturm-Liouville operator

$$L = \frac{d}{dx}\Big\{p(x)\frac{d}{dx}\Big\}+q(x) \tag{4.17}$$

defines second order differential equations by the conditions

$$\begin{aligned} Lu_t &= f_t \in \mathbb{R}_+, \\ Du : u(x_0) &= u_0,\ u'(x_1) = u'_1, \end{aligned} \tag{4.18}$$

with initial values at x_0 and x_1. The linear second order differential operator (4.10) is a Sturm-Liouville operator if and only if

$$\frac{b}{a} = \frac{p'_x}{p}, \quad \frac{c}{a} = \frac{q}{p}.$$

The functions p and q may be written as

$$p(x) = \exp\Big(\int_{x_0}^{x} \frac{b(s)}{a(s)}\, ds\Big),$$

$$q(x) = \frac{c(x)}{a(x)} p(x).$$

By the change of variable $pu'_x = v'_x$, (4.18) is also equivalent to the differential equation

$$v''_{xx} + q_x \int_{x_0}^{x} p_s^{-1}\, dv_s = f_x.$$

Particular solutions of Sturm-Liouville differential equations are well known. For instance, the normal density is unique solution of the homogeneous differential equation $u_t^{(k)} - H_{k,t}u_t = 0$ defined by the Hermite polynomial H_k, for every integer k. This is also the unique solution of the differential equations $\sum_{k=0}^{K} a_k u_t^{(k)} - h_t u_t = 0$ for every function h having an expansion $h = \sum_{k=0}^{K} a_k H_k$.

From (2.11), Legendre's polynomials P_m are solutions of homogeneous mth order linear differential equations $L_m P_m = 0$ defined in $[-1, 1]$ by the differential operators

$$L_m(t) = \frac{d}{dt}\Big\{(1 - t^2)\frac{d}{dt}\Big\} + m(m + 1).$$

The mth polynomial P_{mx} is therefore solution of the implicit equation

$$u_x = u_0 + u'_1 \int_{x_0}^{x} \exp\Big\{\int_{y}^{x_1} \Big(s^2 + m(m + 1)\frac{u_s}{u'_s}\Big)\, ds\Big\}\, dy.$$

For every $k < m$

$$L_m P_k = L_k P_k + \{m(m + 1) - k(k + 1)\}P_k = \{m(m + 1) - k(k + 1)\}P_k$$

then, every function u having an expansion $u = \sum_{k=0}^{m} a_k P_k$ is solution of the inhomogeneous Liouville equation

$$L_{mt}u_t = f_t$$

with a function $f = \sum_{k=0}^{m-1} \alpha_k P_k$ having the coefficients $\alpha_k = \{m(m+) - k(k + 1)\}a_k$. Reciprocally, the second member $f = \sum_{k=0}^{m-1} \alpha_k P_k$ of the equation $L_{mt}u_t = f_t$ defines the coefficients $(a_k)_{k=0,\dots,m-1}$ of the expansion of the solution u and its coefficient a_m is unspecified. This defines a class of equations with the differential operator L_m and with polynomial functions f of degree m.

It is not always possible to write the solution of a second order differential equation in a closed form and the expression of the coefficients of its expansion in a series solution of an implicit equation can be performed in large classes of equations.

Theorem 4.10. *The homogeneous Sturm-Liouville equation (4.18) defined by (4.17) in an interval* $]x_0, x_1[$ *where the function* p *is everywhere different from zero and with initial values* $u_0 = u(x_0)$ *and* $u'_1 = u'(x_1)$ *implies the implicit equation*

$$u_x = u_0 + \int_{x_0}^{x} \frac{1}{p_y}\Big(p_{x_1} u'_1 + \int_y^{x_1} q_s u_s \, ds\Big)\, dy. \tag{4.19}$$

Proof. Let $z_x = u'_x$, it satisfies the first order differential equation $(p_x z_x)' + q_x u_x = 0$. Integrating this equation leads to

$$p_x u'_x = p_x z_x = p_{x_1} u'_1 + \int_x^{x_1} q_s u_s \, ds$$

and the implicit expression of the solution is obtained by integrating this equation over $]x_0, x]$, for every $x < x_1$. □

Let u and v be two solutions of the homogeneous equation (4.18), according to Sturm (1833) it follows that $v(pu')' = u(pv')'$ and using an integration by parts

$$u_x p_x v'_x - u'_x p_x v_x = k,$$

a constant. On every interval (x_0, x_1) where p^{-1} is integrable, we deduce

$$\begin{aligned} k \int_{x_0}^{x} p^{-1} &= \int_{x_0}^{x} u_x^2 \, d\Big(\frac{v_x}{u_x}\Big) + c \\ &= u_x v_x - 2 \int_{x_0}^{x} u'_x v_x + c, \end{aligned}$$

with a constant c. Moreover, the knowledge of a solution u provides another solution proportional to u, $v_x = u_x k \int_{x_0}^{x} p_x^{-1} u_x^{-2}$ and a general solution is $u_x\{1 + k \int_{x_0}^{x} p_x^{-1} u_x^{-2}\}$.

Let $F_x = -\int_x^{x_1} f_s \, ds$ be a primitive of a continuous and integrable functions f, null at x_1.

Theorem 4.11. *The inhomogeneous Sturm-Liouville equation (4.18) defined by (4.17) in an interval* $]x_0, x_1[$ *where the function* p *is everywhere different from zero has a solution* $y_x = u_x + v_x$ *such that* u_x *is solution of the homogeneous differential equation and* $v_x = u_x \int_{x_0}^{x} p_s^{-1} u_s^{-2} \int_{x_0}^{s} f_t u_t^{-1} \, dt \, ds$ *is a the particular solution of the inhomogeneous equation.*

As special cases of the equation (4.18), let us consider the case of a constant $p = 1$ with a function q_x.

Theorem 4.12. *The solutions of the homogeneous equation (4.18) defined by* $L_x = \frac{d^2}{dx^2} + q_x$ *in an interval* $]x_0, x_1[$ *and with initial values* $u_0 = u(x_0)$ *and* $u'_1 = u'(x_1)$ *satisfy the implicit equation*

$$u_x = u_0 + u'_1(x - x_0) + \int_{x_0}^{x} Q_s u_s \, ds - \int_{x_0}^{x} \int_{t}^{x_1} Q_s \, du_s \, dt. \tag{4.20}$$

where $Q_x = \int_x^{x_1} q_s \, ds$.

Proof. Let $z_x = u'_x$, it satisfies the first order differential equation $z'_x + q_x u_x = 0$. Integrating this equation leads to

$$u'_x = z_x = u'_1 + \int_x^{x_1} q_s u_s \, ds.$$

An integration by parts implies

$$u'_x = u'_1 + Q_x u_x - \int_x^{x_1} Q_s \, du_s,$$

and a second integration on $]x_0, x]$ gives the result, for every $x < x_1$. □

Integrating again by parts, (4.20) is equivalent to

$$\begin{aligned} u_x &= u_0 + u'_1(x - x_0) + \int_{x_0}^{x} Q_s u_s \, ds \\ &\quad - (x - x_0) \int_x^{x_1} Q_s u_s \, ds + \int_{x_0}^{x} (s - x_0) Q_s \, du_s. \end{aligned}$$

From Theorems 4.7 and 4.8, the inhomogeneous Sturm-Liouville differential equation defined by (4.17) with the initial conditions y_0 and y'_0 at x_0, y_1 and y'_1 at x_1 such that y_x is different from zero in (x_0, x_1), has an unique solution in (x_0, x_1). It is written as $y_x = u_x(v_x + \varphi_x)$ such that u_x is the solution of the homogeneous Sturm-Liouville differential equation, v_x is solution of the equation $u_x v''_x + (2u'_x + p_x^{-1} p'_x u_x) v'_x = 0$ and φ_x is defined by (4.14) from u_x and f_x. Then v_x is defined from u_x and p_x as

$$v_x = 1 + v'_1 u_1^2 \int_{x_0}^{x} u_t^{-2} p_t \, dt.$$

As a consequence, the inhomogeneous Sturm-Liouville defined by $p_x = x^2$ has an unique solution $y_x = u_x\{1 + v'_1 u_1^2 (x - x_0) + \varphi_x\}$ in every interval

(x_0, x_1) such that the solution u_x of the homogeneous equation is different from zero, with φ_x defined by (4.14). The implicit equation for u_x becomes

$$u_x = u_0 + u_1'(x - x_0) + \frac{1}{3}\int_{x_0}^{x} (s - x_0)s^3\, du_s\, dt$$
$$+ \frac{1}{3}\int_{x}^{x_1} (x - x_0)s^3\, du_s\, dt + \frac{1}{3}\int_{x_0}^{x} s^3 u_s\, ds.$$

Let u_1 and u_2 be two solutions of (4.17) and (4.18) satisfying, and let $v = u_2 u_1^{-1}$ then the coefficients of the differential equations are expressed according to u_1 and u_2

$$p(x) = -\frac{v'(x)}{v(x)} - 2\frac{u_1'(x)}{u_1(x)},$$
$$q(x) = \frac{u_1''(x)}{u_1'(x)} - p(x)\frac{u_1'(x)}{u_1(x)}. \qquad (4.21)$$

The function $f(u_2, u_1) = u_2(qu_1')' - u_1(qu_2')'$ has the form

$$f(u_2, u_1) = u_1 u_2 \frac{(pq)'}{p}\Big(\frac{u_1'}{u_1} - \frac{u_2'}{u_2}\Big).$$

If u_1 and u_2 are linearly independent solutions of the differential equation, linear combinations of u_1 and u_2 are also solutions.

4.6 Applications

The second order derivative $a(x)u_{xx}'' + b(x)u_x' + c(x)u_x$ is strictly positive or negative if the equation $a(x) + b(x)r + r^2 c(x) = 0$ has no real root i.e. $b^2(x) - 4a(x)c(x) < 0$. This is equivalent to $F_{xu}''^2 - F_{xx}'' F_{uu}'' < 0$ for the derivatives of an implicit function $F(x, u)$ of x and u_x. Under these conditions, the second order differential equation $a(x)u_{xx}'' + b(x)u_x' + c(x)u_x = 0$ has no solution.

The second order differential equations with constant functions p and q are equations for harmonic oscillators (Proposition 4.13), if $q_x = 0$ they can be solved using a Sturm-Liouville operator.

Example 24. Let $a > 0$, the differential equation

$$xy'' - ay' = f_x,$$

with initial values y_0 and y_1 at non null x_0 and x_1, has the solution

$$y_x = y_0 + \frac{x^{a+1}}{a+1}\Big\{\frac{F_{x_0}}{x_0^{a+2}} - \frac{F_x}{x^{a+2}} + \int_{x_0}^{x} \frac{f_s}{x^{a+1}}\, ds\Big\},$$

where $F(x) = F(x_1) - \int_x^{x_1} f_s\, ds$ is a primitive of f.

Proof. A solution of the homogeneous equation $x^{-(a+1)}(xy''-ay')=0$ is solution of $(x^{-a}y'_x)'=0$ and its primitive is $x^{-a}y'_x=x_0^{-a}y'_0$. Its solution is

$$y_x=y_0+y'_0(a+1)^{-1}x_0^{-a}x^{a+1}.$$

A solution of the inhomogeneous equation is written as $y_x=y_0+A_xx^{a+1}$ with a function A_x solution of the differential equation

$$\begin{aligned}f_x&=x^{a+2}A''_x+(a+2)x^{a+1}A'_x=(x^{a+2}A'_x)',\\ F_x&=x^{a+2}A'_x+k_1\end{aligned}$$

with a primitive $F(x)=F(x_1)-\int_x^{x_1}f_s\,ds$ of f. Integrating the differential equation for A'_x

$$\begin{aligned}A_x&=\int_{x_0}^x s^{-(a+2)}F_s\,ds\\ &=\frac{1}{a+1}\Big\{x_0^{-(a+2)}F_{x_0}-x^{-(a+2)}F_x+\int_{x_0}^x s^{-(a+1)}f_s\,ds\Big\}.\end{aligned}$$ □

If the function p is a constant, the Sturm-Liouville second order differential equations reduce to $y''_x+q_xy_x=f$.

Example 25. The differential equation

$$y''_x+(1-x^2)y=0$$

has a solution $y(x)$ proportional to $e^{-\frac{x^2}{2}}$. A function $u(x)e^{-\frac{x^2}{2}}$ is also solution if $u''_x-2xu_x=0$, this entails $u'_x=c_1e^{x^2}$ and $u(x)=c_1\int_0^x e^{t^2}\,dt+c_2$ with arbitrary constants c_1 and c_2.

A necessary condition for the existence of an exponential solution $ke^{-b(x)}$ of the differential equation $y''_x+q_xy_x=0$ is $q_x=b''_x-b'^2_x$. This entails an implicit equation for b

$$b(x)=\int_0^x e^{b(s)}\Big\{\int_0^s q_te^{-b(t)}\,dt+c\Big\}\,ds$$

with arbitrary constants k and c_j. The solutions of the inhomogeneous equations $y''_x+q_xy_x=f$ are then deduced from Theorems 4.7 and 4.8.

4.7 Nonlinear differential equations

There is no standard way to solve nonlinear differential equations. Their variables may be separated by a reparametrization. A homogeneous equation including several variables may be normalized with respect to one of them and solved with respect to one variable. The differential equation is reduced to the question of integrating the ratio of polynomials or to the ratio of functional polynomials.

Sturm established that every second order linear differential equation $y''_x + p_x y'_x + q_x y_x = 0$ is equivalent to a first order nonlinear equation with functional coefficients such as (4.24) by the reparametrization $wy = xy'_x$

$$w'_x + \frac{xp_x - 1}{x} w_x + \frac{1}{x} w_x^2 + xq_x = 0. \tag{4.22}$$

So a differential equation $y''_x - ax^{-2}y_x^2 = f(x)$ is equivalent to

$$u''_x + 2u'_x - au_x^2 = f(x)$$

by the reparametrization $y_x = xu_x$.

A second order differential equation $y''_x - q_x y_x = 0$ is equivalent, by the reparametrization $y_x = e^{A_x}$ with a function A_x such that $A'_x = a_x$, to the nonlinear differential equation

$$a'_x + a_x^2 = q_x.$$

The differential equation $y''_x + 2x^{-1}y'_x - y_x = 0$ is equivalent to $u''_x - u_x = 0$ by the reparametrization $y_x = x^{-1}u_x$, and it has the solutions $y_x = x^{-1}(c_1 e^x + c_2 e^{-x})$.

First and second order order nonlinear differential equations (4.5) and (4.7) defined by continuous locally Lipschitz functions f may be locally aprroximated by linear equations and solved.

Bernoulli's equation

$$y'_x = ay - by^2, \tag{4.23}$$

with constants or functions $a(x)$ and $b(x)$, is a nonlinear differential equation. By the change of variable $u(x) = y^{-1}(x)$ it becomes linear and it is equivalent to

$$u'_x + au - b = 0$$

and Theorem 4.1 implies the existence and the unicity of a solution for this equation.

With non zero constants a and b, its solution with initial value $u_0 = y_0^{-1}$ at x_0 is $u(x) = y^{-1}(x) = e^{-ax}\{y_0^{-1} + a^{-1}b(e^{ax} - e^{ax_0})\}$ and the solution of (4.23) is the logistic equation

$$y(x) = \frac{ay_0e^{ax}}{a + by_0(e^{ax} - e^{ax_0})}.$$

With non null functional coefficients, its solution is expressed by the mean of the primitive function A of a with value zero at zero, as

$$y(x) = y_0e^{A(x)}\Big\{1 + y_0\int_{x_0}^{x} b(s)e^{A(s)}\,ds\Big\}^{-1}.$$

Adding a continuous function $f(x)$ as second member to (4.23), the differential equation

$$y'_x = a(x)y - b(x)y^2 + f(x) \tag{4.24}$$

is equivalent to an equation of the same form $u'_x + au - b + fu^2 = 0$, by the reparametrization $u(x) = y^{-1}(x)$. If the coefficient b is a constant, the second member becomes constant. The nonlinear first order Bernoulli equation with a second member cannot be solved like the linear equations by (4.1). Equation (4.24) has a solution if is equivalent to a nonlinear implicit equation $F(x, y) = 0$ with partial derivatives satisfying

$$y'_x = -\frac{F'_x}{F'_y} = ay - by^2 + f.$$

Proposition 4.15. *Let y_1 be a solution of (4.23) and let y_2 satisfy the differential equation $y'_x = (a - 2by_1)y - by^2 + f$, then $y = y_1 + y_2$ is solution of (4.24).*

The solutions of (4.24) are not necessarily additive, however they cannot be the product of y_1 and another function.

Example 26. The function $y = y_1 + k_1\{a + by_0(e^{ax} - 1)\}^{-1}$ is solution of the differential equation (4.24) with $f_x = k_2\{a + by_0(e^{ax} - 1)\}^{-2}$ and $k_2 = k_1(bk_1 + aby_0 - a^2)$.

An implicit solution may be written as $y(x) = u^{-1}(x)$ with

$$u(x) = u_f^{-1}(x)\Big\{u_0 + \int_{x_0}^{x} u_f(s)b(s)\,ds\Big\} \tag{4.25}$$

where u_f is solution of the implicit differential equation $u_f' = au_f + fuu_f$

$$u_f(x) = e^{A(x)}\Big[1 + \int_{x_0}^{x} f_s\Big\{u_0 + \int_{x_0}^{s} u_f(t)b(t)\,dt\Big\}\,ds\Big]$$
$$= e^{A(x)}\Big\{1 + \int_{x_0}^{x} f_s u_s u_{f,s} e^{-A(x)}\,ds\Big\}$$

and $u_x' = -(au + fu^2) + b$.

It may be directly written using the implicit equations

$$y(x) = y_b^{-1}(x)\Big\{y_0 + \int_{x_0}^{x} y_b(s)f(s)\,ds\Big\},$$
$$y_b(x) = e^{-A(x)}\Big\{1 + \int_{0}^{x} b(s)y(s)y_b(s)e^{A(s)}\,ds\Big\},$$

their derivatives are $y_b' = -ay_b + byy_b$ and $y'(x) = -y_b'(x)y_b^{-1}(x)y(x) + f(x) = a(x)y(x) - b(x)y^2(x) + f(x)$.

Equation (4.24) may be written in several equivalent forms by reparametrizations. Denoting $auy = u_x'$, it becomes

$$u_x'' = \frac{u_x'^2}{u}\Big(1 - \frac{b}{a}\Big) + au_x' + buf. \tag{4.26}$$

Let $wu = xu_x'$ with derivative $w_x'u + u_x'w = u_x' + xu_x''$, the equation is equivalent to

$$w_x' = \frac{w^2}{x^2}\Big(1 - x - \frac{b}{a}\Big) + \frac{w}{x}(1 + a) + bf.$$

It is also equivalent to the implicit equation of Theorem 4.6 or Proposition 4.10 for second order homogeneous differential equations with a variable or constant coefficients.

Proposition 4.16. *If $a = b$ in the inhomogeneous equation (4.24), a sufficient condition for the existence of a solution is $a_x' = af$, then $y = 1$ and $a_x = a_0e^{F_x - F_0}$ with a primitive F of the function f.*

Proof. If $a = b$, the differential equation reduces to the linear equation $u_x'' = u_x' + auf$. Denoting $u_x = e^{A_x}$ with a primitive A of a null at t_0, the equation is equivalent to the condition $a_x' = af$ and $auy = u_x'$ implies $y = 1$. □

The solutions of (4.24) may be explicitly calculated from the expression of the function f, for instance the differential equation

$$y_x' + ay + by^2 = (1 - 2ax)e^{-ax} + bx^2e^{-2ax}$$

has the solution $y(x) = xe^{-ax} - ab^{-1}$. With $b(x) = (x+c)^{-2}ke^{ax}$, the function $y(x) = (x+c)e^{-ax}$ is solution of the differential equation

$$y'_x + ay + b(x)y^2 = (k+1)e^{-ax}.$$

If f is a polynom, a polynomial solution may be calculated by identity of the coefficients of $y'_x + ay + by^2$ and f.

In the same way, Bernoulli's differential equation

$$y'_x = a(x)y - b(x)y^m \tag{4.27}$$

is equivalent to the first order linear differential equation

$$\frac{u'_x}{m-1} + a(x)u = b(x)$$

by the change of variable $u = y^{1-m}$. It has a solution

$$u(x) = \exp\{-(m-1)A(x)\}\Big[u_0 + \exp\Big\{\int_0^x e^{(m-1)A(s)}b(s)\,ds\Big\}\Big]$$

according to Theorem 4.1, and

$$y_H(x) = \exp\{A(x)\}\Big[u_0 + \exp\Big\{\int_0^x e^{(m-1)A(s)}b(s)\,ds\Big\}\Big]^{\frac{1}{1-m}}$$

is solution of the differential equation (4.27).

Riccati's equations with real coefficients are

$$\begin{aligned} y'_x &= ay^2 + bx + cx^2, \\ y'_x &= ay^2 + bx^m, \end{aligned} \tag{4.28}$$

with functions $g(x) = (bx + cx^2)$ or $g(x) = x^m$ as second members. Let $g(x) = bx^m$, by the change of variable $y = x^m u$ the differential equation (4.28) becomes

$$xu'_x + mu = ax^{m+1}u^2 + bx$$

this is a Bernoulli equation (4.24) with $f(x) = b(x)$ and with the coefficients $a(x) = -mx^{-1}$ and $b(x) = -ax^m$. The change of variable $y = y_1 u$, with the solution $y_1 = -(ax+k_1)^{-1}$ of the homegeous equation $y'_x = ay^2$, leads to another Bernoulli differential equation $u'_x + ay_1u - ay_1u^2 = -(ax+k_1)bx^m$.

Proposition 4.17. *A power function is solution of (4.28) if and only if $m = -2$. In that case, the solution of (4.28) is $y = -kx^{-1}$ with a constant k such that $b = k - ak^2$.*

Proposition 4.18. *The equations $y'_x + ay^2 = a^{-1}bx^m$ and $u''_x - bx^m u = 0$, are equivalent for all constants a, b and m.*

Proof. Let $u = e^{aY}$ be solution of $u''_x - bx^m u = 0$, where Y is a primitive of a function y and a is an arbitrary constant. Its derivatives are $u'_x = ayu$ and $u''_x = (ay'_x + a^2y^2)u = bx^m u$. □

Proposition 4.19. *Assuming that $a > 0$, $b > 0$ and $y_x = (k_x - ax)^{-1}$ with a function k_x of $C_1(\mathbb{R}_+)$, the differential equation (4.28) reduces to $k'_x + (k_x - ax)^2 bx^m = 0$. The function y_x is increasing and it tends to infinity as x tends to infinity.*

Let $g(x) = bx + cx^2$, with the reparametrization $y(x) = g(x)u(x)$ the differential equation (4.28) becomes

$$gu'_x + ug' = ag(x)u^2 + g(x)$$

it still has the same form as (4.24), the reparametrization $y = y_1 u$ leads to an equation $u'_x + ay_1 u - ay_1 u^2 = g(x)$ and in both cases the question is to solve a inhomogeneous Bernoulli equation with variable coefficients.

By differentiation of an exponential $u_t = e^{\int_0^t y_s\,ds}$, we obtain the equivalence of the following differential equations

$$\begin{aligned} y'_t + y_t^2 + at^m &= 0, \\ u''_t + aut^m &= 0 \end{aligned}$$

and $\log u_t = y_t + c$, with an arbitrary constant. With the parameters $a = m = -1$, the second equation is the Airy differential equation and its solution u is the special function A_t of Figure (4.1), the solution of Riccati's equation (4.28) $y'_t + y_t^2 + at^m = 0$ is then

$$y_t = \log A_t + c.$$

The existence of a solution for other values of m and of the constants does not allow us to express it by the means of other functions.

The differential equation $y'_x = ay^2 + g_x$ has a solution $y_x = (k - ax - h_x)^{-1}$ with a function $g_x = h'_x(k - ax - h_x)^{-2} = y_x^2 h'_x$, for all constants a and k and for every function h_x of C_1. This is not immediatly integrable. The change of variable $u = y^{-1}$ leads to an equation equivalent to (4.28)

$$u'_x + g_x u_x^2 + a = 0$$

with the functional coefficient g_x and a constant second member a. The homogeneous equation has the solution $u_x = (k + G_x)^{-1}$ with a primitive G_x of g_x and an arbitrary constant k.

Proposition 4.20. *A function $y_x = k_x + G_x$ defined with a function k_x of $C_1(\mathbb{R}_+)$ is solution of the differential equation $y'_x = ay^2 + g_x$ if and only if k_x is solution of the differential equation $k'_x = a(k_x + G_x)^2$.*

Proposition 4.21. *A function $y = u + v$ such that u and y are solutions of Riccati's equation (4.28) is defined by a function v solution of the differential equation $v'_x = av^2 + 2auv$.*

When the function u is known, v is solution of a homogeneous Bernoulli equation and its solution is known. Riccati's equations with other functions g may be easily solved.

Example 27. The equation $y'_x = ay_x^2 + a$ has the solution $y_x = \tan(ax)$. Let $y = u + v$, such that y and $u = \tan(ax)$ are solutions of the differential equation $y'_x = ay_x^2 + a$, then $v'_x = av_x^2 + 2a\tan(ax)v_x$. The function v is solution of a Bernoulli equation (4.23) and its solution is

$$v^{-1}(x) = e^{-A(x)}\Big\{v_0^{-1} - \int_0^x ae^{A(s)}\,ds\Big\} = \cos^2 x\Big\{v_0^{-1} - a\int_0^x \frac{ds}{\cos^2 s}\Big\},$$

with $A(x) = 2a\int_0^x \tan(at)\,dt = -\log\cos^2 ax$. Let $z_s = \cos^{-2} as$ and let $X = \cos ax$, the integral becomes

$$\int_0^x \frac{ds}{\cos^2 as} = \frac{1}{a}\int_1^X z(z^2-1)^{-\frac{1}{2}}\,dz = \frac{1}{a}\{\cos^2(ax) - 1\}^{\frac{1}{2}}$$

and

$$v^{-1}(x) = \cos^2 x[v_0^{-1} - \{\cos^2(ax) - 1\}^{\frac{1}{2}}].$$

Example 28. The differential equation $y'_x = a_x(y_x^2 + 1)$ has the solution $y_x = \tan(\int_0^x a_s\,ds)$.

From Proposition 4.21 and the following example, the solutions of nonlinear first order equations are not unique. This was expected from the equivalence of the second order differential equation $y''_x - g_xy_x = 0$ with Riccati's equation.

4.8 Differential equations of higher orders

A linear differential equation of order n for a function u of $C_n(\mathbb{R})$

$$u_x^{(n)} + \sum_{k=0}^{n-1} A_{k,x} = f_x \tag{4.29}$$

is written as a system of $n-1$ differential equation

$$\frac{du_x}{dx} = v_{1x}, \quad \frac{dv_{1,x}}{dx} = v_{2,x}, \ \ldots,$$
$$u_x^{(n)} + \sum_{k=0}^{n-1} A_{k,x} v_{k,x} = f_x.$$

Let $U(x)$ be the vector of $E^{\times n}$ with components $u(x), u'_x(x), \ldots, u_x^{(n-1)}(x)$, the homogeneous linear differential equation

$$\frac{d^n u}{dx^n}(x) = \sum_{k=0}^{n-1} a_{k,x} u_x^{(k)}$$

is written as

$$\frac{dU}{dx}(x) = A_x U(x)$$

with the matrix

$$A_x = \begin{pmatrix} 0 & Id & & \cdots & 0 \\ 0 & 0 & Id & \cdots & 0 \\ 0 & 0 & \cdots\cdots & & Id \\ a_{0,x} & a_{1,x} & a_{2,x} & \cdots & a_{n-1,x} \end{pmatrix}.$$

If the functions $A_{k,x}$ are constant, the differential equation has n independent solutions satisfying the initial conditions and every linear combination of these solutions is a general solution of the differential equation. Let $u_1, \ldots, u_n$ be solutions of the differential equation (4.29), they are independent if and only if the determinant of the matrix A is different from zero.

Proposition 4.22. *The homogeneous equation $U'(x) = AU(x)$ with constant coefficients has solutions $u(x) = \sum_{k=0}^{n-1} e^{\lambda_k x} p_k(x)$ where the parameters λ_k are the eigen values of the matrix A and the functions $p_k(x)$ are polynomials with degree the multiplicity of $lambda_k$.*

Let

$$Y_n = y_x^{(n)} + a_1 y_x^{(n-1)} + \ldots + a_n y = f_x$$

be a nth order non homogeneous linear differential equation. Lagrange' theorem proved that if $n - m$ solutions of the equation $Y_n = 0$ are known, the differential equation $Y_n = f_x$ may be reduced to a mth order linear differential equation. The proof uses recursively the method of variation of the constant. Knowing a solution y_1 of $Y_n = 0$, a function $y = y_1 \int_{x_0}^{x} y_2(s)\, ds$ is solution of a $(n-1)$th order linear differential equation in y_2, $Y_{n-1} = f_x$ depending on the derivatives $y_{2x}^{(n-1)}, \ldots, y_{2x}^{(1)}$ and y_2, with modified coefficients.

A nth order homogeneous linear equation with constant coefficients and with a characteristic polynomial having $2k$ distinct conjugate real and/or complex roots has a solution sum of k sine and cosine hyperbolic functions and/or sine and cosine functions and of $n - 2k$ exponential functions.

If the characteristic polynomial has multiple roots, the number of solutions related to these roots equals their multiplicity. Let m_i be the multiplicity of a root z_i, then $e^{z_i x}, xe^{z_i x}, x^{m_i - 1} e^{z_i x}$ are solutions of the differential equation.

Example 29. The third order linear differential equation

$$y_x^{(3)} - 3y_x^{(1)} + 2y_x = 0$$

has exponential solutions $y_{a,x} = e^{ax}$ and its characteristic polynomial is $a^3 - 3a + 2 = (a-1)^2(a+2) = 0$. The function $y_x = c_1 e^x + c_2 x e^x e^x + c_3 e^{-2x}$ with arbitrary constants c_i is then a solution of the equation. Solutions of the inhomogeneous equation

$$y_x^{(3)} - 3y_x^{(1)} + 2y_x = x$$

are obtained from the solution of the equation homogeneous as the sum $y_x = c_1 e^x + c_2 e^{-2x} + c_3 x e^x + \frac{1}{2}x + \frac{3}{4}$.

Let $U_{n,x}$ be the vector with components $(u_x, \ldots, u_x^{(n-1)})$, let $V_{n,x}$ be the vector with components $(u'_x, \ldots, f_x)$ and let $B_{n,x}$ be the $n \times n$ matrix with diagonal 1, upper triangular matrix zero and lower triangular matrix the vector $(A_{1,x}, \ldots, A_{n,x})$. The nth order differential equation is also written as a system of differential equations

$$U'_{n,x} = B_{n,x} V_{n,x}.$$

A nth order linear differential equation with constant coefficients

$$\sum_{k=0}^{n} a_k y^{(k)} = f(x)$$

is written in a vectorial form with a constant matrix B_n with last line $-(a_{n-1}, \dots, a_0)$ and zeros everywhere except the sub-diagonal one. The real or complex eighen roots λ_k of the matrix B_n determine solutions of the homogeneous differential equation as linear combinations $y(x) = y(x_0) + \sum_{k=1}^{n} c_k e^{\lambda_k(x-x_0)}$. The inhomogeneous differential equation has solutions $y = y_H + y_p$ where $y_H = \sum_{k=1}^{n} c_k y_k$ is solution of the homogeneous equation with arbitrary constants c_k and $y_1(x) = \sum_{k=1}^{n} b_k(x) y_k(x)$ is a particular solution of the inhomogeneous equation. Its derivatives are

$$\begin{aligned}
y_1^{(1)} &= \sum_{m=1}^{n} (b'_m y_m + b_m y'_m),\\
y_1^{(2)} &= \sum_{m=1}^{n} (b''_m y_m + 2b'_m y'_m + b_m y''_m),\\
y_1^{(3)} &= \sum_{m=1}^{n} \sum_{j=0}^{3} \binom{3}{j} b_m^{(j)} y_m^{(3-j)}, \dots\\
y_p^{(n)} &= \sum_{m=1}^{n} \sum_{j=0}^{n} \binom{n}{j} b_m^{(j)} y_m^{(n-j)},
\end{aligned}$$

since y_H is solution of the homogeneous equation, the functions b'_m satisfies a $(n-1)$th order linear differential equation

$$f = \sum_{k=1}^{n} a_k \sum_{m=1}^{n} (b_m y_m)^{(k)} = \sum_{k=1}^{n} a_k \sum_{m=1}^{n} \sum_{j=0}^{k} \binom{k}{j} b_k^{(j)} y_k^{(k-j)}.$$

It is solved by adding the constraints of null linear combinations of the functions b'_m and the derivatives of the functions y_m. The Wronskian is defined as the determinant

$$W = \begin{vmatrix} y_1 & y_2 & \dots y_n \\ y_1^{(1)} & y_2^{(1)} & \dots y_n^{(1)} \\ & & \dots \\ y_1^{(n-1)} & y_2^{(n-1)} & \dots y_n^{(n-1)}. \end{vmatrix}$$

The vector $(y_H, y_H^{(1)}, \dots, y_H^{(n)})^t$ is the product of $(c_1, c_2, \dots, c_n)^t$ and the matrix of the derivatives of the functions y_k therefore W is different from zero if and only if the functions $y_1, y_2, \dots, y_n$ are linearly independent, then y_H is non null as soon as the constants are not all equally zero. For a second order homogeneous differential equation with linearly independent solutions y_1 and y_2, for every x such that a_2 is non null

$$y_1y_2\Big(\frac{y_1''}{y_1}-\frac{y_2''}{y_2}\Big)+\frac{a_1}{a_2}y_1y_2\Big(\frac{y_1'}{y_1}-\frac{y_2'}{y_2}\Big)=0$$

therefore the determinant $\left|\begin{pmatrix} y_1 & y_2 \\ y_1'' & y_2'' \end{pmatrix}\right|$ is different from zero.

Let W_k be the determinant of the matrix obtained by replacing in W the kth colomn of the derivatives of y_k by a vector of zeros except the function f as the last term

$$W_k=\begin{vmatrix} y_1 & y_2 & \ldots 0 \ldots y_n \\ y_1^{(1)} & y_2^{(1)} & \ldots 0 \ldots y_n^{(1)} \\ & & \ldots \\ y_1^{(n-1)} & y_2^{(n-1)} & \ldots f \ldots y_n^{(n-1)}. \end{vmatrix}$$

Theorem 4.13. *A nth order linear differential equation with constant coefficients* $\sum_{k=0}^n a_k y^{(k)} = f(x)$ *has the solutions* $y = \sum_{k=1}^n \{c_k + b_k(x)\} y_k$ *where the functions* y_k *are the solutions of the homogeneous equation and with arbitrary constants* c_k *and*

$$b_k(x)=\int_{x_0}^x \frac{W_k}{W}(s)\,ds.$$

This solution is identical to the solution obtained by varying the constant. The form of the second member may also be used directly with a varying constant to find solutions by identification of the coefficients and yields roots of the characteristic polynomial. Other properties of the solutions of n-th order linear differential equations have been studied (Trench, 2001).

The linear differential equation

$$\frac{dU}{dx}(x)=A_xU(x)$$

with initial conditions $U(x_0)=U_0$ has a unique solution

$$U(x)=U_0\exp\{\int_{x_0}^x A_s\,ds\}$$

where the exponentiel of a matrix $B_x=\int_{x_0}^x A_s\,ds$ is defined by the expansion

$$e^{B_x}=Id+\sum_{k=1}^{\infty}\frac{B_x^k}{k!},$$

with the initial value $e^{B_0} = 1$, its derivative is

$$\frac{de^{B_x}}{dx}(x) = A_x e^{B_x}.$$

If there exists an orthonormal matrix P such that the matrix B is changed as a diagonal matrix $D = P^T BP$, then the exponential of B has the same transform $e^B = Pe^D P^T$.

The inhomogeneous linear differential equation

$$\frac{dU}{dx}(x) = A_x U(x) + f(x)$$

has a solution

$$U(x) = e^{B_x}\{U_0 + \int_{x_0}^{x} f_s B_s \, ds\}.$$

If the matrix A is constant, the solution becomes

$$U(x) = \exp\{(x - x_0)A\}\{U_0 + \int_{x_0}^{x} e^{(s-x_0)A} f(s) \, ds\}.$$

Proposition 4.22 is proved writing

$$e^{xA} = e^{x\lambda_i}\{Id + \sum_{k=1}^{k_i-1} \frac{x^k}{k!}(A - \lambda_i)^k$$

for each eigen value λ_i of A with multiplicity k_i.

More general differential equations of order n are written

$$\begin{aligned} \frac{d^n u_x}{dx^n} &= F(x, U_x), \\ U_0 &= f(x_0, U_0) \end{aligned} \tag{4.30}$$

in an open real interval I, where u is $C^1(I)$ with value in a real set E, U is $C^1(I)$ with value in a subset E_n of $\mathbb{R}^n$, and F is a continuous function on $I \times E_n$.

If F has the local Lipschitz property

$$\sup_{t \in I} |F(x, U) - F(x, V)| \leq k\|U - V\|$$

for every U in E_n and V in a neighborohhod of U, then Theorem 4.2 extends to establish the existence and the unicity of a solutions of (4.30).

Theorem 4.14. *Let I be bounded then there exists an unique solution of equation (4.30) with a local Lipschitz function F.*

The existence of an approximated solution u_ε on a partition $(I_i)_{i\leq n}$ of I, with a sufficiently small path $\delta > 0$, is established by an n oder expansion of a solution u of the n oder differential equation on each subinterval I_i. It satisfies

$$\sup_{t\in I} |U'_{\varepsilon,x}) - f(x, U_{\varepsilon,t})| \leq \varepsilon$$

and it converges to a solution U as ε tends to zero. The unicity under initial conditions is proved like for Theorem 4.4.

Example 30. The differential equation $u'' + a^2u = 0$ is written as $U' = AU$ where U is the vector with components u and u', and $A = \begin{pmatrix} 0 & 1 \\ -a & 0 \end{pmatrix}$ is such that $A^2 = -a^2Id$. The eigen values of A are $\pm ia$, where $a > 0$, and the exponential matrix is

$$e^{xA} = \begin{pmatrix} \sin(ax) & \cos(ax) \\ \cos(ax) & -\sin(ax) \end{pmatrix}.$$

Example 31. The differential equation $u''_x + a_x^2 u_x = 0$, with a non constant function a_x, is written as $U'_x = A_x U_x$ and the exponential matrix is

$$e^{\int_{0^x} a_s\, ds} = \begin{pmatrix} \sin(\int_{0^x} a_s\, ds) & \cos(\int_{0^x} a_s\, ds) \\ \cos(\int_{0^x} a_s\, ds) & -\sin(\int_{0^x} a_s\, ds) \end{pmatrix}.$$

4.9 Differential equations in $\mathbb{C}$

A function f defined on an open set Ω of $\mathbb{C}$, with values in $\mathbb{C}$, is analytic at z_0 if there exists an open disc $D_0(r) = D(z_0, r)$ centered at z_0 and with radius r, where f is differentiable, then it has a convergent series expansion

$$f(z) = \sum_{j=0}^{\infty} a_j(z - z_0)^j,$$

and the derivatives of f are analytic in $D_0(r)$. This expansion is unique and also written as

$$f(z) = \sum_{j=0}^{\infty} \frac{(z - z_0)^j}{j!} f^{(j)}(z_0).$$

A complex function with a k order pole in a neighborhood of z_0 has the form

$$f(z) = \sum_{j=-k}^{\infty} a_j(z - z_0)^j = \frac{h(z)}{(z - z_0)^k}$$

where the function h is analytic in $D_0(r)$.

Let f be a function having an isolated singulatity at z_0, then there exist discs D_1 and $D_2 \subset D_1$ centered at z_0 such that, in $D_1 \setminus D_2$, f has Laurent's expansion

$$f(z) = \sum_{k=1}^{\infty} c_k (z - z_0)^{-k} + \sum_{k=0}^{\infty} c_{-k} (z - z_0)^k.$$

Solutions of differential equations are obtained from such expansions and identification of the coefficients.

Example 32. The differential equation $w'_z + w_z = (1-z)^{-1}$ in $D = \{z \in \mathbb{C} : |z| < 1\}$ has a solution $w(z) = u(z)e^{-z}$ such that $u'_z = e^z(1-z)^{-1}$. Another form of the solution is obtained by expansions of $w(z) = \sum_{j=0}^{\infty} \frac{1}{j!} w_j z^j$ and its derivative and from $(1-z)^{-1} = \sum_{j=0}^{\infty} z^j$ with coefficients w_j such that $w_j + (j+1)w_{j+1} = j!$ for every integer $j \geq 0$.

The solutions of first order linear differential equations in $\mathbb{C}$ are similar to the solutions of first order linear differential equations in $\mathbb{R}$. The solution of the equation $f'(z) = z^{-1}$ in $\mathbb{C} \setminus \{0\}$, with initial value $f(1) = 0$ is the complex logarithm function

$$f(z) = \log(z) = |z| + i \arg(z), \tag{4.31}$$

with the norm $|z| = (x^2 + y^2)^{\frac{1}{2}}$ for $z = x + iy = |z|e^{i\theta}$ and

$$\arg(z) = \theta \pm 2k\pi = \arctan \frac{y}{x},$$

for every integer k.

Proposition 4.23. *A first order linear differential equation $u'(z) - \zeta u(z) = f(z)$, with initial value $u(z_0) = u_0$, has the unique solution*

$$u(z) = u_0 \exp\{\zeta(z - z_0)\} + \int_{z_0}^{z} \exp\{\zeta(z - \xi)\} f(\xi)\, d\xi.$$

Proof. The homogeneous differential equation $u'(z) - \zeta u(z) = 0$ has uniquely the logarithm solution

$$\log \frac{u(z)}{u_0} = \zeta(z - z_0)$$

so $u(z) = u_0 \exp\{\zeta(z - z_0)\}$. A solution of the differential equations $u'(z) - \zeta u(z) = f(z)$ is deduced in the form $u(z) = \omega(z) \exp\{\zeta(z - z_0)\}$ with a complex function ω such that $\omega'(z) = \exp\{-\zeta(z - z_0)\} f(z)$, with initial value $\omega(z_0) = u_0$. The unicity of this solution is deduced from the unicity of this solution for the homogeneous differential equation. □

Example 33. The differential equation $z' + z = \omega$ in $\mathbb{C}$ is equivalent to the equations

$$u'_x + u = \cos x, \quad v'_x + v = \sin x,$$

with the notations $z = u + iv$ and $\omega = \cos x + i \sin x$. They have the solutions $u(x) = \frac{1}{2}(\sin x + \cos x)$ and $v(x) = \frac{1}{2}(\sin x - \cos x)$ then

$$z(x) = \frac{1}{2}\{(1+i)\cos x + (1-i)\sin x\} = \frac{(1-i)\omega}{2}.$$

Example 34. The complex logarithm function of $z = x+iy$ has the partial derivatives

$$f'_x(z) = \frac{x}{|z|^2} + i\theta'_x, \quad f'_y(z) = \frac{y}{|z|}^2 + i\theta'_y$$

where the derivatives of $\arctan u$ and $\theta = \arctan \frac{y}{x}$ are

$$\begin{aligned} \frac{\partial \arctan u}{\partial u} &= \frac{1}{\tan' u} = \frac{1}{1 + \tan^2 u}, \\ \theta'_x &= \frac{\partial(\frac{y}{x})}{\partial x}\frac{x^2}{|z|^2} = -\frac{y}{|z|^2}, \\ \theta'_y &= \frac{\partial(\frac{y}{x})}{\partial y}\frac{x^2}{|z|^2} = \frac{x}{|z|^2} \end{aligned}$$

which imply

$$\begin{aligned} f'_x(z) &= \frac{x}{|z|^2} - i\frac{y}{|z|^2}, \\ f'_y(z) &= \frac{y}{|z|^2} + i\frac{x}{|z|^2}, \end{aligned}$$

they satisfy Cauchy's equalities (3.1).

Example 35. For $z = x + iy$, the partial derivatives of $w(z) = z(\log z - 1)$ with respect to x and y are

$$\begin{aligned} w'_x(z) &= (x+iy)\Big(\frac{x}{|z|^2} - i\frac{y}{|z|^2}\Big) + \log z - 1 = \log z, \\ w'_y(z) &= (x+iy)\Big(\frac{y}{|z|^2} + i\frac{x}{|z|^2}\Big) + i(\log z - 1) = i\log z \end{aligned}$$

so the complex derivative of w is $w'(z) = \log z$.

Proposition 4.23 generalizes to first order linear differential equations on an open subset Ω of $\mathbb{C}^n$, for vectors $u = (u_1, \ldots, u_n)^t$ of differentiable functions on Ω, for $\zeta = (\zeta_1, \ldots, \zeta_n)^t$ in $\mathbb{C}^n$.

Let u be an analytic function solution of a first order linear differential equation

$$u'(z) + p(z)u(z) = f(z), \tag{4.32}$$

with initial value $u(0) = u_0 \neq 0$, then the function p and f are analytic in a neighborhood of zero. The unique solution of the homogeneous equation $u'(z) + p(z)u(z) = 0$ is $u(z) = u_0 \exp\{-P(z)\}$ where $P(z) = \int_0^z p(s)\,ds$.

Proposition 4.24. *A first order linear differential equation (4.32) with initial value $u(z_0) = u_0$ has the unique solution*

$$u(z) = u_0 \exp\{-P(z)\} + \exp\{-P(z)\} \int_0^z \exp\{P(\xi) f(\xi)\, d\xi,$$

it is analytic in a neighborhood of zero if $u_0 \neq 0$

Proposition 4.25. *A first order linear differential equation (4.32) with initial value $u(0) = u_0$ has the unique solution of the form*

$$u(z) = z^a \exp\{-P(z)\}\Big\{u_0 + \int_0^z \xi^{-a} \exp\{P(\xi)\} f(\xi)\, d\xi\Big\},$$

if and only if the function p has a first order pole at zero, $p(z) = az^{-1} + h(z)$ where h is analytic near zero.

Let p have a convergent expansion in a disc near z_0, for $z \neq z_0$

$$p(z) = \sum_{j=-k}^{\infty} a_j (z - z_0)^j,$$

then its integral is

$$P(z) = P(z_0) + \sum_{j=-k-1}^{-1} \frac{a_{j-1}(z - z_0)^j}{j}$$
$$+ a_{-1} \log(z - z_0) + \sum_{j=0}^{\infty} a_j (z - z_0)^j$$

and the unique solution of the homogeneous equation with initial value $u(z_0) = u_0$ is

$$u(z) = u_0 (z - z_0)^{a-1} \exp\Big\{ \sum_{j=-k-1}^{-1} \frac{a_{j-1}(z - z_0)^j}{|j|} \Big\} \exp\Big\{ -\sum_{j=0}^{\infty} a_j (z - z_0)^j \Big\}$$

and the inhomogeneous differential equation has the form given by Proposition 4.25.

The following generalizes Section 4.2 to complex functions on $\Omega \times E$, where Ω and E are open set of $\mathbb{C}$. Let f be a continuous function with values in a complex set V, satisfying the local Lipschitz property

$$\sup_{z\in\Omega} \|f(z,u_1) - f(z,u_2)\|_V \leq k\|u_1 - u_2\| \tag{4.33}$$

for every complex u_1 in E and every u_2 in a neighborhood of u_1. and with a complex function f defined on $\Omega \times E$. The general first order differential equations for complex z and u have the form

$$\begin{aligned} \frac{du(z)}{dz} &= f(z, u(z)), \\ u(z_0) &= u_0 \neq 0. \end{aligned} \tag{4.34}$$

Theorem 4.15. *Let I be a bounded interval and let f be locally a Lipschitz function, then there exists an unique solution of the first order differential equation (4.34).*

Proof. The existence of a solution is proved by construction on a partition of Ω in sufficiently small squares, of an approximated solution u_ε which converges to a solution as ε tends to zero, like in the proof of Theorem 4.2.

To prove the unicity of the solution, let u and v be solutions of the equation with initial values u_0 and respectively v_0 at z_0, then for every z in Ω

$$\|u'_z - v'_z\| = \|f(t, u_z) - f(t, v_z)\| \leq k\|u_z - v_z\|,$$

the unique solution of the real valued equation $\|u'_z - v'_z\| = k\|u_z - v_z\|$ is

$$\|u_z - v_z\| = \|u_0 - v_0\| \, \|e^{k(z-z_0)}\|,$$

and for the inequality, $\|u_z - v_z\| \leq \|u_0 - v_0\| \, \|e^{k(z-z_0)}\|$ where the bound is zero if $u_0 = v_0$. □

Theorem 4.15 generalizes to complex functions on $\Omega \times E$, where Ω is an open subset of $\mathbb{C}$ and E an open subset of $\mathbb{C}^n$, under the local Lipschitz property (4.33).

Second order linear differential equations in $\mathbb{C}$ have the form (4.11) with an operator L_z given by (4.10)

$$u''(z) + p(z)u'(z) + q(z)u(z) = f(z) \tag{4.35}$$

with initial values $u(z_0) = u_0$ and $u'(z_0) = u'_0$, for complex functions u, p,q and f.

Theorem 4.16. *The second order differential equation (4.35) has a unique solution if p and q are analytic.*

Proof. Let $u(z) = v(z)+iw(z)$, $p(z) = p_1(z)+ip_2(z)$, $q(z) = q_1(z)+iq_2(z)$ and $f(z) = f_1(z) + if_2(z)$. The second order differential equation (4.35) is equivalent to a set of two differential equations, for the real and for the imaginary parts of u

$$\begin{aligned} v''(z) + p_1(z)v'(z) + q_1(z)v(z) &= f_1(z) + p_2(z)w'(z) + q_2(z)w(z) \\ &:= f_1(z) + h_z(w, w'), \\ w''(z) + p_1(z)w'(z) + q_1(z)w(z) &= f_2(z) - \{p_2(z)v'(z) + q_2(z)v(z)\} \\ &= f_2(z) - h_z(v, v'), \end{aligned}$$

with the initial conditions for v and w. Applying Cauchy's equalities (3.1) for the first order partial derivatives of v and w, these differential equations are equivalent to a systems of partial derivatives with respect to x and respectively y

$$\begin{aligned} v''_x(z) + q_1(z)v(z) &= g_{1z}(w, w') := f_1(z) + h_z(w, w') - p_1(z)w'_y(z), \\ v''_y(z) + q_1(z)v(z) &= g_{2z}(w, w') := p_1(z)w'_x(z) + f_1(z) + h_z(w, w'), \\ w''_x(z) + q_2(z)w(z) &= g_{3z}(v, v') := p_2(z)v'_y(z) + f_2(z) - h_z(v, v'), \\ w''_y(z) + q_2(z)w(z) &= g_{4z}(v, v') := f_2(z) - h_z(v, v') - p_2(z)v'_x(z). \end{aligned}$$

The complex differential equation $u''(z) + p(z)u'(z) + q(z)u(z) = f(z)$ is therefore equivalent to a set of partial differential equations for real functions and Proposition 4.12 applies. □

Though Proposition 4.12 provides explicit solutions of real second order linear differential equations in $\mathbb{R}_+$, it yields implicit solutions for complex functions on $\mathbb{C}$ where the expression of v depends on w and reciprocally.

Proposition 4.26. *The complex differential equation $u''(z) + pu'(z) + qu(z) = f(z)$ with constant coefficients p and q has unique solutions v and w defining u on $\mathbb{C}$, under the initial conditions.*

Proof. Let $u(z) = v(z) + iw(z)$, $p = p_1 + ip_2$ and $q = q_1 + iq_2$, the second order differential equation with constant coefficients is equivalent to the set of differential equations

$$\begin{aligned} v''(z) + p_1v'(z) + q_1v(z) &= p_2w'(z) + q_2w(z) := h_z(w, w'), \\ w''(z) + p_1w'(z) + q_1w(z) &= -\{p_2v'(z) + q_2v(z)\} = -h_z(v, v'), \end{aligned}$$

with the initial conditions for v and w. Applying Cauchy's equalities (3.1) for the first order partial derivatives of v and w, these differential equations are equivalent to a systems of partial derivatives with respect to x and respectively y

$$\begin{aligned}
v''_x(z) + q_1 v(z) &= h_z(w, w') - p_1 w'_y(z),\\
v''_y(z) + q_1 v(z) &= h_z(w, w') + p_1 w'_x(z),\\
w''_x(z) + q_2 w(z) &= -h_z(v, v') + p_2 v'_y(z),\\
w''_y(z) + q_2 w(z) &= -h_z(v, v') - p_2 v'_x(z).
\end{aligned}$$

The solutions of the homogeneous differential equations depend only on a real variable and they are given by Proposition 4.11, the general solutions of the above inhomogeneous differential equations are deduced by the method of variation of the constants like in Proposition 4.12. □

If the function p has a first order pole and the and if the function q has a second order pole, the three terms of the second order differential equation (4.35) are homogeneous in z and the solutions are modified, like in Proposition 4.23

Proposition 4.27. *A second order linear differential equation*

$$u''(z) + \frac{p}{z} u'(z) + \frac{q}{z^2} u(z) = f(z)$$

with initial values $u(0) = u_0$ *and* $u'(0) = u'_0$, *has the unique solution.*

Proof. Denoting $z = e^\xi$ and $v(\xi) = u(e^\xi)$, we have

$$\begin{aligned}
v'(\xi) &= \frac{\partial z}{\partial \xi} u'(z) = z u'(z),\\
v''(\xi) &= \frac{\partial z}{\partial \xi} u'(z) + z^2 u''(z)\\
&= v'(\xi) + z^2 u''(z)
\end{aligned}$$

and the diffferential equation is equivalent to a second order diffferential equation for v

$$v''(\xi) - v'(\xi) + p v'(\xi) + q v(\xi) = e^{2\xi} f(e^\xi)$$

which has a unique solution, by Proposition 4.26. □

Example 36. A second order differential equation (4.35) with coefficients $p(z) = z^{-1} h(z)$ and $q(z) = z^{-2} g(z)$, where h and g are analytic have unique solutions. By the change of variable $z = e^\xi$ and $v(\xi) = u(e^\xi)$, we have $v'(\xi) = z u'(z)$ and $v''(\xi) = v'(\xi) + z^2 u''(z)$ and the differential equation is equivalent to

$$v''(\xi) - v'(\xi) + v'(\xi)h_2(\xi) + v(\xi)g_2(\xi) = f_2(\xi)$$

with $h_2(\xi) = h(e^\xi)$, $g_2(\xi) = g(e^\xi)$ and $f_2(\xi) = e^{2\xi} f(e^\xi)$. Theorem 4.16 applies.

Example 37. The solution u of a second order differential equation (4.35) with analytic functions $p(z) = \sum_{j\geq 0} p_j z^j$ and $q(z) = \sum_{j\geq 0} q_j z^j$ is analytic, denoted $u(z) = \sum_{j\geq 0} u_j z^j$ and such that

$$\sum_{j\geq 0}(j+1)(j+2)u_{j+2}z^j + \Big(\sum_{j\geq 0}(j+1)u_{j+1}z^j\Big)\Big(\sum_{k\geq 0} p_k z^k\Big) + \Big(\sum_{j\geq 0} u_j z^j\Big)\Big(\sum_{k\geq 0} q_k z^k\Big) = \sum_{j\geq 0} f_j z^j$$

then for every $j \geq 0$

$$f_j = (j+1)(j+2)u_{j+2} + \sum_{k,\ell:k+\ell=j} \{(\ell+1)u_{\ell+1}p_k + u_\ell q_k\}.$$

Second order complex differential equations in an open subset Ω of $\mathbb{C}$ have the general form

$$u''_z = f(z, u_z, u'_z), \qquad u_0 = u_{z_0}, \quad u'_0 = u'(z_0), \tag{4.36}$$

for u in $C_b^2(E_1)$ with values in a subset V_1 of $\mathbb{C}$, and u' in $C_b(E_2)$ with values in a subset V_1 of $\mathbb{C}$, where E_1 and E_2 are open subsets of $\mathbb{C}$, and f is a continuous function on $\Omega \times E_1 \times E_2$. Let f have the local Lipschitz property

$$\sup_{z\in\Omega} |f(z,u,u') - f(t,v,v')| \leq k_1|u-v| + k_2|u'-v'|$$

for all u and v in V_1, u' and v' in V_2.

Theorem 4.17. *Under the conditions, there exists an unique solution of equation (4.36).*

4.10 Exercises

4.7.1. Solve the equation $xy'' + 2y' + xy = 0$.

4.7.2. Solve $y'' + ma\sin(ax)y'_x + ma^2\cos(ax)y = 0$.

4.7.3. Solve $x^2y'' + x^2y' - (x+2)y = e^{-x}(x^{-1} + 6x^{-2})$.

4.7.4. Solve $y'' - (a+b)y' + aby = abx$.

4.7.5. Solve $x^2y'' + xy' + y = x^{-1}$.

4.7.6. Solve $y'' - y' + 4y = 2\sin(2x)$.

4.7.7. Solve $y'' + 4axy' + 4a^2x^2y = f(x)$.

4.7.8. Solve $x^3y''' + ax^2y'' + bxy' + cy = 0$.

4.7.9. Solve $(a-x)^2y''_x + y\frac{P_1(x)}{P_2(x)} = 0$ with 2nd degree polynomials P_1 and P_2.

4.7.10. Solve $y''_x + x^2y = \cos\frac{x^2}{2}$.

4.7.11. Solve $y''_x + y = 3\cos(3x)$.

4.7.12. Solve $y''_x + 2ay'_x + a^2y = e^{-ax}h(x)$.

4.7.13. Solve $y - xy'_x + x = 0$ and $(n+1)y - xy'_x + \frac{x^{n+1}}{n+1} = 0$.

4.7.14. Solve $y'_x + \frac{xy}{1-x^2} = xy^{\frac{1}{2}}$.

4.7.15. Solve $xy^2y'_x - y^3 = \frac{a^3}{x}$.

4.7.16. Solve $yy'_x + \frac{ay^2}{x^2} = \frac{b}{x^3}$.

4.7.17. Solve $xyy'_x - y^2 = f(x)$ with $y(0) = y_0$.

4.7.18. Solve $x^3y'_x - x^2y + y^3 - xy^2 = 0$.

4.7.19. Solve $ay + y'_x = 2xe^{-ax}$.

4.7.20. Solve $y'_x + \cos^{-2}xy^2 = \cos^2 x$.

4.7.21. Solve $ay + bxy'_x + x^my^n(my + nxy'_x) = 0$.

4.7.22. Solve $y'_x - (1-a)x^{-1}y - y + 1 = 0$.

4.7.23. Solve $x^2 y''_x - 2axy'_x + a(a-1)y = x^a h(x)$.

4.7.24. Solve $x(y'_x + y^3) - y^2 = 0$.

4.7.25. Solve $xy'' - 2yy' = 2x$.

4.7.26. Solve $a_x y_x^2 + y'_x + \frac{b}{xa_x} = 0$.

4.7.27. Solve $xu'_x - u^2 + u = 0$.

4.7.28. Solve $x^2 y'_x y''_x - xy'_x + y = 0$.

4.7.29. Solve $x^3 y''_x - (y - xy'_x)^2 = 0$.

4.7.30. Solve $(1 + y'^2_x)^{\frac{3}{2}} = \frac{a^2}{2x} y''_x$.

4.7.31. Solve $xy'_x = (a^2 + x^2)^{\frac{1}{2}}(xy''_x - y'_x)$.

4.7.32. Solve $xyy''_x + yy'_x + xy'^2_x + \frac{nx^2}{(a^2 - x^2)^{\frac{1}{2}}} = 0$.

4.7.33. Solve $y'_x = (x + a)y''_x + xy'^2$.

4.7.34. Solve $yy''_x + \frac{y'^2_x}{2} + \frac{y^2}{4(x+1)^2}\left\{3 - \frac{1}{(x+1)^2}\right\} = \frac{1}{x+1}$.

4.7.35. Solve $y' + \frac{xy}{1-x^2} - xy^{\frac{1}{2}} = 0$.

4.7.36. Solve $x^2 yy' + ay^2 = b$.

Chapter 5

Linear differential equations in $\mathbb{R}^p$

This chapter studies first and second order spatial and spatio-temporal differential equations. We give exact and implicit solutions of first order equations, Laplace's equation, heat conduction, wave, potential and elasticity equations. Several solutions of the equations with constant coefficient are presented and the differential equations with functional coefficients are also studied.

5.1 Introduction

A first order linear differential equation in $\mathbb{R}^p$ is defined by a vector operator L_x with components

$$L_{k,x} = \frac{\partial}{\partial x_k} + \gamma_k(x), \tag{5.1}$$

as a system of p first order partial differential equations having a vector second member f in $L^2(\mathbb{R}^p)$ or by a linear combination of these operators

$$L_{k,x} = \sum_{k=1}^{p} \beta_k(x) L_k(x)$$

as $L_x u_x = f(x)$ with boundary conditions $u(x_0)$ for functions u of $C_1(\mathbb{R}^p)$ and for all x in a domain Ω of $\mathbb{R}^p$ and $u(x_0)$ in $\partial\Omega$. There is no general explicit solution of a linear combination of partial differential equation with functional coefficients.

In $\mathbb{R}^p$, let $\nabla_x u = (u'_{k,x})_{k=1,\dots,p}$ be the vector of the partial derivatives of u with respect to the components of x. Let u be a function u of $C_1(\mathbb{R}^p \times \mathbb{R})$, a differential equation $L_x u(x) = (\nabla_x + \gamma)u = 0$, with a vector function

$\gamma(x) = \{\gamma_k(x_k)\}_{k=1,\dots,p}$, has the solution

$$u_x = u(x_0) \exp\Big(-\sum_{k=1}^{p} \int_{x_{0,k}}^{x_k} \gamma_k(s)\,ds\Big) \tag{5.2}$$

and a solution of the inhomogeneous equation $L_{k,x}u_x = f_{k,x}$ under the same conditions is

$$\begin{aligned} u_x &= u(x_0) \exp\Big(-\sum_{k=1}^{p} \int_{x_{0,k}}^{x_k} \gamma_k(s)\,ds\Big) \\ &\quad + \sum_{k=1}^{p} \int_{x_{0,k}}^{x_k} f_k(t_k) \exp\Big(\sum_{k=1}^{p} \int_{t_k}^{x_k} \gamma_k(s)\,ds\Big)\,dt_k\,. \end{aligned} \tag{5.3}$$

The boundary conditions on a domain strictly included in a rectangle of $\mathbb{R}^2$ are two marginal functions, they may be constant.

A nonlinear first order differential equation $\nabla_x u = f(u_x, x)$ in a subset Ω of $\mathbb{R}^p$, with boundary value $u(x_0)$ at x_0 in $\partial\Omega$, has a solution

$$u_x = u(x_0) + \sum_{k=1}^{p} \int_{x_{0,k}}^{x_k} f_k(u_{s_{x,k}}, s_{x,k})\,ds_{x,k} \tag{5.4}$$

where $s_{x,k}$ is equal to the vector x except its kth component s. It is well known that a small variation of the initial condition at x_0 implies a small variation of the solutions of the differential equations in $\mathbb{R}$. This result is still valid in $\mathbb{R} \times \mathbb{R}^p$.

Proposition 5.1. *For every real function u of $C_1(\mathbb{R}^p)$, let $f(u,x)$ be a continuous function of $L^1(\mathbb{R} \times \mathbb{R}^p)$ with values in $\mathbb{R}^p$ and locally Lipschitz with respect to u, for all x in $\mathbb{R}^p$, u_1 in $C_1(\mathbb{R}^p)$ and u_2 in a neighborhood of u_1*

$$\|f(u_1,x) - f(u_2,x)\|_2 \leq C|u_1 - u_2|.$$

Let u be a solution of the differential equation $\nabla_x u(x) = f(u_x, x)$ with boundary conditions $u(x_0)$, and let u_η be a solution of the same differential equation with boundary conditions $u_\eta(x_0) = u(x_0) + \eta > u(x_0)$, for every x_0 in $\partial\Omega$. Then

$$|u(x) - u_\eta(x)| \leq \eta e^{C \sum_{k=1}^{p}(x - x_{0,k})}.$$

Proof. From (5.4), $|u(x) - u_\eta(x)|$ has the bound

$$\begin{aligned} |u(x) - u_\eta(x)| &\leq \eta + \sum_{k=1}^{p} \int_{x_{0,k}}^{x_k} |f_k(u_{\eta,s_{x,k}}, s) - f_k(u_{s_{x,k}}, s)|\,ds \\ &\leq \eta + C \sum_{k=1}^{p} \int_{x_{0,k}}^{x_k} |u_s - u_{\eta,s}|\,ds \end{aligned}$$

from the Lipschitz condition for f. The result is then a consequence of Gronwall's inequality

$$\varphi_t \le c \exp\{a \int_{t_0}^t \psi_s \, ds\}$$

for all positive functions φ and ψ and all positive constants a and c such that $\varphi_t \le c + a \int_{t_0}^t \varphi_s \psi_s \, ds$. □

For a function u of $C_1(\mathbb{R}^p \times \mathbb{R})$, the differential equation $du(x, y) = 0$ is equivalent to a vector of k partial differential equations

$$(u'_{k,x} + y'_{k,x} u'_y)_{k=1,\dots,p} = 0.$$

For example, in $\mathbb{R}^2$ the differential equation

$$x \frac{\partial}{\partial x} u(x, y) + y \frac{\partial}{\partial y} u(x, y) = 0$$

has the solutions

$$u(x, y) = c_1 e^{\alpha_1 \frac{y}{x}} + c_2 e^{-\alpha_1 \frac{y}{x}} + c_3 e^{\alpha_2 \frac{y}{x}} + c_4 e^{-\alpha_2 \frac{x}{y}},$$

for all constants $\alpha_1, \alpha_2, c_1, \dots, c_4$. With initial values $u(0, y) = u_0$ for every $y > 0$, a solution is

$$u(x, y) = u_0 (c e^{\alpha \frac{x}{y}} + (1 - c) e^{-\alpha \frac{x}{y}}).$$

The differential equation

$$ax \frac{\partial}{\partial x} u(x, y) + y \frac{\partial}{\partial y} u(x, y) = 0$$

has the solutions

$$u(x, y) = c_1 e^{\alpha_1 \frac{x^a}{y}} + c_2 e^{-\alpha_1 \frac{x^a}{y}} + c_3 e^{\alpha_2 \frac{y^a}{x}} + c_4 e^{-\alpha_2 \frac{y^a}{x}},$$

moreover, $a y'_x = x^{-1} y$ hence $y(x) = k x^{\frac{1}{a}}$. The equation

$$x \frac{\partial}{\partial x} u(x, y) + ay \frac{\partial}{\partial y} u(x, y) = 0$$

has the solutions

$$u(x, y) = c_1 e^{\alpha_1 \frac{y}{x^a}} + c_2 e^{-\alpha_1 \frac{x}{y^a}} + c_3 e^{\alpha_2 \frac{y^a}{x}} + c_4 e^{-\alpha_2 \frac{y^a}{x}}$$

for all constants $\alpha, c_1, \dots, c_4$. $\alpha > 0$ and c in $[0, 1]$ such that the constraints of the initial values are fulfilled, and $y(x) = k x^a$.

The solutions of such equations are not necessarily exponential, more generally, evry function $h(\frac{x}{y})$ is solution of the differential equation

$xu'_x + yu'_y = 0$. Denoting $u(x,y) = f(x)\,g(y)$ with functions $f \neq 0$ and $g \neq 0$, the differential equation

$$ax\frac{\partial}{\partial x}u(x,y) + by\frac{\partial}{\partial y}u(x,y) = 0$$

is equivalent to

$$ax\frac{f'(x)}{f(x)} + by\frac{g'(x)}{g(x)} = 0.$$

As they do not depend on the same variable, each term of this sum is a constant $\pm k$ which yields

$$f(x) = f_0\Big(\frac{x}{x_0}\Big)^{\frac{k}{a}}, \quad g(x) = g_0\Big(\frac{y_0}{y}\Big)^{\frac{k}{b}}.$$

In the same way, the function $u(x,y) = e^{H(x)K(y)}$ of $C_1(\mathbb{R}^2)$ defined by functions H and K with derivatives $h = H'$ and, respectively, $k = K'$, is solution of the differential equation

$$\varphi(x)\frac{\partial}{\partial x}u(x,y) + \psi(y)\frac{\partial}{\partial y}u(x,y) = 0 \tag{5.5}$$

with $\varphi(x) = H(x)h^{-1}(x)$ and $\psi(y) = K(y)k^{-1}(y)$, for all x and y in intervals where the functions h and k are non null and under conditions for initial values. In $C_1(\mathbb{R}^p \times \mathbb{R})$, the differential equation

$$\varphi(x).\nabla_x u(x,y) + \psi(y).\nabla_y u(x,y) = 0$$

defined by continuous functions φ, ψ and f with values in $\mathbb{R}^p$, has the solution $u(x,y) = e^{\sum_{m=1}^{p} H_m(x)K_m(y)}$.

The function $u(x,y) = e^{H(x)+K(y)}$ is solution of the same differential equation defined by the functions $\varphi(x) = h^{-1}(x)$ and $\psi(y) = k^{-1}(y)$, for all x and y in intervals where h and k are non null and under conditions for initial values. It is also solution of the differential equation

$$k(y)\frac{\partial}{\partial x}u(x,y) + h(x)\frac{\partial}{\partial y}u(x,y) = 0 \tag{5.6}$$

without constraint on the functions h and k. These examples are not exhaustive and the functions φ and ψ may depend on (x,y).

The solutions of the second order differential equations in Hilbert spaces on convex subsets of $\mathbb{R}^p$ depend on their values on the frontier of the domains of the equations and on the Hilbert spaces. A periodic function with period $\omega\pi$ satisfies a periodicity condition $u(0) = u(\omega\pi)$ at the frontier of the interval $[0,\omega\pi]$, so u constant on the frontier of the domain.

The equations can be expressed in several equivalent forms. For example, the second order partial derivatives of the function $u(x,y) = e^{H(x)+K(y)}$ satisfy the differential equation

$$(k' + k^2)(y)\frac{\partial^2}{\partial x^2}u(x,y) + (h' + h^2)(x)\frac{\partial}{\partial y}u(x,y) = 0$$

and $u(x,y) = e^{H(x)K(y)}$ is solution of the differential equation

$$H(x)\{k'(y) + H(x)k^2(y)\}\frac{\partial^2}{\partial x^2}u(x,y)$$
$$+ K(y)\{h'(x) + h^2(x)K(y)\}\frac{\partial}{\partial y}u(x,y) = 0.$$

Let u be a function of $C_2(\mathbb{R}^p)$ and let f be a function of u, $\nabla_x u$ and x with values in $(\mathbb{R}^p)^{\otimes 2}$ endowed with its l_2-norm. The second order differential equation $\nabla_x\nabla_x u = f(u, u'_x, x)$ in a subset Ω of $\mathbb{R}^p$, with boundary value u_0 at x_0 in $\partial\Omega$, has a solution

$$u(x) = u(x_0) + \sum_{k=1}^p\sum_{l=1}^p \int_{x_{0,k}}^{x_k}\Big[u'_{x,l}(x_{0,l})$$
$$+ \int_{x_{0,l}}^{x_l}\{f_k(u(v_{x,k,l}(s,t))) - f_k(u_\eta(v_{x,k,l}(s,t)))\}\,ds\Big]\,dt, \quad (5.7)$$

where $v_{x,k,l}(s,t)$ is identical to x but its (k,l)th components equal to (s,t). This differential equation only includes the second order partial derivatives independently of each other. The Sturm-Liouville operator is generalized in $\mathbb{R}^p$ as

$$L_a u(x) = \sum_{i=1}^p \frac{\partial}{\partial x_i}\Big\{a_i(x)\frac{\partial u}{\partial x_i}\Big\}. \quad (5.8)$$

Proposition 5.2. *For every real function u of $C_1(\mathbb{R}^p)$, let $f(u, u'_x, x)$ be a continuous function of $L^1(\mathbb{R}\times\mathbb{R}^{2p})$ with values in $\mathbb{R}^{p^2}$, such that*

$$\|f(u_{1x}, u'_{1x}, x) - f(u_{2x}, u'_{2x}, x)\|_2 \le C_1(|u_{1x} - u_{2x}| + \|u'_{1x} - u'_{2x}\|_2)$$

and $\|u'_{1x} - u'_{2x}\|_2 \le C_2(|u_{1x} - u_{2x}|$. Let u be a solution of the differential equation $\nabla_x\nabla_x u(x) = f(u, u'_x, x)$ with boundary conditions $u(x_0)$ and $u'_x(x_0)$, and let $u_{\eta,\varsigma}$ be a solution of the same differential equation with boundary conditions $u_{\eta,\varsigma}(x_0) = u(x_0) + \eta$ and $u'_{x,\eta,\varsigma}(x_0) = u'_x(x_0) + \zeta$, for every x_0 in $\partial\Omega$, with strictly positive constants η and ζ. Then

$$|u(x) - u_\eta(x)| \le \eta\exp\Big\{C\sum_{k=1}^p\sum_{l=1}^p(x_l - x_{0,l})(x_k - x_{0,k})\Big\} - C^{-1}\zeta.$$

Proof. From (5.7) and the assumptions $|u(x) - u_\eta(x)|$ is bounded by the sum $S(x)$ of $\eta + \zeta \sum_{k=1}^p (x_k - x_{0,k})$ and

$$\sum_{k=1}^p \sum_{l=1}^p \int_{x_{0,k}}^{x_k} \int_{x_{0,l}}^{x_l} |f_k(u, u'_{x_l}, s) - f_k(u_\eta, u'_{\eta,x_l}, s)| \, ds \, dt$$

$$\leq C \sum_{k=1}^p \sum_{l=1}^p \int_{x_{0,k}}^{x_k} \int_{x_{0,l}}^{x_l} |u - u_\eta| \, ds \, dt$$

with $C = C_1(1 + C_2)$. The derivative of $CS(x) + \zeta$ with respect to x_k is

$$\zeta + \sum_{l=1}^p \int_{x_{0,l}}^{x_l} \{f_k(u(v_{x,l}(s))) - f_k(u_\eta(v_{x,l}(s)))\} \, ds$$

and its ratio with $S(x) + C^{-1}\zeta$ is lower than $C \sum_{l=1}^p (x_l - x_{0,l})$. From Gronwall's inequality $|u(x) - u_\eta(x)|$ is bounded by $\eta \exp\{C \sum_{k=1}^p \sum_{l=1}^p (x_l - x_{0,l})(x_k - x_{0,k})\} - C^{-1}\zeta$. □

5.2 Laplace's differential equation

Laplace's differential equation $\Delta u = 0$ on $\mathbb{R}^2$ has the complex solution $u(z) = e^z$, for every $z(x, y) = x + iy$, and trigonometric solutions in a convex subset Ω of $\mathbb{R}^2$,

$$u(x) = \int_\Omega h(\lambda_1 + \lambda_2)\{c_1 \cos\{\alpha(\lambda_1 - x_1)$$
$$+ c_2 \sin\{\alpha(\lambda_1 - x_1)\} e^{-\alpha(\lambda_2 - x_2)} \, d\lambda_1 \, d\lambda_2,$$

where Ω is considered as a subset of $\mathbb{R}\times] - k\pi, k\pi]$, for every function h such that the integral is finite on Ω. The boundary values $u(x_1, 0)$ and $u(0, x_2)$ depend on h. The limiting behaviour at infinity of these solution are different but their boundary properties on bounded sets are not always different and the unicity cannot be established.

Example 38. The function $u(x, y) = \log(x^2 + y^2)$ is solution of Laplace's equation $\Delta_{x,y} u(x, y) = 0$ on a strictly positive open set of $\mathbb{R}^2$, its second order derivatives are $u''_{xx} = 2(y^2 - x^2)(x^2 + y^2)^{-2}$ and $u''_{yy} = 2(x^2 - y^2)(x^2 + y^2)^{-2}$.

For a function u of $C_2(\mathbb{R}^p)$, with $p \geq 1$, provided with the $l_{2,p}$-norm $r(x_1, \cdots, x_p) = \{\sum_{k=1}^p x_k^2\}^{\frac{1}{2}}$, the functions r and $\log r$ have the Laplacian

$$\Delta r = \frac{p-1}{r},$$

$$\Delta \log r = \frac{p-2}{r^2}.$$

Proposition 5.3. *Laplace's differential equation $\Delta u = 0$ in $\mathbb{R}^p$, $p > 2$, has the solution*

$$u(r) = r^{-(p-2)}.$$

Proof. Let $u(r) = r^k(x)$, for x in $\mathbb{R}^p$, its partial derivatives with respect to x_i are

$$u_i'(r) = kr_i'(x)r^{k-1}(x) = kx_i r^{k-2}(x),$$
$$u_{ii}''(r) = kr^{k-2}(x) + k(k-2)x_i^2 r^{k-4}(x)$$

$\Delta u(r) = k(p + k - 2)r^{k-2}(x)$ is zero for $k = 2 - p$. □

Poisson's second order differential equation is

$$\Delta u = f \tag{5.9}$$

for a function u of $C_2(\Omega)$ where Ω a convex subset of $\mathbb{R}^p$, with initial conditions on the frontier $\partial\Omega$ of Ω. The Dirichlet problem is to find out solutions of Poisson's equation in the Sobolev space

$$H_{1,0}(\Omega) = \{v \in C_2(\Omega) : v \in L^2(\Omega), v' \in L^2(\Omega), v_{|\partial\Omega} = 0\}$$

of the equation $u''(x) = f(x)$ in a subset Ω of $\mathbb{R}^p$ with $\partial\Omega$ in $\mathbb{R}^2$. Green's formula for real functions u and v of $C^2 \cap H_{1,0}(\Omega)$ states that

$$\int_\Omega (u\Delta v - v\Delta u) + \int_{\partial\Omega} \{u(v_x' + v_y') - v(u_x' + u_y')\} = 0. \tag{5.10}$$

Theorem 5.1. *A solution in $H_{1,0}(\Omega)$ of Laplace's equation $u''(x) = f(x)$ is solution of differential equation*

$$\sum_{i=1}^{2} \int_\Omega \frac{\partial}{\partial x_i} u(x) \frac{\partial}{\partial x_i} v(x)\, dx = \int_\Omega f(x)v(x)\, dx,$$

for every v of $H_{1,0}(\Omega)$.

A solution of the Dirichlet equation $u''(x) = \delta_y(x)$ in $\Omega =]-1, 1[$, defined by the Dirac measure δ_y is

$$u(x) = \int_\Omega E(x, y) f(y)\, dy$$

with the second order primitive of the Dirac measure

$$E(x, y) = (x - y)1_{\{-1<y\leq x<1\}}.$$

Its values on the frontier $\partial\Omega$ are $u(-1) = 0$ and $u(1) = \int_{-1}^{1}(1-y)f(y)\,dy$. In the subset Ω^p of $\mathbb{R}^p$, the second order primitive of the Dirac measure is

$$E_p(x,y) = \prod_{i=1}^{p}(x_i - y_i)1_{\{y_i \le x_i\}}.$$

Theorem 5.2. *The function* $u(x) = \int_\Omega E_p(x,y)f(y)\,dy$ *is solution of Laplace's differential equation* $\Delta u(x) = f(x)$ *in* $\Omega =]-1,1[^{\otimes p}$, *with the initial conditions* $u(x_0) = u'(x_0) = 0$ *for every* x_0 *having a component equal to* -1.

For a function u of $C_2(\mathbb{R}^p)$, $p \ge 1$, depending on the coordinates $(x_1, \cdots, x_p)$ through the $l_{2,p}$-norm $r = \{\sum_{k=1}^p x_k^2\}^{\frac{1}{2}}$, the Poisson equation with an arbitrary second member $f(r)$ is solved using only the expression of Δu. Its solution depends on the dimension p of the space.

Theorem 5.3. *For a function* u *of* $C_2(\mathbb{R}^p)$ *depending on the* $l_{2,p}$*-norm* r *on* $\mathbb{R}^p$, *the Poisson equation with initial conditions*

$$\Delta u(r) = f(r),\ r \in [r_0, r_1],$$
$$u(r_0) = u_0, \quad u'(r_1) = u'_1,$$

has the solution

$$u_p(r) = u_0 + \frac{1}{p-1}\Big(\frac{1}{r_0^{p-1}} - \frac{1}{r^{p-1}}\Big)\Big\{u'_1 r_1^p + \int_{r_0}^{r_1} s^p f(s)\,ds\Big\} + \int_{r_0}^{r} sf(s)\,ds,$$

if $p > 1$ *and*

$$u_1(r) = u_0 + u'_1 r_1 \log\frac{r}{r_0} + \int_{r_0}^{r} y^{-1}\int_{y}^{r_1} sf(s)\,ds, \quad \textit{if } p = 1.$$

Proof. From (1.9), the Poisson equation is equivalent to the differential equation $pu'(r) + ru''(r) = rf(r)$ also written as $(r^p u'_r)' = r^p f(r)$. By integration on $(r, r_1]$

$$r^p u'(r) = r_1^p u'_1 - \int_r^{r_1} s^p f(s)\,ds$$

and the unique solution is obtained as the primitive of $u'(r)$ with initial value u_0 at r_0 and using an integration by parts. □

Example 39. The equation

$$\Delta_{x,y} u(x,y) = \frac{4(x+y)}{x^2+y^2}$$

has the solution $u(x,y) = (x+y)\log(x^2+y^2)$. Its second derivatives are $u''_{xx} = (6x+2y)(x^2+y^2)^{-1} - 4x^2(x-y)(x^2+y^2)^{-2}$ and $u''_{yy} = (6y+2x)(x^2+y^2)^{-1} - 4y^2(x-y)(x^2+y^2)^{-2}$.

Example 40. In $\mathbb{R}^p$, the function $u(x,r) = x^p \log(r)$ is solution of the differential equation

$$\Delta u(x,r) - p(p-1)\frac{u(x,r)}{x^2} + \frac{2px^p}{r^2} = 0.$$

where the Laplacian is considered with respect to the p coordinates of r.

Proposition 5.4. *In* $\mathbb{R}^p$, *the function* $u(r) = r^{-\frac{p-1}{2}}\cos(kr)$ *is solution of the differential equation*

$$\Delta u(r) - \frac{(p-1)(p-3)}{4}\frac{u(r)}{r^2} - k^2 u(r) = 0.$$

Proof. The second derivative of $u(r) = r^n \cos(kr)$ with respect to x_i and its Laplacian are

$$\begin{aligned} u_i''(r) &= \cos(kr)\{nr^{n-2} + n(n-2)x_i^2 r^{n-4} - k^2 x_i^2 r^{n-2}\} \\ &\quad - k\sin(kr)\{r^{n-1} + nx_i^2 r^{n-3} + (n-1)x_i^2 r^{n-3}\}, \\ \Delta u(r) &= r^{n-2}\cos(kr)\{np + n(n-2) - k^2 r^2\} \\ &\quad - kr^{n-1}\sin(kr)(p+2n-1) \end{aligned}$$

and the coefficient of the sine function is zero as $n = -\frac{p-1}{2}$. □

For $p = 3$, the differential equation reduces to $\Delta u(r) - k^2 u(r) = 0$.

Proposition 5.5. *On a symmetric set* Ω *of* $\mathbb{R}^p$ *having the width* d, *the function* $u(r) = v(r)w(r)$ *defined by* $v(r) = r^{-\frac{p-1}{2}}\cos(kr)$ *and*

$$w(r) = w(r_0) + \frac{1}{pd^{p-1}}\int v^{-2}(s)\int_\Omega v(\rho)f(\rho)1_{\{\rho_i \le s_i\}}\,d\rho\,ds_i$$

is solution of the differential equation

$$\Delta u(r) - \frac{(p-1)(p-3)}{4}\frac{u(r)}{r^2} - k^2 u(r) = f(r).$$

Proof. Let $u(r) = v(r)w(r)$ be a solution with $v(r) = r^{-\frac{p-1}{2}}\cos(kr)$ solution of the homogeneous equation, then w is solution of the differential equation

$$2\sum_{i=1}^{p} v_i'(r)w_i'(r) + v(r)\Delta w(r) = f(r).$$

Assuming that $w_i'(r) = v^{-2}(r)h(r)$ for every i yields

$$w_i''(r) = \frac{h_i'(r)}{v^2(r)} - \frac{2v_i'(r)h(r)}{v^3(r)}$$

and the differential equation for w becomes

$$v(r)f(r) = \sum_{i=1}^{p} h_i'(r).$$

Integrating this equation on Ω implies

$$h(r) = \frac{1}{pd^{p-1}} \int_\Omega v(s)f(s)1_{\{s_i \leq x_i\}}\, ds,$$

$$w(r) = w(r_0) + \int v^{-2}(s)h(s)1_{\{s_i \leq x_i\}}\, ds_i. \qquad \square$$

Proposition 5.6. *Let u and v in $C_2(\Omega)$ where Ω is a convex subset of $\mathbb{R}^p$, such that $\Delta u(x) = f(x)$ and $u\Delta v + 2(\nabla u)^t\nabla v = 0$. Then Laplace's differential equation $\Delta y(x) = f(x)$ has the solution $y = vf$*

Laplace's differential equation has therefore solutions $y = U + uv$ where v is a general solution of the first order equation for ∇v of Proposition 5.6 and u is the solution of Theorem 5.3. The integral of

$$u\Delta v + 2(\nabla u)^t\nabla v = 0$$

is equivalent to $u\sum_i v'_{x_i}|_{\partial\Omega} + \int_\Omega(\nabla u)^t\nabla v = 0$ and

$$u\sum_i v'_{x_i}\Big|_{\partial\Omega} - v\sum_i u'_{x_i}\Big|_{\partial\Omega} = \int_\Omega u\Delta v - \int_\Omega vf.$$

A periodic function u of $L^2([0,T])$ having the frequencies $\omega_k = T^{-1}2k\pi$, $k \geq 0$, has a convergent Fourier series $u(x) = \sum_{j\geq 0}^\infty\{a_j\cos(w_jx) + b_j\sin(w_jx)$. Let f be a periodic function of $L^2([0,T])$ with a Fourier series $f = \sum_{j\geq 0}^\infty\{A_j\cos(w_jx) + B_j\sin(w_jx)\}$ such that $\Delta u(x) = f$, then for every x in $[0,T]$

$$\sum_{j=0}^{\infty}\{(w_j^2a_j - A_j)\cos(w_jx) + (w_j^2b_j - B_j)\sin(w_jx)\} = 0.$$

Solving this equation defines the constants of the expansion of u according those of the expansion of f, the frequencies of f and u must be identical and $a_j = w_j^{-2}A_j$, $b_j = w_j^{-2}B_j$.

5.3 Potential equations

The potential theory concerns mainly the solutions in $H^2(\Omega) = C_2 \cap L^2(\Omega)$, of differential equations

$$\Delta u = f(u) \text{ on } \Omega,$$
$$\sum_{i=1}^{n} u'_{x_i} v_i = 0 \text{ on } \partial\Omega,$$

where $(v_i)_{i=1,\dots,n}$ is a basis of vectors orthogonal to $\partial\Omega$, and Ω is a domain of $\mathbb{R}^n$, $n = 1, 2$ or 3. The problem is also formulated as

$$\sum_{i=1}^{n} \int_\Omega u''_{x_i} v_i dx = \sum_{i=1}^{n} \int_\Omega f v_i dx,$$
$$\sum_{i=1}^{n} \int_{\partial\Omega} u'_{x_i} v_i = 0, \ i = 1, \dots, n,$$

and they imply

$$\sum_{i=1}^{n} \int_\Omega u'_{x_i} v'_{x_i} dx = \int_\Omega f v_i dx.$$

The problems in physics provide various potential models.

In a bounded connex domain Ω of $\mathbb{R}_+\mathbb{R}^3$, with surface $S = \partial\Omega$, the solutions of the differential equation

$$\Delta V(t,x,y,z) - P(x,y,z)V(t,x,y,z) = f(t,x,y,z), \tag{5.11}$$
$$\sum_{i=1}^{3} V'_{x_i} v_i = 0 \text{ on } \partial\Omega$$

are the sum of a solution of the homogeneous equation and a solution of (5.11). Let

$$V_1(t,x,y,z) = \varphi(x,y,z)(\alpha \sin t + \beta \cos t),$$
$$\Delta\varphi(x,y,z) = -P(x,y,z)\varphi(x,y,z),$$

then V_1 is solution of the homogeneous differential equation (5.11) if the initial conditions are satisfied by φ.

On $\mathbb{R}$, a function $\varphi(x) = e^{A(x)}$, with a twice continuously differentiable functions A having a first derivative a, is solution if

$$\Delta\varphi(x) = \Big\{a'(x) + a^2(x)\Big\}\varphi(x,y,z) = -P(x)\varphi(x),$$

hence $a'(x) + a^2(x) = -P(x)$. A solution of the differential equation (5.11) has the form $V(t,x) = u(x)V_1(t,x)$, with $V_1(t,x) = \varphi(x)(\alpha \sin t + \beta \cos t)$ and u satisfies the second order differential equation

$$u''_{xx}(x)V_1(t,x) + 2u'_x(x)V'_{1x}(t,x) - u(x)P(x)V_1(t,x) = f(t,x),$$

a solution is provided by Proposition (4.9).

The Poisson-Boltzmann equation for the ionic potential of a fluid in $\mathbb{R}^3$ with a constant load q and at the Kelvin temperature T is

$$\Delta V \pm a \exp\Big\{-\frac{qV}{k_B T}\Big\} = 0. \tag{5.12}$$

A solution of (5.12) with $\pm a = 1$ is a function u such that

$$v(x,y,z) = \frac{2}{b}\exp\Big\{\frac{bu(x,y,z)}{2}\Big\} = \frac{\sqrt{2}}{\sqrt{3b}}(x+y+z),$$

with $b = qV(k_B T)^{-1} > 0$. The partial derivatives of u and v satisfy

$$v'_x = u'_x \exp\Big\{\frac{bu}{2}\Big\} = \frac{\sqrt{2}}{\sqrt{3b}},$$
$$u'^2_x = \frac{2}{3b}e^{-bu}$$

hence $u''_x = -\frac{1}{3}e^{-bu}$. Since u''_y and u''_z are identical to u''_x, the function u is solution of the differential equation $\Delta u + e^{-bu} = 0$. From the expressions of the functions v, a solution of (5.12) is

$$V(x,y,z) = \pm au = \pm\frac{a}{b}\log\Big\{\frac{b(x+y+z)^2}{6}\Big\}$$

and

$$ae^{-bV} = \Big\{\frac{b(x+y+z)^2}{6}\Big\}^{-\frac{a}{b}}.$$

If the temperature T is variable t, let $p = tb$. The spatial derivatives of the function u are not changed and its derivative with respect to t is

$$u'_t = \frac{u}{t} - \frac{1}{p},$$

the function V is solution of the differential equation

$$\Delta V - V'_t + \frac{V}{T} \pm ae^{-bV} = \frac{1}{p}.$$

More generally, the Poisson-Boltzmann equation for n kinds of ions with respective loads q_i, $i = 1, \ldots, n$, is a mean differential equation

$$\Delta V \pm \sum_{i=1}^{n} a_i \exp\Big\{-\frac{q_i V}{k_B T}\Big\} = 0.$$

It satisfies the equations

$$V_x' V_x'' \pm \frac{1}{3} \sum_{i=1}^{n} a_i V_x' e^{-b_i V} = 0,$$

$$V_x'^2 \mp \frac{2}{3} \sum_{i=1}^{n} \frac{a_i}{b_i} e^{-b_i V} = C,$$

with a constant C. Taking $C = 0$, the equations become

$$V_x' = i^k \Big\{2 \sum_{i=1}^{n} \frac{a_i}{b_i} e^{-b_i V}\Big\}^{\frac{1}{2}}, \ k = 0, 1,$$

and it cannot be integrated as in the case $n = 1$.

The distance r from the earth of a solid with a mass m follows the differential equation $r_t'' = -Mmg(r)$ under the gravitational attraction, with

$$g(r) = \frac{\gamma}{(R+r)^2}$$

where M and R are the mass and the radius of the earth. Multiplying this equation by r_t' and integrating, its velocity is such that

$$\frac{r_t'^2}{2} = a + \frac{mM\gamma}{R+r}$$

with $2a = r_0'^2$. It follows that

$$r_t'(R+r)^{\frac{1}{2}} = \{2a(R+r) + 2mM\gamma\}^{\frac{1}{2}}.$$

Approximating the right-hand term by the constant $K = (aR + mM\gamma)^{\frac{1}{2}}$ as r is small implies

$$R + r_t \approx \Big(\frac{3}{2} K t\Big)^{\frac{2}{3}}.$$

The distance r becomes zero after a duration approximated by

$$T = \frac{2}{3}(aR + mM\gamma)^{-\frac{1}{2}} R^{\frac{3}{2}}.$$

If $r_0' = 0$, $T = \frac{2}{3}(mM\gamma)^{-\frac{1}{2}} R^{\frac{3}{2}}$ and the quantity $m^{\frac{1}{2}} T$ must be a constant.

The potential of attraction of a disk having the radius r has been defined in polar coordinates as $V(r) = M(c^2 + r^2)^{-\frac{1}{2}}$ and a function u on $\mathbb{R}^2$ such that $\Delta_r u = -V(r)$ has a velocity $v = u'_r$ satisfying the first order differential equation

$$v'(r) + \frac{2}{r}v(r) + V(r) = 0,$$

a solution $v(r)$ is given by theorem (4.1). The function G is $G(r) = ar^{-2}$ with the constant $a = G(r_0)r_0^2$ at an arbitrary r_0, and

$$\begin{aligned} v(r) &= \frac{a}{r^2} + \frac{1}{r^2}\int_0^r s^2 V(s)\,ds \\ &= \frac{a}{r^2} + \Big(\frac{c^2}{r^2} + 1\Big)^{\frac{1}{2}} - \frac{1}{r^2}\int_0^{\frac{r}{c}} (1 + y^2)^{\frac{1}{2}}\,dy. \end{aligned}$$

The function u is deduced as

$$\begin{aligned} u(r) &= -\frac{a}{r} + (c^2 + r^2)^{\frac{1}{2}} - \frac{1}{r}\int_0^r s^2 V(s)\,ds \\ &= -\frac{a}{r} + \Big(1 - \frac{1}{r}\Big)(c^2 + r^2)^{\frac{1}{2}} + \frac{1}{r}\int_0^{\frac{r}{c}} (1 + y^2)^{\frac{1}{2}}\,dy. \end{aligned}$$

The last term depends on the primitive of $h(x) = (1 + x^2)^{\frac{1}{2}} = (1 + x^2)u'_x$ where $u(x) = \arg\sinh x$ denoting $x^2 = \sinh^2 u$. The function h becomes $h(x) = (1 + \sinh^2 u_x)u'_x$ and its primitive is $H(x) = u(x) + \frac{1}{3}\sinh^3 u(x)$.

From (1.9), the Laplacian of $V(r)$ in $\mathbb{R}^2$ is

$$\Delta_r V = \frac{2}{r}V'(r) + V''(r) = -\frac{3}{M^2}V^3\Big(1 - \frac{r^2}{M^2}V^2\Big).$$

The equation $\Delta_r V = 0$ is equivalent to $MV = r$ which has two solutions

$$r^2 = \frac{1}{2}\{\pm(c^4 + 4r^4M^4)^{\frac{1}{2}} - c^2\}.$$

In the polar coordinates of the circle in $\mathbb{R}^2$, the partial derivatives of a function $u(r, \theta)$ are

$$\begin{aligned} u'_x &= u'_\theta \theta'_x + u'_r r'_x = \frac{u'_\theta}{y} + \frac{x u'_r}{r}, \\ u''_{xx} &= \frac{u''_{\theta\theta}}{y^2} + 2\frac{x u''_{\theta r}}{ry} + \frac{x^2 u''_{rr}}{r^2}, \end{aligned} \tag{5.13}$$

it follows that

$$\Delta_{x,y} u = \Big(\frac{1}{x^2} + \frac{1}{y^2}\Big)u''_{\theta\theta} + \frac{2r}{xy}u''_{\theta r} + u''_{rr}.$$

The derivatives of the potential $V(r)$ with respect to θ are zero and its Laplacien with respect to x reduces to $\Delta_r V$.

The coordinates $x_t = r_t \cos\theta_t$ and $y_t = r_t \sin\theta_t$ of a system varying with a strictly positive parameter t have the derivatives

$$x'_t = r'_t \cos\theta_t - r_t \sin\theta_t \theta'_t = \frac{r'_t}{r} x - y\theta'_t,$$
$$y'_t = r'_t \sin\theta_t + r_t \cos\theta_t \theta'_t = \frac{r'_t}{r} y + x\theta'_t.$$

The derivatives of the polar coordinates $\theta_t = \arctan x_t^{-1} y_t$ and the norm $r_t = (x_t^2 + y_t^2)^{\frac{1}{2}}$ are written as

$$r'_t = r'_x x'_t + r'_y y'_t = \frac{x x'_t + y y'_t}{r},$$
$$\theta'_x = \frac{r'_x \cos\theta - 1}{r \sin\theta} = -\frac{\sin\theta}{r}, \quad \theta'_y = \frac{\cos\theta}{r},$$
$$\theta'_t = \frac{x y'_t - y x'_t}{r^2}.$$

The cartesian Laplacian of $r^2 u$ is calculated from the derivatives

$$\frac{d}{dx}(r^2 u) = 2 r r'_x u + r^2 u'_x,$$
$$\frac{d^2}{dx^2}(r^2 u) = 2 r r''_{xx} u + 2 r'^2_x u + 4 r r'_x u'_x + r^2 u''_{xx},$$

they depend on the derivatives of r

$$r'_x = \frac{x}{r}, \quad r''_{xx} = \frac{1}{r} - \frac{x^2}{r^3},$$

which implies

$$\begin{aligned}\Delta_{x,y}(r^2 u) &= 2r(r''_{xx} + r''_{yy})u + 2(r'^2_x + r'^2_y)u \\ &\quad + 4r(r'_x u'_x + r'_y u'_y) + r^2(u''_{xx} + u''_{yy}) \\ &= 3u + 4(x u'_x + y u'_y) + r^2 \Delta_{x,y} u.\end{aligned}$$

In the polar coordinates, $\Delta_{x,y}(r^2 u)$ is mapped by (5.13) into

$$\begin{aligned}\Delta_{x,y}(r^2 u) &= 3u + 4r\Big(\frac{r}{xy} u'_\theta + u'_r\Big) + r^2 u''_{rr} \\ &\quad + \frac{2 r u''_{\theta r}}{\cos\theta \sin\theta} + \frac{u''_{\theta\theta}}{\cos^2\theta \sin^2\theta} \\ &= \Big\{3 + 4r\Big(\frac{r}{\cos\theta\sin\theta}\frac{d}{d\theta} + \frac{d}{dr}\Big) + \Big(r\frac{d}{dr} + \frac{1}{\cos\theta\sin\theta}\frac{d}{d\theta}\Big)^2\Big\} u\end{aligned}$$

whereas $\Delta_{r,\theta}(r^2 u) = 2u + 2 r u'_r + r^2 u''_{rr} + r^2 u''_{\theta\theta}$.

With the potential $u(r) = V(r)$ related to a circle, the cartesian differential equation for r^2u depends on $u'_x = u'_r r'_x = -xu^3$ and $u'_y = u'_r r'_y = -yu^3$, in polar coordinates it reduces to the terms including derivatives with respect to r

$$\Delta_{x,y}(r^2u) = 3u - 4r^2u^3 + M^{-4}r^2(2r^2 - c^2)u^5.$$

5.4 Heat conduction equations

The second order differential operator describing the heat conduction in any solid has the form

$$L_{x,t} = \Delta - \alpha(x,t)\frac{\partial}{\partial t}, \tag{5.14}$$

for every x in a domain Ω of $\mathbb{R}^p$, $p = 1, 2, 3$, and for every t in a time interval $[0, T]$. A function u is solution of a heat equation with second member f if

$$L_{x,t}u(x,t) = f(x,t), \quad \text{in } \Omega \times [0,T], \tag{5.15}$$

with initial conditions $u(x, 0) = \psi(x)$ and $u'_x(x, 0) = \varphi(x)$ for every x of the frontier of Ω where the temperature is measured. The heat diffusion in a parallelogram has the boundary condition $u(x_k, t) = \varphi_k(t)$ at points x_k of the edges.

A multiplicative solution of a heat equation along a line was defined by (1.10). In $\mathbb{R}^2 \times \mathbb{R}_{+*}$, a product function $u(x, y, t) = e^{-kt}v(x)w(y)$ solution of the differential equation $\Delta_{x,y}u - \alpha u'_t = 0$, with a positive constant α, may be chosen such that $k = \alpha^{-1}\gamma^2$ and

$$\frac{v''}{v}(x) + \frac{w''}{w}(y) + \gamma^2 = 0,$$

then the functions $v(x) = v_0 \sin\left(\frac{\gamma}{\sqrt{2}}x + \omega_1\right)$ and $w(y) = w_0 \sin\left(\frac{\gamma}{\sqrt{2}}y + \omega_2\right)$ provide a solution. The sums of multiplicative functions

$$\begin{aligned} u(x,y,t) = \sum_{j=1}^{\infty} \{ & A_j \sin(\theta_1 x + \omega_1)\sin(\theta_2 y + \omega_2) \\ & + B_j \cos(\theta_1 x + \omega_1)\cos(\theta_2 y + \omega_2)\}e^{-\gamma t} \end{aligned}$$

with constants satisfying $(\theta_1^2 + \theta_2^2) - \alpha\gamma = 0$ are also solutions. The sums

$$\begin{aligned} u(x,y,t) = \sum_{j=1}^{\infty} \{ & A_j \sin(\theta_1 x + \omega_1) + B_j \sin(\theta_2 y + \omega_2) \\ & + C_j\{\cos(\theta_1 x + \omega_3) + D_j \cos(\theta_2 y + \omega_4)\}e^{-\gamma t} \end{aligned}$$

or any linear combinaison of product of trigonometric functions of x and y, with constants θ_1 and θ_2 satisfying the same equation are solutions of the differential equation.

The functions $u(x,y) = \sum_{j=1}^{\infty}\{A_j \sin(ax - by) + B_j \cos(ax - by)\}e^{-\gamma t}$ have the partial derivatives $u''_{xx} = -a^2 u$ and $u''_{yy} = -b^2 u$, they are solutions of the differential equation with a constant parameter α such that $(a^2 + b^2) - \alpha\gamma = 0$.

In $\mathbb{R}^3 \times \mathbb{R}_{+*}$, a product function $u(x,y,t) = e^{-kt}v(x)w(y)g(z)$ solution of the differential equation $\Delta_{x,y,z}u - \alpha u'_t = 0$, with a positive constant α, may be chosen such that $k = \alpha^{-1}\gamma^2$ and

$$\frac{v''}{v}(x) + \frac{w''}{w}(y) + \frac{g''}{g}(z) + \gamma^2 = 0.$$

A product of trigonometric functions in x, y and respectively z are solution. Solutions of the differential equation may be defined as sums of multiplicative functions

$$\begin{aligned} u(x,y,z,t) = \sum_{j=1}^{\infty}\{&A_j \sin(\theta_1 x + \omega_1)\sin(\theta_2 y + \omega_2)\sin(\theta_3 z + \omega_3) \\ &+ B_j \cos(\theta_1 x + \omega_1)\cos(\theta_2 y + \omega_2)\cos(\theta_3 z + \omega_3)\}e^{-\gamma t} \end{aligned}$$

or multiplicative functions including the same sine and cosine functions, and as sums

$$\begin{aligned} u(x,y,z,t) = \sum_{j=1}^{\infty}\{&A_{1j}\sin(\theta_1 x + \omega_1) + A_{2j}\sin(\theta_2 y + \omega_2) \\ &+ A_{3j}\sin(\theta_3 z + \omega_3) + B_{1j}\cos(\theta_1 x + \omega_4) \\ &+ B_{2j}\cos(\theta_2 y + \omega_5) + B_{3j}\cos(\theta_3 z + \omega_6)\}e^{-\gamma t} \end{aligned}$$

with constants satisfying $\theta_1^2 + \theta_2^2 + \theta_3^2 - \alpha\gamma = 0$. Additive functions are other form of the solutions

$$\begin{aligned} u(x,y,z,t) = \sum_{j=1}^{\infty}\{&A_j \sin(\theta_1 x \pm \theta_2 y \pm \theta_3 z + \omega) \\ &+ B_j \cos(\theta_1 x \pm \theta_2 y \pm \theta_3 z + \omega)\}e^{-\gamma t}, \end{aligned}$$

with $\alpha\gamma = \theta_1^2 + \theta_2^2 + \theta_3^2$. It is not always possible to distinguish the marginal functions under the boundary conditions of this solution from those of the previous ones.

In solids, the differential heat equation may be written as

$$a^2 u''_{xx} + b^2 u''_{yy} + c^2 u''_{zz} = \alpha u'_t = 0$$

where the constants a, b, c are length of the object and the condition for the constants of the functions u sum of the previous trigonometric functions are modified as $a^2\theta_1^2 + b^2\theta_2^2 + c^2\theta_3^2 - \alpha\gamma = 0$.

From Lemma (3.1), the Laplacian of a function u defined on a disk is expressed in terms of the radius of the disk and the differential equation with a constant coefficient α is equivalent to

$$u''_{rr} + \frac{u'_r}{r} - \alpha u'_t = 0.$$

A multiplicative solution $u(r,t) = e^{-kt}v(r)$ has a function v solution of the second order differential equation $v''_{rr} + r^{-1}v'_r + k\alpha v = 0$. Particular solutions may be searched in the form $v(r) = e^{g(r)}$, the derivatives of $v(r)$ are $v'_r = g'v$ and $v''_{rr} = (g''_{rr} + g'^2)v$, so the function g is solution of the equation

$$r(g'' + g'^2) + g' + k\alpha r = 0.$$

The function g with primitive G is solution of the implicit equation obtained by integration

$$-k\alpha\frac{r^2}{2} = rg'(r) + \int_0^r sg'^2(s)\,ds + c_1$$

$$-k\alpha\frac{r^3}{6} = \int_0^r s\,dG(s) + r\int_0^r sg'^2(s)\,ds - \int_0^r s^2g'^2(s)\,ds + c_1r + c_2,$$

with arbitrary constants. From Lemma (3.5), replacing r^{-1} by $2r^{-1}$, a similar differential equation applies to the heat conduction from the surface of an object of $\mathbb{R}^3$.

The heat equation with the functional coefficient $\alpha(x,t) = k(x - x_0)^{-1}$ has the solutions

$$u(x,t) = \exp\{-\frac{k(x - x_0)}{t}\}$$

for all constant $k > 0$ and real variables x in an interval centered at x_0 and t in $[0,T]$. In a bounded subset of $\mathbb{R}^2$ centered at (x_0, y_0), the heat equation with the functional coefficient $\alpha(x,y,t) = k(x - x_0 + y - y_0)^{-1}$ has the solution

$$u(x,y,t) = \exp\Big\{-\frac{k(x - x_0 + y - y_0)}{t}\Big\},$$

for t in $[0,T]$.

The function $u(x,t) = \exp\{-\frac{(x-x_0)^2}{2t}\}$, in an interval $I = [a,b]$ centered at x_0, is solution of the heat equation defined with the function

$$\alpha(x,t) = 2\{1 - (x-x_0)^{-2}t\}$$

along a line of length $b-a$, for every x in I and for every $t > 0$. In a plane, the function

$$u(x,y,t) = \exp\Big\{-\frac{(x-x_0)^2 + (y-y_0)^2}{2t}\Big\} = \exp\Big\{-\frac{r^2(x,y)}{2t}\Big\}$$

is solution of the heat equation with the function $\alpha(r,t) = 2\{1 - r^{-2}t\}$, for all x in an interval centered at x_0, y in an interval centered at y_0, and $t > 0$. The same model applies to objects in $\mathbb{R}^3$.

The isotrope differential equations are defined in $\mathbb{R}^p$ with respect to the l^2-norm r of $x = (x_1, \ldots, x_k)$ in $\mathbb{R}^p$. The isotrope heat conduction equations are defined with respect to the l^2-norm r of x in $\mathbb{R}^p$ as time-varying functions of the norm r in $C_2(\mathbb{R}_+^2)$. The second order differential operator has the form

$$L_{r,t} = \Delta - \alpha(r,t)\frac{\partial u}{\partial t}$$

with a strictly positive function α, for all r in an interval (r_0, r_1) and t in a time interval $0, T]$. The initial conditions define the values of the function u and its first partial derivatives on the frontier of the domain of the equation.

The Gaussian density depends only on the radius to its center in $\mathbb{R}^p$, $p \geq 1$, and to the variance t^2, it is the solution of a second order heat differential equation.

Proposition 5.7. *The Gaussian density*

$$u(r,t) = \frac{1}{\sqrt{2\pi t}} \exp\{-\frac{r^2}{2t^2}\}$$

in $\mathbb{R} \times \mathbb{R}_+^*$ is solution of the heat equation $\Delta_r u - t^{-1}u'_t = 0$.

Proof. The partial derivatives of the Gaussian density function u are

$$\begin{aligned} u'_t &= \Big(\frac{r^2}{t^3} - \frac{1}{t}\Big)u, \\ u'_r &= -\frac{r}{t^2}u, \\ u''_{rr} &= -\Big(\frac{1}{t^2} - \frac{r^2}{t^4}\Big)u \end{aligned}$$

and the ratio of u''_{rr} and u'_t gives the result. □

Proposition 5.8. *The function*

$$u(x,t) = \frac{1}{2\sqrt{t}} \exp\left\{-\frac{x^2}{4t}\right\}$$

in $\mathbb{R} \times \mathbb{R}_+^*$ *is solution of the heat equation* $u''_{xx} - u'_t = 0$.

Proposition 5.8 does not generalize to $\mathbb{R}^p \times \mathbb{R}_+^*$ where the function

$$u(r,t) = \frac{1}{2\sqrt{t}} \exp\left\{-\frac{r^2}{4t}\right\}$$

is solution of the inhomogeneous equation $u''_{xx} - u'_t + \frac{p-1}{2t} = 0$.

The differential equation

$$\frac{\partial u(t,x)}{\partial t} = \Delta u(t,x) \tag{5.16}$$

on $\mathbb{R}_+ \times \mathbb{R}$, leads to the expansions in neighborhoods of t and respectively x

$$u(t+h_1,x) = u(t,x) + h_1 \frac{\partial^2 u(t,x)}{\partial x^2} + \cdots + h_1^k \frac{\partial^{2k} u(t,x)}{k!\partial x^{2k}} + \cdots,$$

$$u(t,x+h_2) = u(t,x) + h_2^2 \frac{\partial u(t,x)}{2\partial t} + \cdots + h_2^{2k} \frac{\partial^k u(t,x)}{(2k)!\partial t^k} + \cdots$$

$$+ h_2 \frac{\partial u(t,x)}{\partial x} + h_2^3 \frac{\partial u(t,x)}{3!\partial t \partial x} + \cdots + h_2^{2k+1} \frac{\partial^k u(t,x)}{(2k+1)!\partial t^k \partial x} + \cdots,$$

Poisson wrote a solution of (5.16) in the form

$$u(t,x) = \frac{1}{2\pi\sqrt{t}} \int_{\mathbb{R}} \varphi(\xi) e^{-\frac{(\xi-x)^2}{4t}} \, d\xi.$$

A differential equation

$$\Delta u(t,x) - \alpha(x) u'_t(t,x) = 0 \tag{5.17}$$

on $\mathbb{R}_+ \times \mathbb{R}$, has similar expansions with

$$u(t+h_1,x) = u(t,x) + \frac{h_1}{\alpha(x)} \frac{\partial^2 u(t,x)}{\partial x^2} + \cdots + \frac{h_1^k}{k!\alpha^k(x)} \frac{\partial^{2k} u(t,x)}{\partial x^{2k}} + \cdots,$$

$$u(t,x+h_2) = u(t,x) + h_2^2 \alpha^2(x) \frac{\partial u(t,x)}{2\partial t} + \cdots + h_2^{2k} \alpha^{2k}(x) \frac{\partial^k u(t,x)}{(2k)!\partial t^k} + \cdots$$

$$+ h_2 \alpha(x) \frac{\partial u(t,x)}{\partial x} + h_2^3 \alpha^3(x) \frac{\partial u(t,x)}{3!\partial t \partial x} + \cdots$$

$$+ h_2^{2k+1} \alpha^{2k+1}(x) \frac{\partial^k u(t,x)}{(2k+1)!\partial t^k \partial x} + \cdots.$$

The differential equation

$$\Delta u(r,t) - \alpha_r u'_t(r,t) = 0$$

has a multiplicative solution $u(r,t) = \theta_r\varphi_t$, with $\varphi_t = e^{at}$, if the function θ_r satisfies the differential equations

$$\begin{aligned} f(r,t) &= \theta_r''\varphi_t - \alpha_r\varphi_t'\theta_r, \\ e^{-a_t}f(r,t) &= \theta_r'' - \alpha_r a\theta_r, \end{aligned} \tag{5.18}$$

in the interval $[r_0, r_1]$. The function θ_r satisfies a marginal inhomogeneous second order differential equations for every t. With functions θ_r proportional to r^{-1} and a product function $f(r,t) = f_{1r}f_{2t}$, the differential equation (5.18) reduces to

$$f_{1r}f_{2t} = kr^{-3}(2 - ar^2\alpha_r)e^{a_t}. \tag{5.19}$$

This implies that e^{at} is proportional to f_{2t} and r is proportional to $k\{(2 - ar^2\alpha_r)f_{1r}^{-1}\}^{\frac{1}{3}}$, with $a^2r^2\alpha_r < 2$. The function θ_r depends on f_{1r} and α_r through this expression of r. Under the constraint $ar^2\alpha_r < 2$, the function α_r is bounded by $2a^{-1}r^{-2}$ and

$$u(r,t) = \frac{1}{2}\Big\{kr^{-1} + (2 - ar^2\alpha_r)^{-\frac{1}{3}}f_{1r}^{\frac{1}{3}}\Big\}e^{at}$$

is a solution of (5.19).

The homogeneous heat equation $L_{x,t}u(x,t) = 0$ in a rectangular domain $[x_0, x_1] \times [0, T]$ of $\mathbb{R} \times \mathbb{R}_+$ has the initial conditions $u(x,0) = \psi(x)$ in $[x_0, x_1]$, $u(x_0, t) = \varphi_0(t)$ and $u'(x_1, t) = \varphi_1(t)$ in $[0, T]$. Its solution satisfies the implicit equations

$$\begin{aligned} u(x,t) &= \int_0^t \alpha^{-1}(x,s)u_{x,x}''(x,s)\,ds + \psi(x), \\ u(x,t) &= \varphi_0(t) + \varphi_1(t)\int_{x_0}^x (y - x_1)\,dy - \int_{x_0}^x \int_y^{x_1} \alpha(z,t)u_t'(dz,t)\,dy \\ &= \varphi_0(t) + \varphi_1(t)\int_{x_0}^x (y - x_1)\,dy - \int_{x_0}^x z\alpha(z,t)u_t'(dz,t) \\ &\quad - x\int_x^{x_1} \alpha(z,t)u_t'(dz,t). \end{aligned}$$

It appears from these expressions that a solution of the homogeneous equation cannot generally be written as an integral of the second member with respect to its derivatives, even with a constant coefficient α.

Lemma 5.1. *A sum of marginal functions $u(x,t) = v(x) + w(t)$ is not solution of a homogeneous heat conduction equation (5.14).*

Proposition 5.9. *The function*

$$u_1(x,t) = \sum_{i=1}^{p} a_i \exp\{-k_1(x_i - x_{0i})\} + b \exp\{-k_2 t\}$$

on $\mathbb{R}_+ \times \mathbb{R}^p$, *is solution of the homogeneous second order differential equation*

$$u''_{1,xx}(x,t) + \alpha u'_{1,t}(x,t) - \beta u_1(x,t) = 0,$$

with constant coefficients and with the initial values $u_1(x_0,0) = \sum_{i=1}^{p} a_i + b$ *and* $u'_{1x_i}(x_0,0) = -k_1 a_i + \sum_{j\neq i} a_j + b$, *where* $\beta = k_1^2 > 0$ *and* $k_2 = -\alpha^{-1}\beta$.

Proposition 5.10. *The function*

$$u_1(x,t) = \sum_{i=1}^{p} a_i \exp\{-k_1(x_i - x_{0i})\} + b \exp\{-k_2 t\}$$

on $\mathbb{R}_+ \times \mathbb{R}^p$, *is solution of the homogeneous second order differential equation*

$$u''_{1,xx}(x,t) + \alpha u'_{1,t}(x,t) - \beta u_1(x,t) - \gamma \sum_{i=1}^{p} u'_{1x_i}(x,t) = 0,$$

with constant coefficients and with the initial values $u_1(x_0,0) = \sum_{i=1}^{p} a_i + b$ *and* $u'_{1x_i}(x_0,0) = -k_1 a_i + \sum_{j\neq i} a_j + b$, *where* $\beta = k_1^2 + \gamma$ *and* $k_2 = -\alpha^{-1}\beta$.

Proposition 5.11. *The function*

$$u_2(x,t) = a \exp\{-k_1(x - x_0)\} + b \exp\{-k_2(t)\}$$

on $\mathbb{R}_+ \times \mathbb{R}$, *is solution of the differential equation*

$$u''_{2,xx}(x,t) - \alpha(x,t) u'_{2,t}(x,t) + \beta(x) u_2(x,t) = 0$$

depending on functions k_1 *and* k_2 *such that* $k_1(x - x_0) = 0$ *and* $k_2(0) = 0$, *with initial values* $u_2(x_0,0) = a + b$ *and with functions* α *and* β *such that* $\alpha(x,t)k_2'(t) + \beta(x) = 0$ *and* $\beta(x) = k_1''(x - x_0) - k_1'^2(x - x_0)$, *depending only on* x.

The following cases are solutions of the homogeneous equation as products of functions of x and, respectively, t. The function

$$u_1(x,t) = u_0 \exp\{-k_1(x - x_0) + k_2 t\}$$

satisfies the homogeneous differential equation with a constant coefficient

$$u''_{1,xx} - \frac{k_1^2}{k_2} u'_{1,t} = 0,$$

for all constants k_1 and k_2 such that $k_1^2 = \alpha k_2$. The functions

$$u(x,t) = u_0[\sin\{k_1(x-x_0)\} + \cos\{k_1(x-x_0)\}]\exp\{-kt\}$$

are also solutions of the equation $u''_{xx} - \alpha u'_t = 0$ for all constants k_1 and k_2 such that $\alpha = k_1^2 k^{-1}$. Under the constraints of initial conditions

$$u(x_j,t) = \psi_j(t), \quad u(x,0) = \varphi(x)$$

with functions φ and ψ_j such that

$$\begin{aligned}\varphi(x) &= u_0\sin\{k_1(x-x_0)\} + u_0\cos\{k_1(x-x_0)\},\\ \psi_j(t) &= u_j\exp\{-kt\},\end{aligned}$$

with the constants $u_j = u_0[\sin\{k_1(x_j-x_0)\}+\cos\{k_1(x_j-x_0)\}]$, for $j = 1, 2$. So the product of the boundary functions defines solutions of the differential equation up to a constant.

Lemma 5.2. *All functions $u(x,t) = v(x)w(t)$ with marginal functions such that $v''_{xx}(x) = \alpha\lambda v(x)$ and $w'(t) = \lambda w(t)$ are solutions of the differential equation $u''_{xx} - \alpha u'_t = 0$.*

A linear combination of functions $u_{0j}\exp\{-(k_{1j}(x-x_0) + k_{2j}t)\}$ is not solution of a heat conduction equation if their coefficients $\alpha_j = k_{1j}^2 k_{2j}^{-1}$ differ. The function $w(t)$ cannot be a sine and cosine functions or a hyperbolic function sinh and cosh, so the solutions of heat equations are not oscillatory functions of t.

Theorem 5.4. *Let $u_j(x,t) = v_j(x)w_j(t)$ be solutions of the heat equation $u''_{xx} - \alpha u'_t = 0$ with marginal functions such that $v''_{j,xx}(x) = \alpha\lambda_j v_j(x)$ and $w'_j(t) = \lambda_j w_j(t)$, for every integer j, then $u(x,t) = \sum_{j\geq 0} u_j(x,t)$ is solution of the same equation.*

A well known solution of the heat equation $u''_{xx} - \alpha u'_t = 0$ is the exponential-trigonometric series

$$\begin{aligned}u(x,t) = \exp\{-k^2 t\}\sum_{k=1}^{+\infty}[A_k\sin\{\sqrt{\alpha}k(x-x_0)\}\\ + B_k\cos\{\sqrt{\alpha}k(x-x_0)\}].\end{aligned} \tag{5.20}$$

A linear combination of exponential-exponential series or the hyperbolic functions like

$$u(x,t) = \sum_{k=1}^{+\infty}[A_k\sinh\{\sqrt{\alpha}k(x-x_0)\} + B_k\cosh\{\sqrt{\alpha}k(x-x_0)\}]\exp\{-\alpha k^2 t\}$$

is also solution of this equation.

Theorem 5.5. *The functions*

$$u(x,t) = \sum_{j=1}^{+\infty} u_j[\sin\{k_j(x-x_0)\} + \cos\{k_j(x-x_0)\}]\exp\{-2k_j^2\alpha^{-1}t\}$$

are solutions of the homogeneous heat equation $\Delta u - \alpha u'_t = 0$ *in* $\mathbb{R} \times \mathbb{R}_+^*$.

Theorem 5.6. *The functions*

$$u(x,t) = \sum_{j=1}^{+\infty}\Big[u_j \sin\Big\{\sum_{i=1}^{p} k_{ij}(x_i - x_{0i})\Big\} + v_j \cos\Big\{\sum_{i=1}^{p} k_{ij}(x_i - x_{0i})\Big\}\Big]\exp\Big\{-\alpha^{-1}t\sum_{i=1}^{p} k_{ij}^2\Big\}$$

and

$$u(x,t) = \sum_{j=1}^{+\infty} u_j \exp\Big\{-\sum_{i=1}^{p} k_{ij}(x_i - x_{0i})\Big\}\exp\Big\{\alpha^{-1}t\sum_{i=1}^{p} k_{ij}^2\Big\}$$

are solutions of the homogeneous heat equation $\Delta u - \alpha u'_t = 0$ *in* $\mathbb{R} \times \mathbb{R}_+$.

Replacing the constants by functions, the parameter α of the differential equation becomes a funtion is related to functions k_j. Let I be a bounded real interval centered at x_0, for all functions $k_1(x-x_0)$ of $C_2(I)$ and $k_2(t)$ of $C_2([0,T])$, the function

$$u_2(x,t) = u_0 \exp\{-k_1(x-x_0) - k_2(t)\}$$

is solution of the homogeneous equation $\Delta u_2(x,t) - \alpha_2(x,t)u'_{2t}(x,t) = 0$ with the function

$$\alpha_2(x,t) = \frac{k_1''(x-x_0) - k_1'^2(x-x_0)}{k_2'(t)}.$$

Proposition 5.12. *For every functions f having an expansion on Laguerre's polynomial basis, there exists solution of the differential equation*

$$\Delta u(x,t) - \frac{at}{x^2}u'_t(x,t) = f(x,t) \tag{5.21}$$

where f is a square integrable function having an expansion on Laguerre's polynomial basis.

Proof. Considering the Laguerre polynomials with the derivatives

$$xL_n'(x) = nL_n(x) - nL_{n-1}(x),\ n \geq 1$$
$$x^2L_n''(x) = n(n-1)L_n(x) - n(2n-1)L_{n-1}(x) + n^2L_{n-2}(x),\ n \geq 2,$$

and function f and u having the expansions

$$f(x,t) = \sum_{n\geq 0, m\geq 0} f_{n,m}L_n(x)L_m(t),$$
$$u(x,t) = \sum_{n\geq 0, m\geq 0} u_{n,m}L_n(x)L_m(t),$$

the differential equation (5.21) is expressed as

$$\begin{aligned} tx^2f(x,t) = t\sum_{m\geq 0} L_m(t)\Big\{\sum_{n\geq 2} n(n-1)u_{n,m}L_n(x) \\ - \sum_{n\geq 1}(n+1)(2n+1)u_{n+1,m}L_n(x) + \sum_{n\geq 0}(n+2)^2u_{n+2,m}L_n(x)\Big\} \\ - x^2\alpha(x,t)\sum_{n\geq 0} L_n(x)\Big\{\sum_{m\geq 1} mu_{n,m}L_m(t) \\ - \sum_{m\geq 0}(m+1)u_{n,m+1}L_m(t)\Big\}, \end{aligned}$$

with $\alpha(x,t) = atx^{-2}$. The values of the coefficients $u_{n,m}$ are deduced by identification with the expansion of $x^2f(x,t)$. □

A trigonometric-exponential solution of heat equation with a constant coefficient α

$$u''_{xx}(x,t) - \alpha u'_t(x,t) = f(x) \tag{5.22}$$

with a known second member f_x has been defined by Liouville using the Fourier transform. By an expansion of the function f as a Fourier series

$$f(x) = f(0) + \sum_{k=1}^{+\infty}\{\kappa_k \sin(k\omega x) + \zeta_k \cos(k\omega x)\}$$

with the constants ω and

$$\kappa_k = \frac{2}{\pi}\int_{-\frac{\pi}{2k\omega}}^{\frac{\pi}{2k\omega}} f(x)\sin(k\omega x)\,dx,$$
$$\zeta_k = \frac{2}{\pi}\int_{-\frac{\pi}{2k\omega}}^{\frac{\pi}{2k\omega}} f(x)\cos(k\omega x)\,dx.$$

A particular solution of the differential equation (5.22) is specified with constants ω, A_k and B_k such that

$$\omega = \sqrt{\alpha}, \quad \kappa_k = (\alpha - \omega^2)k^2 A_k, \quad \zeta_k = (\alpha - \omega^2)k^2 B_k. \tag{5.23}$$

Theorem 5.7. *A solution of the inhomogeneous heat equation (5.22) is the sum of the solution (5.22) of the homogeneous equation and of a particular solution of the same form, with coefficients A_k and B_k satisfying (5.23).*

Boundary conditions $u(x,0) = \psi(x)$, $u(0,t) = \varphi_0(t)$ and $u(l,t) = \varphi_l(t)$, with ψ, φ_0 and φ_l such that

$$\Delta\psi(x) - \alpha\frac{\varphi_k'(t)}{\varphi_k(t)}\psi(x) = 0, \; k = 0, l, \; x \in \partial\Omega,$$

determine the form of the solutions of the homogeneous equation with coefficient α. Since the function φ does not depend on t, the ratio of $\varphi_k'(t)$ and $\varphi_k(t)$ must be a constant hence the trigonometric or exponential functions are the unique bounded solutions. In the same way, the ratio of $\Delta\psi(x)$ and $\psi(x)$ is constant and negative, so that ψ is an exponential function on $\mathbb{R}_-$. Theorems 5.5 and 5.6 give therefore the unique multiplicative solutions.

The solutions of Theorem (5.7) extend to a thin rectangular plane like the surface of the ocean and to objects in the space. The equations and the solutions for a plane are similar to (5.15) with the operator (5.14) replacing the real x by (x, y) in $\mathbb{R}^2$.

Theorem 5.8. *The functions*

$$u(x,y,t) = \sum_{j=1}^{+\infty} u_j[\sin\{k_j(x - x_0)\}\cos\{k_j(y - y_0)\}]\exp\{-k_j^2\sqrt{2}\alpha^{-1}t\}$$

and

$$u(x,y,t) = \sum_{j=1}^{+\infty} u_j[\cos\{k_j(x - x_0)\}\sin\{k_j(y - y_0)\}]\exp\{-k_j^2\sqrt{2}\alpha^{-1}t\}$$

are solutions of the homogeneous heat equation $\Delta u - \alpha u_t' = 0$ in a subset Ω of $\mathbb{R}^2 \times \mathbb{R}_+^*$.

Boundary conditions $u(x,y,0) = \psi(x,y)$ and $u(x_k, y_k, t) = \varphi_k(t)$ at (x_k, y_k) on the frontier $\partial\Omega$, with ψ and φ_k such that

$$\Delta\psi(x_k, y_k) - \alpha\frac{\varphi_k'(t)}{\varphi_k(t)}\psi(x_k, y_k) = 0$$

imply that the ratio of $\varphi_k'(t)$ and $\varphi_k(t)$ must be a constant for every $t > 0$, the exponential functions is still the unique solutions. For multiplicative solutions of the homogeneous heat equation with a constant coefficient α the function ψ must be sine, cosine or exponential, like in Theorem 5.7.

Proposition 5.13. *The Gaussian density on* $\mathbb{R}^2$

$$u(x,y,t) = \frac{1}{2\pi t^2} \exp\{-\frac{ax^2+by^2}{2t^2}\}$$

in $\mathbb{R}^2 \times \mathbb{R}_+^*$ *is solution of the heat equations*

$$\frac{u''_{xx}}{a^2} + \frac{u''_{yy}}{b^2} - t^{-1}u'_t = 0.$$

Proof. The partial derivatives in $\mathbb{R}^2$ of the Gaussian density function u are

$$u''_{xx} = -\Big(\frac{a^2}{t^2} - \frac{a^4x^2}{t^4}\Big)u,$$
$$u''_{yy} = -\Big(\frac{b^2}{t^2} - \frac{b^4y^2}{t^4}\Big)u.$$

A Gaussian density for depend variables has a covariance matrix

$$\Sigma = \begin{pmatrix} \sigma_1^2 & \sigma_{12} \\ \sigma_{12} & \sigma_2^2 \end{pmatrix}$$

with components proportional to t^2 and its determinant is proportional to t^4. Its partial derivatives are

$$u''_{xx} = -\Big(\frac{a^2}{t^2} - \frac{a^4x^2}{t^4}\Big)u,$$
$$u''_{yy} = -\Big(\frac{b^2}{t^2} - \frac{b^4y^2}{t^4}\Big)u.$$

Then $u(x,y,t)$ is proportional to

$$\frac{1}{2\pi t^2} \exp\{-\frac{a^2x^2+b^2y^2+2c^2xy}{2t^2}\}$$

with constants a, b and c, and the differential equation of the Gaussian density follows. $\square$

From Proposition 5.7, the differential equations for the Gaussian densities on $\mathbb{R}^2 \times \mathbb{R}_+^*$ and $\mathbb{R} \times \mathbb{R}_+^*$ differ due to their different normalizations.

The surface of solid in $\mathbb{R}^3$ satisfies a similar heat conduction equation. A solid Ω with a hot point on its surface diffuses the heat inside and outside Ω and on its surface $\partial\Omega$. Three heat equations are then defined with constants depending on the densities of each domain.

5.5 Wave differential equations

Poisson's second order differential differential operator of vibratory strings or planes is defined with a strictly positive constant α as

$$L_{x,t} = \Delta - \alpha \frac{\partial^2}{\partial t^2}, \tag{5.24}$$

for every $x = (x_1, \dots, x_p)$ in a domain Ω of $\mathbb{R}^p$, $p = 1, 2, 3$ and for every t in a time interval $[0, T]$. The differential equation $L_{x,t}u(x,t) = 0$ may be obtained from a system of spatial differential equations with constant parameters a_i such that $\sum_{i=1}^p a_i^2 = \alpha$

$$u''_{x_i,x_i} = a_i^2 u''_{t,t},\ i = 1, \dots, p.$$

Boundary conditions define u, u'_x and u'_t at time $t_0 = 0$ and on the frontier $\partial\Omega$ of the domain.

Let

$$r(x,t) = \Big(\sum_{k=1}^p x_k^2 - t^2\Big)^{\frac{1}{2}}. \tag{5.25}$$

The function $u(x,t) = r^2(x,t)$, on $\mathbb{R} \times \mathbb{R}_+$, is solution of the differential equation

$$\Delta u(x,t) + u''_{tt}(x,t) = 0.$$

The function $u(x,t) = e^{r(x,t)}$, on $\mathbb{R}^p \times \mathbb{R}_+$, is solution of the differential equation

$$\Delta u(x,t) - u''_{tt}(x,t) - u(x,t) - \frac{pu(x,t)}{r(x,t)} = 0.$$

Theorem 5.9. *The wave differential equation $\Delta u(x,t) - u''_{tt}(x,t) = 0$ on $\mathbb{R}^p \times \mathbb{R}_+$, $p > 1$, has the solution $u(x,t) = r^{-(p-1)}(x,t)$.*

Proof. The function $u(x,t) = r^{-k}(x,t)$ has the partial derivatives with respect to x_i and t

$$u'_i(x,t) = -\frac{kr'_i}{r^{k+1}}(x,t) = -\frac{kx_i}{r^{k+2}(x,t)},$$

$$u''_{ii}(x,t) = -\frac{k}{r^{k+2}(x,t)} + \frac{k(k+2)x_i^2}{r^{k+4}(x,t)},$$

$$u'_t(x,t) = -\frac{kr'_t}{r^{k+1}}(x,t) = \frac{tk}{r^{k+2}(x,t)},$$

$$u''_{tt}(x,t) = \frac{k}{r^{k+2}(x,t)} + \frac{k(k+2)t^2}{r^{k+4}(x,t)},$$

so

$$\Delta u(x,t) = -\frac{kp}{r^{k+2}(x,t)} + \frac{k(k+2)(r^2+t^2)}{r^{k+4}(x,t)},$$

$$\Delta u(x,t) - u''_{tt}(x,t) = \frac{k(k+1-p)}{r^{k+2}(x,t)}$$

and it is zero as $k = p-1$. □

Theorem 5.10 (Liouville 1839). *The wave differential equation $L_{x,t}u(x,t) = f(x,t)$ on $\mathbb{R}^p \times \mathbb{R}_+$ has a solution for every f belonging to a Hilbert space provided with an orthonormal basis.*

Defining a weighted distance

$$r_h(x,t) = \Big(\sum_{k=1}^{p} h_k(x_k^2) - t^2\Big)^{\frac{1}{2}}, \tag{5.26}$$

with twice continuously differentiable functions h_k the partial derivatives of $u_h(x,t) = r_h^{-k}(x,t)$ with respect to x_i are modified as

$$u'_{hi}(x,t) = -\frac{kr'_{hi}}{r_h^{k+1}}(x,t) = -\frac{kx_i h'_i(x_i^2)}{r_h^{k+2}(x,t)},$$

$$u''_{hii}(x,t) = -\frac{kh'_i(x_i^2)}{r_h^{k+2}(x,t)} - \frac{2kx_i^2 h''_i(x_i^2)}{r_h^{k+2}(x,t)} + \frac{k(k+2)x_i^2 h'^2_i(x_i^2)}{r_h^{k+4}(x,t)},$$

so its Laplacian is

$$\Delta u_h(x,t) = -\frac{k\sum_{i=1}^{p}\{h'_i(x_i^2) + 2x_i^2 h''_i(x_i^2)\}}{r_h^{k+2}(x,t)} + \frac{k(k+2)\{\sum_{i=1}^{p} x_i^2 h'^2_i(x_i^2)\}}{r_h^{k+4}(x,t)}.$$

The differential equation $\Delta u_h(x,t) - u''_{htt}(x,t) = f(x,t)$ is then equivalent to

$$f(x,t) = -\frac{k\{\sum_{i=1}^{p} h'_i(x_i^2) + 2x_i^2 h''_i(x_i^2) + 1)\}}{r_h^{k+2}(x,t)} + \frac{k(k+2)\{\sum_{i=1}^{p} x_i^2 h'^2_i(x_i^2) - t^2\}}{r_h^{k+4}(x,t)}.$$

According to Sturm (1861), the wave differential equation defined by (5.24) on $\mathbb{R} \times \mathbb{R}_+$ has the solutions

$$u(x,t) = \varphi\Big(x + \frac{t}{a}\Big) + \psi\Big(x - \frac{t}{a}\Big), \tag{5.27}$$

for x in $[0,l]$ and t in $[0,T]$, with boundary conditions $u^{(k)}(x,0)$ and $u^{(k)}(0,t)$, for $k=0,1$. A necessary condition for the functions φ and ψ is the consistency of the initial conditions. In particular

$$\begin{aligned} u(x,0) &= \varphi(x)+\psi(x) \\ u(0,ax) &= \varphi(x)+\psi(-x) \end{aligned} \tag{5.28}$$

and the boundary functions are equal if ψ is symmetric. The derivatives of u satisfy the equations

$$\begin{aligned} u'_x(x,0) &= \varphi'(x)+\psi'(x), \\ u'_t(0,ax) &= a^{-1}\{\varphi'(x)-\psi'(-x)\}. \end{aligned}$$

The functions φ and ψ are deduced by integration of the differential equation for φ' and ψ' according to u'_x and u'_t

$$\begin{aligned} \psi(x)-\psi(-x) &= u(x,0)-u(0,0)-a\int_0^x u'_t(0,as)\,ds \\ &= u(x,0)-u(0,ax), \end{aligned}$$

and

$$\begin{aligned} 2\varphi(x) &= u(x,0)-u(0,0)+a\int_0^x u'_t(0,as)\,ds-\{\psi(x)+\psi(-x)\} \\ &= u(x,0)+u(0,ax)-\{\psi(x)+\psi(-x)\}. \end{aligned}$$

If the function ψ is symmetric, $\psi'(x)+\psi'(-x)=0$ and $u'_x(x,0)-au'_t(0,ax)=0$. These properties are fulfilled by the function $\psi(x)=\cos x$ with $u(0,0)=1$ but the function $\varphi(x)=\sin x$ is antisymmetric and $u(x,t)=\varphi(x+\frac{a}{t})$ satisfies the same properties with $u(0,0)=0$.

Theorem 5.11. *The wave differential equation $L_{x,t}u(x,t)=0$ on $[0,l]\times[0,T]$ has the solutions (5.27) for all functions φ and ψ of $C_2(\mathbb{R}_+)$ such that $\psi(x)=-\psi(-x)$, and they are defined as*

$$\begin{aligned} \varphi(x) &= \frac{1}{2}\{u(x,0)+u(0,ax)\}, \\ \psi(x) &= \frac{1}{2}\{u(x,0)-u(0,ax)\} \end{aligned}$$

with the initial values $u'_x(x,0)$ and $u'_t(0,t)$ given by (5.28).

The condition $u'_x(x,0)=0$ is verified if and only if $\varphi'(x)+\psi'(x)=0$ and $u'_t(0,t)=0$ if and only if $\varphi'(x)=\psi'(-x)$.

Under the conditions of Theorem 5.11, the functions

$$u(x,t)=\sum_{j=1}^{\infty} a_j\varphi\Big\{\omega_j\Big(x+\frac{t}{a}\Big)\Big\}+b_j\varphi\Big\{\omega_j\Big(x-\frac{t}{a}\Big)\Big\}$$

are solutions of the wave differential equation, for all constants a_j, b_j and ω_j. Other linear combinations of exponential functions are also solutions.

Example 41. The differential equation $\Delta y - a^2 y_t'' = 0$ has the solutions

$$y(t) = \sum_{n \geq 0} c_n e^{k_n(x+at)}$$

with arbitrary constants c_n and k_n, and

$$y(t) = \sum_{n \geq 0} \{A_n \sin(k_n x) + B_n \cos(k_n x)\} e^{ak_n t}$$

with arbitrary constants A_n, B_n and k_n.

Example 42. For every function u of $C_2([0,T] \times \Omega)$ such that $u(0) = 0$, the function $y(x,t) = \sinh u(x_1 + x_2 - vt)$, defined for x in a subset Ω of $\mathbb{R}^2$ and t in $[0,T]$, is solution of the differential equation $\Delta y - 2v^{-2} y_t'' = 0$ with boundary condition $y(0) = 0$. It is solution of the differential equations

$$\begin{aligned} y_t' &= -vy_{x_1}' = -vy_{x_2}', \\ y_{tt}'' &= (1+a)v^2 \Delta y + \frac{a}{2} v(y_{x_1,t}'' + y_{x_2,t}''), \end{aligned}$$

for every a in $[0,1]$. This function y is not oscillatory but decreasing. The function $y(x,t) = \frac{1}{2} \sum_{k=1}^{\infty} \{a_k \sinh u(x_1 + x_2 - vt) + b_k \cosh u(x_1 + x_2 - vt)\}$ is solution of the same wave differential equation. The function $1 - y^2$ with partial derivatives $-2(yy'' + y'^2)$ is also solution of this equation and this is a model for a single symmetric wave.

Example 43. The periodic functions $u(x,t) = \sin(\omega x)\sin(\nu\omega t)$ and $u(x,t) = \cos(\omega x)\cos(\nu\omega t)$ are solutions of the homogeneous wave equation $u_{xx}''(x,t) - \nu^{-2} u_{tt}''(x,t) = 0$ on a rectangle of $\mathbb{R} \times \mathbb{R}_+$, for every real ω. Adding conditions for the initial values $u(x,0) = 0$ for every real x and $u(0,t) = 0$ for every positive t. The functions $u(x,t) = \sum_{j=1}^{\infty} A_j \sin(\omega_j x)\sin(\nu\omega_j t)$ are also solutions, with constants A_j proportional to

$$k_j = \int_{-\pi}^{\pi} \int_{-\pi}^{\pi} u(x,t) \sin(\omega_j x) \sin(\nu\omega_j t)\, dx\, dt$$

and such that $S_0 = \sum_{j=1}^{\infty} A_j$ and $S_2 = \sum_{j=1}^{\infty} A_j \omega_j^2$ are finite. Since

$$\int_{-\pi}^{\pi} \sin^2(\omega x)\, dx = \frac{1}{2} \int_{-\pi}^{\pi} \Big\{1 - \cos(2\omega x)\Big\}\, dx = \pi - \frac{1}{2\omega} \sin(2\omega\pi)$$

for every ω, $k_j = \Big\{\pi - \frac{1}{2\omega_j} \sin(2\omega_j \pi)\Big\}\Big\{\pi - \frac{1}{2\nu\omega_j} \sin(2\nu\omega_j \pi)\Big\}$.

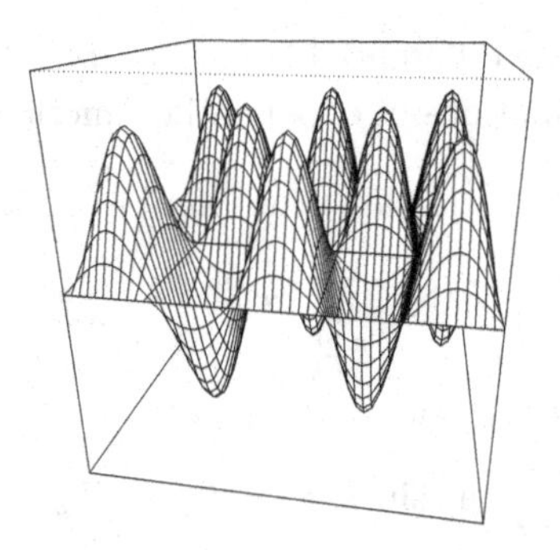

Fig. 5.1 A multiplicative wave function $u(x,t)$.

Example 44. The differential equation

$$u''_{xx}(x,t) + \nu^{-2}u''_{tt}(x,t) = 0, \ x \text{ in } [0,l], \ t \text{ in } [0,T]$$

with initial conditions $u(x,0) = A\sin(\omega x)$, and $u(0,t) = 0$ has the solution $u(x,t) = \{A\sin(\theta x + \omega) + B\cos(\theta x + \omega)\}e^{\pm\nu\gamma t}$, with arbitrary constants, under the condition $\theta^2 - \gamma^2 = 0$.

Example 45. In $\mathbb{R}^p \times \mathbb{R}_+$, a function $u(r,t) = f(r + a^{-1}t)$ is solution of the differential equation

$$\Delta u - a^2 u''_{tt}(x,t) = (p-1)\,\frac{f'(r + a^{-1}t)}{r}$$

with initial conditions $u(r,0) = f(r)$ and $u(0,t) = f(a^{-1}t)$.

Theorem 5.12. *The functions*

$$u(x,t) = \sum_{j=1}^{+\infty} u_j[\sin\{k_j(x - x_0)\} + \cos\{k_j(x - x_0)\}]\{\sin(k_j\nu t) + \cos(k_j\nu t)\}$$

are solutions of the homogeneous wave equation $\Delta u(x,t) - \nu^{-2}u''_{tt}(x,t) = 0$ *on* $\mathbb{R} \times \mathbb{R}_+$*, for all constants* u_j *and* k_j*.*

The functions $u(x,t) = F(x) + \sum_{j=1}^{+\infty} u_j[\sin\{k_j(x - x_0)\} + \cos\{k_j(x - x_0)\}]\{\sin(k_j\nu t) + \cos(k_j\nu t)\}$ are solutions in $\mathbb{R} \times \mathbb{R}_+^*$ of the inhomogeneous wave equation $u''_{xx}(x,t) - \nu^{-2}u''_{tt}(x,t) = f(x)$ with a function f such that $F''_{xx} = f$, for all constants u_j and k_j.

Theorem 5.13. *The solutions of the inhomogeneous wave equation* $u''_{xx} - a^2 u''_{tt} = f(x - \frac{t}{a})$ *on* $[0, l] \times [0, T]$ *are sums of solutions (5.27) and a function* F *with second derivative* f.

Let $\xi(x,t) = x_1 + x_2 - vt$, x in a subset Ω of $\mathbb{R}^2$ and t in $[0,T]$. The function $y(x,t) = u \circ \xi(x,t)$ is a solution of the wave equation with the constant coefficient $\alpha = 2v^{-2}$, for every function u of $C_2(\Omega \times [0,T])$.

In a subset Ω of $\mathbb{R}^3$, the differential equation

$$a^2 u''_{xx}(x,y,t) + b^2 \Delta_{yz} u(x,y,z,t) - u''_{tt}(x,y,t) = 0,$$

for (x,y,z) in Ω, t in $[0,T]$, has solutions $u(x,y,z,t) = U(x,t) + V(y,z,t)$ such that

$$a^2 U''_{xx} = U''_{tt}(x,y,t), \quad b^2 \Delta_{yz} V(y,z,t) = V''_{tt}(x,y,t)$$

and these equations may be solved separately.

Theorem 5.14. *The wave differential equation* $\Delta u - ka^2 u''_{tt} = 0$ *with boundary conditions* $u(0,t)$ *and* $u(x,0)$ *on* $[0,l_1] \times \ldots \times [0,l_k] \times [0,T]$, $k \geq 2$, *and such that*

$$u(x,0,\ldots,0) - u(0,\ldots,0,ax) + u(-x,0,\ldots,0) - u(0,\ldots,0,-ax) = 0$$

has the solutions

$$u(x,t) = \varphi\Big(x_1 + \ldots + x_k + \frac{t}{a}\Big) + \psi\Big(x_1 + \ldots + x_k - \frac{t}{a}\Big)$$

with the functions

$$\varphi(x) = \frac{1}{2}\{u(x,0,\ldots,0) + u(0,\ldots,0,ax)\},$$
$$\psi(x) = \frac{1}{2}\{u(x,0,\ldots,0) - u(0,\ldots,0,ax)\}.$$

In particular, the exponential models of waves on a plane with a constant coefficient are multiplication functions of separate variables, they include the products of sine and cosine functions like in Theorem 5.12.

Theorem 5.15. *The functions*

$$u(x,t) = \sum_{j=1}^{+\infty}\Big[A_{ij}\sin\Big\{\sum_{i=1}^{p} k_{ij}(x_i - x_{0i})\Big\} + B_{ij}\cos\Big\{\sum_{i=1}^{p} k_{ij}(x_i - x_{0i})\Big\}\Big]$$
$$\times\Big[A_{ij}\sum_{i=1}^{p}\Big\{\sin\Big(\frac{t}{\alpha}k_{ij}\Big) + B_{ij}\cos\Big(\frac{t}{\alpha}k_{ij}\Big)\Big\}\Big]$$

is solution of the equation $\Delta u - \alpha^2 u'_t = 0$ *in* $\mathbb{R}^p \times \mathbb{R}^*_+$.

Replacing the sum of the trigonometric time functions by a sum of exponential functions with the same parameters in Theorem 5.15, we get solutions of the equation $\Delta u + \alpha^2 u'_t = 0$.

In $\mathbb{R}^2$, $u(x,y,t) = \sum_{j=1}^{+\infty} u_j \cos(k_j x)\sin(k_j y)\cos(k_j\omega\sqrt{2}t)$ has the partial derivatives

$$\begin{aligned}
u''_{xx}(x,y,t &= -k_j^2 \cos(k_j x)\sin(k_j y)\cos(k_j\omega\sqrt{2}t),\\
u''_{yy}(x,y,t &= -k_j^2 \cos(k_j x)\sin(k_j y)\cos(k_j\omega\sqrt{2}t),\\
u''_{tt}(x,y,t &= -2k_j^2\omega^2 \cos(k_j x)\sin(k_j y)\cos(k_j a\sqrt{2}t),
\end{aligned}$$

which satisfy the differential equation

$$\Delta u - a^2 u''_{tt} = 0.$$

According to the boundary conditions

$$u(x,y,t) = \sum_{j=1}^{+\infty}\{A_j\cos(k_j x)\sin(k_j y) + B_j\sin(k_j x)\cos(k_j y)\}$$
$$\{C_j\cos(k_j\omega\sqrt{2}t) + D_j\sin(k_j\omega\sqrt{2}t)\}$$

with arbitrary constants, is solution of this equation.

The solutions on $\mathbb{R}^p \times \mathbb{R}_+^*$ of the inhomogeneous equation

$$u''_{xx}(x,t) + \nu^{-2}u''_{tt}(x,t) = f(x,t)$$

are sums of the general solution of Theorem 5.15 and a particular solution of the inhomogeneous equation. If $f(x,t)$ has an expansion as a sum of product of sine and cosine functions, the function u has the same form with coefficients depending on those of the expansion of f, like in Theorem 5.7. The solutions may be defined on $\mathbb{R}^p$ instead of $\mathbb{R}$. For every function F of $C_2(\mathbb{R}^p)$, $u(x,t) = F(x_1 \pm \nu t, \ldots, x_p \pm \nu t)$ is a solution of the wave equation $\Delta u(x,t) - \nu^{-2}u''_t(x,t) = 0$ with the conditions of initial values $u(x_0,t) = F(x_{01} \pm \nu t, \ldots, x_{0p} \pm \nu t)$ and $u(x,0) = F(x)$. The function $u(x,t) = F(x_1 \pm \nu_1 t, \ldots, x_p \pm \nu_p t)$ is a solution of the wave equation $\Delta u(x,t) - (\sum_{j=1}^p \nu_j)^{-2}u''_t(x,t) = 0$ with the conditions of initial values $u(x_0,t) = F(x_{01} \pm \nu_1 t, \ldots, x_{0p} \pm \nu_p t)$ and $u(x,0) = F(x)$.

The equation is more generally written by replacing the constant of the differential equation by a function $\alpha_{x,t}$. Sufficient conditions for the existence of a multiplication solution $u(x,t) = \varphi_1(x)\varphi_2(t)$ of the differential equation

$$\Delta u(x,t) - \alpha_{x,t}u''_{tt}(x,t) = f(x,t) \tag{5.29}$$

is the factorization of the function f with a constant α or the factorization of the functions f and α. In the homogeneous equation with a coefficient $\alpha_{x,t} = \phi_1(x)\phi_2^{-1}(t)$ with the derivatives $\phi_1(x) = \varphi''_{1,xx}(x)$ and $\phi_2(t) = \varphi''_{2,tt}$, on a line. In a plane, a function $u(x_1, x_2, t) = \varphi_1(x_1)\varphi_2(x_2)\varphi_3(t)$ is solution of this equation if $\varphi_1^{-1}\varphi''_{1,xx} + \varphi_2^{-1}\varphi''_{2,xx} = \alpha_{x,t}\varphi_3^{-1}\varphi''_{3,tt}$.

The isotrope wave equations are defined with respect to the l^2-norm r of x in $\mathbb{R}^2$ for time-varying functions of the norm r in $C_2(\mathbb{R}^2_+)$. Their second order differential operator is written as

$$L_{r,t} = \frac{\partial^2}{\partial r^2} - \alpha(r,t)\frac{\partial^2}{\partial t^2}$$

with a strictly positive function α, for all r in an interval (r_0, r_1) and t in a time interval $[0, T]$. The coefficient of this equation is necessarily a function of r. The initial conditions define the values of the function u and its first partial derivatives on the frontier of the domain of the equation. The equation is equivalent to

$$(ru'_r)_r - r\alpha(r,t)u''_{tt} = 0.$$

Like in the Poisson equation (1.9), $\Delta_x u(r,t) = \frac{2}{r}u'(r,t) + u''(r,t)$ and the differential equation $L_{x,t}u = 0$ in a plane is equivalent to

$$\frac{2}{r}u'_r(r,t) + u''_{rr}(r,t) = \alpha(r,t)u''_{tt}(r,t).$$

This equation is also equivalent to a wave equation in a rectangle with an operator (5.24)

$$v''_{rr}(r,t) = r\alpha(r,t)u''_{tt}(r,t),$$

by the reparametrization $v'_r(r,t) = r^{-1}\{r^2u(r,t)\}'_r$.

The differential equation

$$\Delta u(r,t) - \alpha_r u''_t(r,t) = f(r,t)$$

has a multiplicative solution $u(r,t) = \theta_r\varphi_t$, with $\varphi_t = e^{a_t}$, if the functions θ_r and a_t satisfy the differential equations (5.18)

$$\begin{aligned} f(r,t) &= \theta''_r\varphi_t - \alpha_r\varphi''_t\theta_r, \\ e^{-a_t}f(r,t) &= \theta''_r - \alpha_r\{a''_t + (a'_t)^2\}\theta_r, \end{aligned} \tag{5.30}$$

in the interval $[0, r_1]$ and for t in $[0, T]$. The functions θ_r and a_t satisfy marginal inhomogeneous second order differential equations for every t

and, respectively, r. Their solutions are sums of particular solutions of the equation and a general solution of

$$\theta_r'' = C\alpha_r\theta_r,$$
$$\varphi_t'' = C\varphi_t.$$

The second equation implies $a_t = \pm\sqrt{C}t + k_1$, and the first equation has the implicit solutions θ_r such that $\theta'^2 = C\int_0^r \alpha_s\, d\theta_s^2$, with arbitrary constant k_j. Thus, the exponential function φ has a constant coefficient if the function α depends only on r.

Example 46. With functions θ_r proportional to r^{-1}, a function $a_t = at$ and a product function $f(r,t) = f_{1r}f_{2t}$, the differential equation (5.18) reduces to (5.19)

$$f_{1r}f_{2t} = kr^{-3}(2 - a^2r^2\alpha_r)e^{a_t}.$$

A particular solution is obtained with an exponential e^{at} proportional to f_{2t} and $r = k\{(2 - a^2r^2\alpha_r)f_{1r}^{-1}\}^{\frac{1}{3}}$ and under the constraint $a^2r^2\alpha_r < 2$. The function θ_r depends on f_{1r} and α_r through this expression of r. Under the constraint $a^2r^2\alpha_r < 2$, the function α_r is bounded by $2a^{-2}r^{-2}$ and

$$u(r,t) = \frac{1}{2}\Big\{kr^{-1} + (2 - a^2r^2\alpha_r)^{-\frac{1}{3}}f_{1r}^{\frac{1}{3}}\Big\}e^{at}$$

is solution of (5.19).

At fixed r, Laplace's equation (5.18) is written as a inhomogeneous linear second order time-varying differential equation

$$A_r(t) = \varphi_t'' - k_{1r}\varphi_t + k_{2r}f_t = 0$$

for all r and t, in intervals $[r_0, r_1]$ where θ_r is different from zero. At fixed t, this is an inhomogeneous linear second order differential equation with respect to r

$$B_t(r) = \theta_r'' - e^{-a_t}\varphi_t''\alpha_r\theta_r - e^{-a_t}f(r,t) = 0.$$

With a multiplicative function f, $B_t(r) = \theta_r'' - \{a_t'' + (a_t')^2\}\alpha_r\theta_r - cf_{1r} = 0$. They are Sturm-Liouville partial differential equations with constant functions in $A_r(t)$ and their solutions satisfy the implicit equations of Theorems 4.7 and 4.8. Since the coefficients are time constant in $A_r(t)$, the function φ must be an exponential functions of t or a linear combination of exponential functions of t, with constants depending on r, according to Proposition 4.6. A solution of $B_t(r) = 0$ with α_r proportional to r^{-2} and $\alpha_r = kr^{-1}h_r$ was given in the introduction.

From Theorem 4.12, a solution of the homogeneous equation defined at fixed t by $B_t(r) = 0$ in $]r_0, r_1[$, with initial values $u_t(r_0)$ and $u'_t(r_1)$, satisfies the implicit equation

$$u_t(r) = u_t(r_0) + (r - r_0)u'_t(r_1) + \int_{r_0}^{r} Q_{v,t}u_{v,t}\,dv + \int_{r_0}^{r}\int_{v}^{r_1} Q_{\rho,t}\,du_t(\rho)\,dv,$$

where $Q_{r,t} = e^{-a_t}\varphi''_t \int_r^{r_1} \alpha_v\,dv$. These results provide expressions for the solutions on the frontier of the domain of the differential equation (5.18) from their values at r_0, r_1, 0 and T.

Multiplying (5.18) by $\theta'_r\varphi'_t$ and integrating the homogeneous equation in $[r_0, r] \times [0, t]$ implies

$$2\Big(\int_0^t f_{2s}\,d\varphi_s\Big)\Big(\int_{r_0}^{r} f_{1v}\,d\theta_v\Big) = (\theta_r'^2 - \theta_0'^2)\varphi_t - \varphi_t'^2 \int_{r_0}^{r} \alpha_s\theta_s\,d\theta_s$$

for all r and t, with a function a'_t null at zero. This equality is an inhomogeneous and nonlinear differential equation depending only on the first order derivatives with respect to r and t.

The wave equation (5.29) on $\mathbb{R}_+ \times \mathbb{R}$ with multiplicative functions $f(x,t) = f_1(x)f_2(t)$ and $\alpha(x,t) = \varphi_1(x)\varphi_2(t)$ has multiplicative solutions $u(x,t) = u_1(x)u_2(t)$ such that

$$\frac{u_1''}{\varphi_1 u_1} - \varphi_2\frac{u_2''}{u_2} = \frac{f_1}{\varphi_1 u_1}\frac{f_2}{u_2}$$

and this expression is a constant k such that $u_1'' - k_1\varphi_1 u_1 = k_1$ and $\varphi_2 u_2'' - u_2 = k_2$, with $f_1 = k_3\varphi_1 u_1$, $f_2 = k_4 u_2$ and $k_1 - k_2 = k_3 k_4 = k$. The components of u are primitives of the differential equations $u_1'' = k_3^{-1}k_1 f_1$ and $u_2'' = (k_4\varphi_2)^{-1}k_2 f_2$. However, the factorization of the function f or α is not a sufficient condition for the existence of a multiplication solution u and non multiplicative solutions must be considered.

Example 47. On $[t_0, t_1] \times [x_0, x_1]$, with $t_0 > 0$, the differential equation

$$\Big(\frac{x^2}{t^2} - \frac{2x}{at}\Big)u''_{xx} - u''_{tt} = 0$$

has the solution $u(x,t) = e^{-a\frac{x}{t}}$ under the boundary conditions $u(0,t) = 1$ and $u'_x(0,t) = 1$.

Example 48. On $[t_0, t_1] \times [x_0, x_1]$, with $t_0 > 0$, the differential equation

$$u''_{xx} - \frac{t^2}{x}u''_{tt} + 2a\Big(t - \frac{1}{t}\Big)u = 0$$

has the solution $u(x,t) = xe^{-a\frac{x}{t}}$ under the boundary conditions $u(0,t) = 0$ and $u'_x(0,t) = 1$.

5.6 Parabolic, elliptic and hyperbolic equations

The Poisson-Boltzmann equation (5.12), the heat conduction equations (5.15) and the differential equations of Propositions 5.7 and 5.13 for the Gaussian densities on $\mathbb{R}$ and $\mathbb{R}^p$ are parabolic differential equations. The wave differential equations of Section 5.5 are hyperbolic differential equations. Leray and Lions (1965), Lions (1965a, b), Lions and Strauss (1965) studied the existence and the unicity of first order parabolic and second order elliptic and hyperbolic non-linear equations, and their regularity properties, under monotonicity or coercivity conditions. Here we provide explicit proofs of the existence of solutions for such equations under stronger conditions.

Parabolic differential equations. We first consider parabolic first order non-linear differential equations

$$\Delta u(t,x) + a\frac{\partial u(t,x)}{\partial t} + b(t,u) = f(t), \qquad (5.31)$$
$$u(t_0,x) = u_0(x), \quad u'_x(t_0,x) = u'_0(x),$$

on $I \times \Omega$, where Ω is a bounded subset of $\mathbb{R}^p$ and I is a bounded real interval of $\mathbb{R}_+$, and such that the function u defined on $I \times \Omega$ takes its values in real a Hilbert space V, and it is $C^2(I)$ for every x in Ω, and $C^2(\Omega)$ for every t in I.

Proposition 5.14. *Let b be a locally differentiable function on $I \times V$, then there exist a solution of the homogeneous parabolic differential equation (5.31) with initial values $u(t_0,x_0) = u_0$ and $u'_x(t_0,x_0) = u'_0$.*

Proof. Let $p = 1$, for all $\varepsilon > 0$, there exist $\delta_1 > 0$ and $\delta_2 > 0$ such that

$$|t - t_0| < \delta_1 \text{ implies } \|u(t,x)) - u(t_0,x))\| < \varepsilon, \text{ as } |x - x_0| < \delta_2,$$
$$|x - x_0| < \delta_2 \text{ implies } \|u(t,x)) - u(t,x_0))\| < \varepsilon, \text{ as } |t - t_0| < \delta_1,$$

then, by continuity of u on $\Omega \times I$, we have

$$\|u(t,x) - u_0\| \le \|u(t,x) - u(t_0,x)\| + \|u(t_0,x) - u_0\| < 2\varepsilon,$$

where δ_1 and δ_2 tend to zero as ε tends to zero.

Let $(I_i)_{i \le n}$ be a partition of I starting from t_0 and with a path δ_1, and let $(J_i)_{i \le n}$ be a partition of Ω starting from x_0 and with a path δ_2, an approximated solution $u_{1,\varepsilon}(t,x)$ of a solution for the differential equation is defined on $I_1 \times J_1$ from the differentiation of the function b in a neighborhood of (t_0, x_0) as

$$b(t,u) = b(t_0,u_0) + (t-t_0)\frac{\partial b}{\partial t}(t_0,u_0) + (u-u_0)\frac{\partial b}{\partial u}(t_0,u_0) + o(\varepsilon),$$

on $I_1 \times J_1$, the differential equation has the form

$$\Delta u(t,x) + a\frac{\partial u(t,x)}{\partial t} + c_{00}u(t,x) = h_\varepsilon(t) \tag{5.32}$$

where $h_\varepsilon(t) = f(t) + u_0 b'_u(t_0,u_0) - b(t_0,u_0) - (t-t_0)b'_t(t_0,u_0) + o(\varepsilon)$.

From Proposition 5.9 and according to the sign of c_{00}, a sinusoïdal or hyperbolic function

$$\begin{aligned} u_{00}(x,t) &= \alpha \exp\{-\sqrt{c_{00}}(x-x_0)\} + \beta \exp\{\sqrt{c_{00}}(x-x_0)\} \\ &\quad + \gamma \exp\{-k_2(t-t_0)\} \end{aligned}$$

is solution of the homogeneous differential equation $\Delta u(t,x) + au'_t(t,x) + c_{00}u(t,x) = 0$ on $I_1 \times J_1$, with the initial conditions $\alpha+\beta+\gamma = u_{00}(x_0,t_0)$, $-k_1\alpha + k_1\beta = u'_{x,00}(x_0,t_0)$, $k_1^2 = -c_{00}$ and $\beta k^2 = c_{00}$, for the constant coefficients. A solution $u_{00}(x,t)$ of the inhomogeneous equation (5.31) on $I_1 \times J_1$ is deduced by replacing in the function $v(t) = \beta \exp\{-k_2(t-t_0)\}$ the constant β with a function $w_{00}(t)$ solution of the first order differential equation $aw'_t(t,x) = e^{k_2(t-t_0)}h(t)$, thus

$$w_{00}(t) = u_{00}(t) + \int_{t_0}^{t} h_s e^{k_2(s-t_0)}\,ds.$$

The same arguments prove the existence of an approximated solution $u_{12,\varepsilon}(t,x)$ of (5.31) on $I_1 \times J_2$ and on the following subsets of the partitions of $\Omega \times I$. The approximation solution is $C^2(\Omega)$ for every t and $C^2(I)$ for every x, it converges to a solution u as ε tends to zero.

The proof extends to subsets Ω of $\mathbb{R}^p$ replacing in $u_{00}(x,t)$ the function $\alpha \exp\{-\sqrt{c_{00}}(x-x_0)\} + \beta \exp\{\sqrt{c_{00}}(x-x_0)\}$ by a sum of p functions depending separately on the coordinates of x in Ω. □

Proposition 5.15. *Let $b(t,\cdot)$ be a locally differentiable function on I, then there exist functions $f(x)$ such that the parabolic differential equation*

$$\Delta u(t,x) + a\frac{\partial u(t,x)}{\partial t} + b(t,u) = f(x),$$

with initial values $u(t_0,x) = u_0(x)$ and $u'_x(t_0,x) = u'_0(x)$, has solutions.

Proof. Let Ω be a subset of $\mathbb{R}$. By the local linearization used for the proof of Proposition 5.14, the homogeneous equation has a unique solution $u_{00}(x,t)$ on $I_1 \times J_1$ and the solution of the inhomogeneous equation is deduced for functions $f(x)$ such that $u_1''(x) + c_{00}u_1(x) = f(x)$. The proof extends to subsets Ω of $\mathbb{R}^p$ with sums of n functions depending separately on the coordinates of x in Ω. □

Elliptic differential equations. Simple examples of elliptic differential equations are Poisson's second order differential equation $\Delta u = f$ and the ellipsoïd evolution equations

$$\sum_{i=1}^{p} \frac{x_i^2}{a_i^2} + \frac{t^2}{b^2} = 1,$$

they are solutions of the differential equation $u_{tt}''(t,x) + \Delta u(t,x) = c$ with the constant $c = 2(\sum_{i=1}^{p} a_i^{-2} + b^{-2})$.

According to Proposition 5.3, the homogeneous elliptic differential equation $\Delta u(r) + u_{tt}''(r) = 0$ has the solution $u(r) = r^{-(p-1)}(t,x)$ where $r(t,x) = (\sum_{i=1}^{p} x_i^2 + t^2)^{\frac{1}{2}}$, for $p > 1$.

Linear elliptic differential equations on $\mathbb{R}_+ \times \mathbb{R}^2$ have the form

$$\begin{aligned} &u_{tt}''(t,x,y) + p(x,y)u_{xx}''(t,x,y)(t,x,y) + q(x,y)u_{yy}''(t,x,y) = h(t,x,y),\\ &u(0,x,y) = F(x,y), \quad u_t'(0,x,y) = f(x,y). \end{aligned} \tag{5.33}$$

Theorem 5.16. *Let h be a function of a Hilbert space having an expansion $h(t,x,y) = \sum_{n\geq 0} e^{nt} h_n(x,y)$, then there exist a solution of the elliptic differential equation (5.33).*

Proof. Let ϕ be a solution of the homogeneous equation (5.33) obtained from Proposition 5.3 by replacing x and y by twice differentiable functions of the square of each coordinates, and let $u = \phi + \sum_{n\geq 0} u_n$ such that for every n

$$\begin{aligned} &u_{ntt}''(t,x,y) + p(t,x,y)u_{nxx}''(t,x,y)(t,x,y) + q(t,x,y)u_{nyy}''(t,x,y)\\ &= e^{nt} h_n(x,y), \end{aligned} \tag{5.34}$$

with initial values deduced from the expansion of the solution u and its initial values

$$u_n(0,x,y) = 0, \quad u_t'(0,x,y) = 0, u_t'(0,x,y) = h_n(0,x,y).$$

Like Liouville (1939), let

$$\lambda_n(t,x,y) = u'_{nt}(t,x,y) - mu_n(t,x,y), \tag{5.35}$$

λ_n is solution of the homogeneous equation (5.34) with initial values

$$\lambda_n(0,x,y) = 0, \quad \lambda'_{nt}(0,x,y) = h(0,x,y)$$

and by integration of (5.35)

$$u_n(t,x,y) = \int_0^t e^{n(t-s)} \lambda_n(s,x,y)\, ds$$

the expression of a solution u of (5.33) follows. □

Let t belong to a real interval $I = [0,T]$ and let x in a subset Ω of $\mathbb{R}^p$, defining functions $u(t,x)$ of $C^2(I \times \Omega)$, with values in a real set V and with derivatives in V'. Nonlinear elliptic second order differential equations are written as

$$\begin{aligned} &\Delta u(t,x) + p(x)u''_{tt}(t,x) + b(t,u,u') = f(t,x), \\ &u(t_0,x) = u_0(x),\ u'_x(t_0,x) = v_x(x),\ u'_t(t_0,x) = v_t(x), \end{aligned} \tag{5.36}$$

with a function b defined on $I \times V \times V'$.

Theorem 5.17. *Let the function b be a locally differentiable function on $I \times V \times V'$, then there exists a solution of the elliptic equation*

$$\Delta u(t,x) + au''_{tt}(t,x) + b(t,u,u') = 0$$

with initial values $u(t_0,x_0) = u_0$ and $u'_t(t_0,x_0) = u'_{t0}$.

Proof. Arguing as in the proof of Proposition 5.14 for $p = 1$, for every $\varepsilon > 0$ there exist $\delta_1 > 0$ and $\delta_2 > 0$ such that $\|u(t,x) - u_0\| \leq \varepsilon$ as $|t - t_0| < \delta_1$ and $|x - x_0| < \delta_2$.

Let $(I_i)_{i \leq n}$ be a partition of I starting from t_0 and with a path δ_1, and let $(J_i)_{i \leq n}$ be a partition of Ω starting from x_0 and with a path δ_2, an approximated solution $u_{1,\varepsilon}(t,x)$ of a solution for the differential equation is defined on $I_1 \times J_1$ from the differentiation of the function b for t in a neighborhood of t_0, as

$$\begin{aligned} b(t,u,u') = b(t_0,u_0) &+ (t - t_0)\frac{\partial b}{\partial t}(t_0,u_0) + (u - u_0)\frac{\partial b}{\partial u}(t_0,u_0) \\ &+ (u' - u'_0)\frac{\partial b}{\partial u'}(t_0,u_0) + o(\varepsilon). \end{aligned}$$

On $I_1 \times J_1$, the differential equation then is written as

$$\Delta_x u(t,x) + au''_{tt}(t,x) + a_{00}u(t,x) + b_{00}u'(t,x) = h_\varepsilon(t),$$

where $u'(t,x)$ is the vector of the derivatives with respect to (t,x), it extends equation (5.32) for the parabolic differential equation on $\mathbb{R}_+ \times \mathbb{R}$, including the second order derivative of u with respect to t in the Laplacian. There exists a solution of the homogeneous differential equation on $I_1 \times J_1$, having the same form as the solution of Proposition 5.10, under modified initial conditions for the constant coefficients.

A solution of the inhomogeneous differential equation is deduced, replacing in the function $v(t) = \beta \exp\{-k_2(t-t_0)\}$ the constant β with a solution of the second order differential equation $aw''_{tt}(t) + b_{00,t}w'(t,x) = g(t)h(t)$, where $g(t) = e^{k_2(t-t_0)}$, thus

$$w'_{00}(t) = u'_{t0} + a^{-1}\int_{t_0}^{t} h_s e^{-a^{-1}b_{00,t}(t-s)}\, ds,$$

$$w'_{00}(t) = u_0 + u'_{t0}(t-t_0) + a^{-1}\int_{t_0}^{t}\int_{t_0}^{y} h_s e^{-a^{-1}b_{00,t}(y-s)}\, ds\, dy.$$

The same arguments prove the existence of an approximated solution $u_{12,\varepsilon}(t,x)$ of (5.31) on $I_1 \times J_2$ and on the following subsets of the partitions of $\Omega \times I$. The approximation solution is $C^2(\Omega)$ for every t and $C^2(I)$ for every x, it converges to a solution u as ε tends to zero.

The proof extends to subsets Ω of $\mathbb{R}^p$ replacing in $u_{00}(x,t)$ the local solution by a sum of p functions depending separately on the coordinates of x in Ω. □

Explicit solutions may be expressed using expansions of the functions u, b and f on an orthonormal basis of functions on $\mathbb{R}^p \times \mathbb{R}_+$.

Example 49. Let $u(x,t)$ be a solution of an inhomogeneous differential equation

$$u''_{tt} + a(t,x)u''_{xx} + b(t,x)u'_x = f(t,x)$$

on $\mathbb{R} \times \mathbb{R}_+$ having an expansions on the basis defined by the product of Laguerre's polynomials as $u(x,t) = \sum_{j\geq 0}\sum_{k\geq 0} u_{jk}L_j(x)L_k(t)$, then

$$f(t,x) = \sum_{j\geq 0}\sum_{k\geq 0}\{u_{jk}L_j(x)L''_{k,tt}(t) + a(t,x)L''_{j,xx}(x)L_k(t) \\ + b(t,x)L'_{j,x}(x)L_k(t)\}.$$

The derivatives of Laguerre's polynomials

$$xL'_n(x) = -nL_n(x) - nL_{n-1}(x),\ n \geq 1$$
$$x^2L''_n(x) = n(n-1)L_n(x) - n(2n-1)L_{n-1}(x) - n^2L_{n-2}(x),\ n \geq 2,$$

imply the equivalence of the differential equation and

$$\begin{aligned}
x^2t^2 f(t,x) = \sum_{j\geq 0}\sum_{k\geq 0}\Big[& u_{j,k+2}x^2 L_j(x)\{(k+2)(k+1)L_{k+2}(t) \\
& - (k+2)(2k+3)L_{k+1}(t) - (k+2)^2 L_k(t)\} \\
& + a(t,x)u_{j+2,k}t^2 L_k(t)\{(j+2)(j+1)L_{j+2}(x) \\
& - (j+2)(2j+3)L_{j+1}(x) - (j+2)^2 L_j(x)\} \\
& - b(t,x)u_{j+1,k}xt^2 L_k(t)\{(j+1)L_{j+1}(x) - (j+1)L_j(x)\}
\end{aligned}$$

Let $g(t,x) = x^2t^2 f(t,x)$ have an expansion $g(x,t) = \sum_{j\geq 0}\sum_{k\geq 0} g_{jk}L_j(x)$ $L_k(t)$, the coefficients u_{jk} are deduced recursively by identification from the coefficients g_{jk}.

Hyperbolic differential equations. Wave solutions of the homogeneous hyperbolic differential equation

$$\Delta u(x,t) - \alpha u''_{tt} = 0$$

are defined in Section 5.5. Other solutions on $\mathbb{R}^p \times \mathbb{R}_+$ are the hyperbolic evolution functions

$$\sum_{i=1}^{n} \frac{x_i^2}{a_i^2} - \frac{t^2}{b^2} = 1$$

with $\sum_{i=1}^{n} a_i^{-2} = b^{-2}$. The homogeneous differential equation

$$u''_{tt} - n^{-1}a^2\Delta u(x,t) = 0$$

on $\mathbb{R}^p \times \mathbb{R}_+$ has exponential solutions

$$u(x,t) = e^{\sum_{i=1}^{n}(\pm x_i)}e^{\pm at}.$$

The existence and unicity of solutions for the inhomogeneous hyperbolic differential equations depend on the initial conditions and on conditions about f, as examples proved in Section 5.5 for specific differential equations.

A linear hyperbolic differential equations on $\mathbb{R}_+ \times \mathbb{R}^2$ is written as

$$\begin{aligned}
& p(x,y)u''_{xx}(t,x,y)(t,x,y) + q(x,y)u''_{yy}(t,x,y) - u''_{tt}(t,x,y) = h(t,x,y), \\
& u(0,x,y) = F(x,y), \quad u'_t(0,x,y) = f(x,y). \qquad (5.37)
\end{aligned}$$

Theorem 5.18. *Let h be a function of a Hilbert space having an expansion $h(t,x,y) = \sum_{n\geq 0} e^{nt}h_n(x,y)$, then there exist a solution of the hyperbolic differential equation (5.37).*

The proof is similar to the proof of Theorem 5.16.

Hyperbolic differential equations have the general form

$$\Delta_x u(x,t) - au''_{tt}(t,x) + b(t,u,u') = f(x,t), \tag{5.38}$$

for t in an interval $[0,T]$ and x in a subset Ω of $\mathbb{R}^p$, with u in a Hilbert space V with a norm $\|u\|$ and such that $\|u'\|_2$ is finite. General boundary conditions are defined on $\Sigma = \partial\Omega \times \mathbb{R}_+^*$ as

$$\begin{aligned} \frac{\partial u(x,t)}{\partial t} &\geq 0,\ (x,t) \in \Sigma, \\ \frac{\partial u(x,t)}{\partial \nu} &\geq 0,\ (x,t) \in \Sigma, \\ \frac{\partial u(x,t)}{\partial t}\frac{\partial u(x,t)}{\partial \nu} &= 0,\ (x,t) \in \Sigma, \\ u(x,0) &= 0,\ \frac{\partial u}{\partial t}(x,0) = 0, \end{aligned}$$

where ν is the set of vectors orthogonal to $\partial\Omega$.

Example 50. Let $y(t,x) = e^{\lambda t}\{A\sin(\lambda\sqrt{a}x) + B\cos(\lambda\sqrt{a}x)\}$ and $v = uy$ be solutions of the differential equation $y''_{xx} - ay''_{tt} = 0$, then the function u is solution of the differential equation

$$u''_{xx} - au''_{tt} + 2u'_x\frac{y'_x}{y} - \lambda\sqrt{a}u'_t = 0$$

where $y'_x(t,x) = \lambda\sqrt{a}e^{\lambda t}\{A\cos(\lambda\sqrt{a}x) - B\cos(\lambda\sqrt{a}x)\}$.

Proposition 5.16. *Let the function b be a locally differentiable function on $I \times V \times V'$, then there exist solutions for the homogeneous equation (5.38).*

$$\Delta_x u(x,t) - au''_{tt}(t,x) + b(t,u) = f(x,t)$$

with initial values $u(t_0,x_0) = u_0$ and $u'_t(t_0,x_0) = u'_0$.

Proof. Like in the proof of Proposition 5.14 for $p = 1$, for every $\varepsilon > 0$ there exist $\delta_1 > 0$ and $\delta_2 > 0$ such that $\|u(t,x) - u_0\| \leq \varepsilon$ and $\|u'(t,x) - u'_0\| \leq \varepsilon$ as $|t - t_0| < \delta_1$ and $|x - x_0| < \delta_2$.

Let $(I_i)_{i\leq n}$ be a partition of I starting from t_0 and with a path δ_1, and let $(J_i)_{i\leq n}$ be a partition of Ω starting from x_0 and with a path δ_2, an approximated solution $u_{1,\varepsilon}(t,x)$ of a solution for the differential equation is defined on $I_1 \times J_1$ from a first order expansion of the function b for (t,u) in a neighborhood of (t_0,u_0), like in the proof of Theorem 5.17.

On $I_1 \times J_1$, the nonlinear differential equation has the linear approximation

$$\Delta u(t,x) - a u''_{tt} + c_{00} u(t,x) = h_\varepsilon(t),$$

the homogeneous equation has a solution $u(x,t) = A[\sin\{k(x-x_0)\} + \cos\{k(x-x_0)\}]\{\sin(k\nu t) + \cos(k\nu t)\}$ with $k^2(1-a\nu^2) = c_{00}$. A solution of the inhomogeneous equation is deduced by the same method as in the proof of Theorem 5.17. The same arguments prove the existence of an approximated solutions on the following subsets of the partitions of $I \times J$, and the global solution converges as ε tends to zero. The proof extends to subsets Ω of $\mathbb{R}^p$. □

For the homogeneous equation (5.38) with b depending on the solution and its first order derivatives, a local approximation of b on $I_1 \times J_1$ as

$$\begin{aligned} b(t,u) &= b(t_0,u_0) + (t-t_0)\frac{\partial b}{\partial t}(t_0,u_0) + (u-u_0)\frac{\partial b}{\partial u}(t_0,u_0) \\ &\quad + (u'-u'_0)\frac{\partial b}{\partial u'}(t_0,u_0) + o(\varepsilon). \end{aligned}$$

leads to the linear approximation of (5.38)

$$\Delta u(t,x) - \alpha u''_{tt} + c_{00} u(t,x) + d_{00} u'(t,x) = h_\varepsilon(t)$$

and the solutions of the homogeneous equation cannot be sinusoïdal due to the first order derivatives. Solutions of the form (5.27) under constraints may be locally defined.

Proposition 5.17. *A function u defined on $\mathbb{R} \times \mathbb{R}_+$ by (5.27) with*

$$\begin{aligned} \varphi\Big(x+\frac{t}{a}\Big) &= v\Big(x+\frac{t}{a}\Big)e^{-\frac{ab}{ac+d}(x+\frac{t}{a})}, \\ \psi\Big(x-\frac{t}{a}\Big) &= w\Big(x-\frac{t}{a}\Big)e^{-\frac{ab}{ac-d}(x-\frac{t}{a})} \end{aligned}$$

is solution of a differential equation

$$u''_{xx}(x,t) - a u''_{tt}(t,x) + b u(t,x) + c u'_x(x,t) + d u'_t(t,x) = f(x,t)$$

if $f(x,t) = \frac{ac+d}{a} v'\Big(x+\frac{t}{a}\Big)e^{-\frac{ab}{ac+d}(x+\frac{t}{a})} + \frac{ac-d}{a} w'\Big(x-\frac{t}{a}\Big)e^{-\frac{ab}{ac-d}(x-\frac{t}{a})}$.

Proof. Let $y = x + \frac{t}{a}$ and $z = x - \frac{t}{a}$ and let u defined by (5.27) be solution of homogeneous differential equation with $\alpha = a^2$, then $bu(t,x) + cu'_x(x,t) + du'_t(t,x) = 0$ and

$$b\{\varphi(y) + \psi(z)\} + \frac{ac+d}{a}\varphi'(y) + \frac{ac-d}{a}\psi'(z) = 0,$$

its solutions have the form $\varphi(y) = \varphi_0 e^{-ky}$ and $\psi(z) = \psi_0 e^{-hz}$ with the constants $k = \frac{ab}{ac+d}$ and $h = \frac{ab}{ac-d}$. The inhomogeneous equation has solutions of the form $\varphi(y) = v(y)e^{-ky}$ and $\psi(z) = w(z)e^{-hz}$ with functions v and w such that

$$f\Big(\frac{y+z}{2}, \frac{a(y-z)}{2}\Big) = \frac{b}{k} v'(y)e^{-ky} + \frac{b}{h} w'(z)e^{-hz}$$

so the primitive of f is a mixture of the Laplace transforms of the functions v' and w'. □

5.7 Elasticity equations

The differential equations of elasticity for a line in $\mathbb{R}^3$ have been first studied by Lagrange (1853) and generalized Bertrand as a system of differential equations for its cartesian coordinates, with respect to the variation of the curve s it draws

$$p(y'_s z''_s - z'_s y''_s) = \theta x'_s + gy,$$
$$p(z'_s x''_s - x'_s z''_s) = \theta y'_s - gx,$$
$$p(x'_s y''_s - y'_s x''_s) = \theta z'_s,$$

where x axis has the same direction of the strength of the motion. If $\theta = 0$, $y'_s = cx'_s$ and the curve lies in a plane. Multiplying the equations by x'_s, y'_s and, respectively, z'_s and adding them leads to the equation

$$\theta(x'^2_s + y'^2_s + z'^2_s) + g(yx'_s - xy'_s) = 0. \tag{5.39}$$

In polar coordinates where $r^2 = x^2 + y^2$, $y = x \tan\omega$, $s^2 = r^2 + z^2$ with $r_s = s\cos\varphi$, and $z = s\sin\varphi$ therefore

$$z'_\varphi = r = (s^2 - z^2)^{\frac{1}{2}}$$

and (5.39) is equivalent to

$$\theta + g(yx'_s - xy'_s) = 0. \tag{5.40}$$

From (5.40), the derivative of the tangent of $x_s^{-1} y_s$ is

$$d_s \frac{y_s}{x_s} = \frac{xy'_s - yx'_s}{x^2} ds = \frac{\theta}{gx_s^2} ds.$$

The differential equation for the tangent of ω entails

$$\begin{aligned} y_t &= \frac{y_0 x_t}{x_0} + x_t \frac{\theta}{g} \int_{t_0}^{t} \frac{1}{s^2 \cos^2 \varphi \cos^2 \omega} \, ds \\ &= \frac{y_0 x_t}{x_0} - x_t \frac{\theta}{g \cos^2 \varphi \cos^2 \omega} \Big(\frac{1}{t} - \frac{1}{t_0} \Big) \\ &= \frac{y_0 x_t}{x_0} - \frac{\theta}{g \cos \varphi \cos \omega} \Big(1 - \frac{x_t}{x_0} \Big). \end{aligned}$$

If $\theta = 0$, $y_t = y_0 x_0^{-1} x_t$.

The differential equation $z'_\varphi = r$ implies

$$z_\varphi = z_0 + \int_{\varphi_0}^{\varphi} (x^2 + y^2)^{\frac{1}{2}} \, d\varphi$$

and the integral must be calculated using the expression of y according to x, and according to φ.

In $\mathbb{R}^p$, an elastic distortion is a map u from $C_1(\mathbb{R}^p)$ to $\mathbb{R}^p$ under the operators

$$T_{ij} = \frac{1}{2} \Big(\frac{\partial u_i}{\partial x_j} + \frac{\partial u_j}{\partial x_i} \Big)$$

under shape constraints.

In $\mathbb{R}^3$, let u denote the coordinates of a curve on the surface of a sphere centered at zero, its distortion in $\mathbb{R}^3$ is similar to the distortion of a plane. The motion on the surface of a sphere described by its projections on a grid of orthogonal circles. If the distortion strength is orthogonal to the surface at a point a_0, the motion of the orthogonal axis of the grid intersecting at a_0 are similar and they follow the equation of a string in orthogonal planes, with $\theta = 0$ in (5.40), each of them being orthogonal to the surface. The other curves of the grid have non zero parameters θ. The maximum depth of the distortion is reached as the derivatives of the projections on the grid are zero.

5.8 Exercises

5.7.1. Solve the equation $P''_{m,x} = f$ on $]-1, 1[$ for Legendre's polynomials.

5.7.2. Solve the differential equation $\Delta u(x) = f$ in $L_2([-1, 1])$, using expansions by projections on Chebyshev's polynomials of the functions u and f.

5.7.3. Solve the Sturm-Liouville differential equation in $\mathbb{R}$ using Fourier series for the solution and the coefficients.

5.7.4. Solve $xy''_x + y'_x - y = 0$.

5.7.5. Solve $xy''_x - my'_x - xy = 0$, with an integer $m > 0$.

5.7.6. Solve $u''_{xx}(x,t) - au''_{tt}(t,x) + bu(t,x) + cu'_x(x,t) + du'_t(t,x) = f(x,t)$ using expansions on the basis defined by the product of Laguerre's polynomials.

Chapter 6

Partial differential equations

This chapter focuses on systems of partial differential equations (PDEs), and their solutions. Their equivalence to implicit functions leads to conditions for the existence of exact solutions, for first and second order partial differential equations, and we define classes of solutions for the several equations. By reparametrizations, some systems of time varying PDEs are equivalent to ordinary differential equations where the time variable has disappeared. We present models for population dynamics and their solutions.

6.1 Partial differential equations in $\mathbb{R}^2$

Deriving the straight line equation $u(x, y) = ax + by + c = 0$, its constants are $a = u'_x$ and $b = u'_y$, it satisfies the linear first order partial differential equation

$$u(x, y) = xu'_x + yu'_y + c = 0.$$

The partial derivatives of the parabol curve $u(x, y) = ax^2 + by + c = 0$ are $u'_x = 2ax$, $u''_{xx} = 2a$ and $u'_y = b$, it is defined by the linear second order partial differential equation

$$u(x, y) = \frac{1}{2}x^2 u''_{xx} + yu'_y + c = 0.$$

These differential equations are special cases of polynomials curves with variables x and y, they have the form $u(x, y) = a_j x^j u_x^{(j)} + b_k y^k u_y^{(k)}$ or $u(x, y) = \sum_{j=0}^{J} \sum_{k=0}^{K} c_{jk} x^j y^k u_{x,y}^{(j,k)}$, including only the constant partial derivatives of the curve. A partial differential equation for a function $u(x)$

in $C^k(\mathbb{R}^p)$

$$\sum_{m=1}^{k}\sum_{n=1}^{p} a_{m,n}(x_n)u_{x_n}^{(m)} + u = 0$$

with marginal coefficients has a solution $u(x) = \sum_{n=1}^{p} u_n(x_n)$ with marginal functions $u_n(x_n)$ solutions of the ordinary differential equations

$$\sum_{m=1}^{k} a_{m,n}(x_n)u_{n,x_n}^{(m)} + u_n = 0.$$

The existence of explicit solutions of implicit functions depends on the determinant of their derivatives.

Theorem 6.1. *Let $\varphi_1, \ldots, \varphi_k$ be functions of $C_1(\mathbb{R}^p)$ and let x_0 be in $\mathbb{R}^p$ and such that $\varphi_1(x_0) = 0, \ldots, \varphi_k(x_0) = 0$. Let V_0 be a neighborhood of x_0 such that for every x in V_0, the determinant*

$$\begin{vmatrix} \varphi'_{1,x_1}(a)\ \varphi'_{1,x_2}(a) \ldots \varphi'_{1,x_k}(a) \\ \ldots \\ \varphi'_{k,x_1}(a)\ \varphi'_{k,x_2}(a) \ldots \varphi'_{k,x_k}(a) \end{vmatrix} \neq 0.$$

Then there exist k implicit functions $\psi_1, \ldots, \psi_k$ of $C_1(\mathbb{R}^{n-p})$ defined in a neighborhood W_0 of $(x_{0,k+1}, \ldots, x_{0,n})$ such that $\varphi_1(x) = 0, \ldots, \varphi_k(x) = 0$ in V_0 are equivalent to $x_1 = \psi_1(x_{k+1}, \ldots, x_n), \ldots, x_k = \psi_1(x_{k+1}, \ldots, x_n)$ in W_0.

From Poincaré's theorem, a function u from $C_1(\mathbb{R}^2)$ to $\mathbb{R}^2$ is solution of the differential equation

$$\frac{\partial u_1}{\partial x_1} + \frac{\partial u_2}{\partial x_2} = 0,$$

if and only if there exists a real function ϕ of $C_2(\mathbb{R}^2)$ such that the consistency properties hold

$$u_1 = \frac{\partial \phi}{\partial x_2}, \quad u_2 = -\frac{\partial \phi}{\partial x_1}.$$

Theorem 6.2. *On $\mathbb{R}^3$, a necessary and sufficient condition for the existence of a solution of the differential equation*

$$\frac{\partial u_1}{\partial x_1} + \frac{\partial u_2}{\partial x_2} + \frac{\partial u_3}{\partial x_3} = 0,$$

is the existence of real functions ϕ and ψ of $C_2(\mathbb{R}^3)$ such that

$$\Big(\frac{\partial^2 \phi}{\partial x_1 \partial x_2} + \frac{\partial^2 \phi}{\partial x_1 \partial x_3} + \frac{\partial^2 \phi}{\partial x_2 \partial x_3}\Big)(\phi - \psi) = 0,$$

$$u_1 = \frac{\partial \phi}{\partial x_2} - \frac{\partial \psi}{\partial x_3},$$
$$u_2 = \frac{\partial \phi}{\partial x_3} - \frac{\partial \psi}{\partial x_1},$$
$$u_3 = \frac{\partial \phi}{\partial x_1} - \frac{\partial \psi}{\partial x_2}.$$

The functions ϕ and ψ are not uniquely defined, it is satisfied in particular with $\phi = \psi$. Theorem 6.2 extends straightforwardly to a higher dimensions p including $p - 1$ functions or by alternating sums and differences of the $p-1$ partial derivatives with respect to the coordinates x_j, j different from i, for the variable u_i.

A trigonometric reparametrization may be useful for a homogeneous differential equation of $x = \cos\theta$ and $y = \sin\theta$, with $\theta = \arctan(x^{-1}y)$. A function u has the derivative

$$u'_\theta = x'_\theta u'_x + y'_\theta u'_y = -yu'_x + xu'_y.$$

A linear first order differential equation $u'_\theta = \varphi_\theta$ is therefore equivalent to a differential equation

$$yu'_x - xu'_y = f(x, y)$$

where f is an Euclidean function of $\cos\theta$ and $\sin\theta$. Using these derivatives, the differential equation

$$xu'_x + yu'_y = 0$$

is equivalent to $xu'_\theta - u'_y = 0$ and $yu'_\theta + u'_x = 0$.

When the second member is a trigonometric function, the solution of the differential equation may be a conjugate trigonometric function of a factor in the second member.

Example 51. The function $y(x) = \sin\varphi(x)$ with a function of $C_2(\mathbb{R})$ is solution of the second order differential equation

$$y''_{xx} + \varphi''_{xx}y = \varphi'^2_x \cos\varphi(x).$$

Example 52. For all constants a and b, the functions $u(x, y) = \frac{x^2}{a^2} + \frac{y^2}{b^2}$ satisfy the inhomogeneous first order partial differential equation

$$xu'_x + yu'_y = 2u(x, y),$$

with an initial condition $u(x_0, y_0) = u_0$.

Considering x and y as functions of a variable t, the first order derivative of u with respect to t is $u'_t = (b^2 x x'_t + a^2 y y')a^{-2}b^{-2}$ and the equation $u'_t = 0$ has the solution

$$x_t = a(x_0 \cos t - y_0 \sin t),$$
$$y_t = b(x_0 \sin t + y_0 \cos t),$$

with derivatives

$$x'_t = -\frac{a}{b} y_t, \quad y'_t = \frac{b}{a} x_t,$$

they define an ellipse with

$$\frac{x_t^2}{a^2} + \frac{y_t^2}{b^2} = x_0^2 + y_0^2.$$

Example 53. The function $u(x, y) = (x^2 + y^2)e^{xy}$ satisfies the inhomogeneous first order partial differential equation

$$x u'_x - y u'_y = 2(x^2 - y^2)e^{xy} = 2u(x, y)\frac{x^2 - y^2}{x^2 + y^2}$$

6.2 First order linear partial differential equations

The function $u(x, y) = \arctan(x^{-1} y)$ satisfies the partial differential equation

$$x u'_x + y u'_y = 0.$$

This may be checked denoting

$$v(x, y) = x^{-1} y = \tan u(x, y)$$

its partial derivatives are

$$\tan' u = \frac{d}{u}\frac{\sin u}{\cos u} = 1 + \frac{y^2}{x^2},$$
$$v'_x = \frac{d}{x} \tan u(x, y) = -\frac{y}{x^2} = u'_x \tan' u,$$
$$v'_y = \frac{d}{y} \tan u(x, y) = \frac{1}{x} = u'_y \tan' u$$

then

$$u'_x = -\frac{y}{x^2 + y^2}, \quad u'_y = \frac{x}{x^2 + y^2}.$$

The function $\arctan(x^{-1}y)$ is not the unique solution of the differential equation $xu'_x + yu'_y = 0$. Let $u(x,y) = \arcsin(xy^{-1})$, the partial derivatives of $\sin u = xy^{-1}$ are

$$\frac{d}{dx}\sin u(x,y) = \frac{1}{y} = u'_x \cos u,$$
$$\frac{d}{dy}\sin u(x,y) = -\frac{x}{y^2} = u'_y \cos u,$$

the function u is therefore a solution of $yu'_y + xu'_x = 0$. The homogeneous functions $\arccos(yx^{-1})$, $\arccos(xy^{-1})$ and $\arccos(yx^{-1})$ satisfy the same equation.

Proposition 6.1. *For every function ϕ of $C_1(\mathbb{R})$, the differential equation*

$$(y-b)u'_x(x,y) + (x-a)u'_y(x,y) = 0$$

has for solutions the functions

$$u(x,y) = \phi((x-a)^{-1}(y-b)),$$
$$u(x,y) = \phi((x-a)(y-b)^{-1}),$$
$$u(x,y) = \phi((x-a)^2 - (y-b)^2),$$
$$u(x,y) = \phi((y-b)^2 - (x-a)^2),$$

for all $a \neq a$ and $y \neq b$.

Proof. The function $u(x,y) = \phi((x-a)^{-1}(y-b))$ has the partial derivatives

$$u'_x(x,y) = -\frac{y-b}{(x-a)^2}\phi'((x-a)^{-1}(y-b)),$$
$$u'_y(x,y) = -\frac{1}{x-a}\phi'((x-a)^{-1}(y-b))$$

and the differential equation is satisfied. The function $u(x,y) = \phi((x-a)(y-b)^{-1})$ has the derivatives

$$u'_x(x,y) = \frac{1}{y-b}\phi'((x-a)(y-b)^{-1}),$$
$$u'_y(x,y) = -\frac{x-a}{(y-b)^2}\phi'((x-a)(y-b)^{-1}),$$

the same differential equation is satisfied. The proof is similar for the other functions. □

Note that the equation has no nontrivial exponential solution $e^{\alpha x+\beta y}$ since $\alpha x + \beta y = 0$ reduces u to one. For the sum of two exponential functions $c_1u_1 + c_2u_2$, $c_1(x\alpha_1 + y\beta_1) + c_2(x\alpha_2 + y\beta_2) = 0$ for all x and y therefore $c_1\alpha_1 + c_2\alpha_2 = 0$ and $c_1\beta_1 + c_2\beta_2 = 0$ which provides solutions as linear combinations of exponential function, so the first order partial differential equations have an infinity of solutions. Initial conditions at several points determine the constants. If the function u is not homogeneous in x and y, the partial derivatives are proportional to those of Proposition (6.1).

Proposition 6.2. *The differential equation*

$$\alpha(y-b)u'_x(x,y) + \beta(x-a)u'_y(x,y) = 0$$

with constants α and β has for solutions the functions

$$u(x,y) = \phi((x-a)^{-\alpha}(y-b)^{\beta}),$$
$$u(x,y) = \phi((x-a)^{\alpha}(y-b)^{-\beta}),$$
$$u(x,y) = \phi(\beta(x-a)^2 - \alpha(y-b)^2),$$
$$u(x,y) = \phi(\alpha(y-b)^2 - \beta(x-a)^2)$$

defined by functions ϕ of $C_1(\mathbb{R})$, for all $x \neq a$ and $y \neq b$.

A function of the difference of two polynoms $P(x)$ and $Q(y)$ depending on a single variable, $u(x,y) = \phi(P(x) - Q(y))$ has the partial derivatives $u'_x = P'_x\phi'(P-Q)$ and $u'_y = -Q'_y\phi'(P-Q)$. The function $u(x,y)$ is solution of the differential equation

$$Q'_y(x,y)u'_x(x,y) + P'_x(x,y)u'_y(x,y) = 0$$

for all x and y such that $\phi'(P(x,y) - Q(x,y))$ is non null.

With the ratio of two polynoms $P(x)$ and $Q(y)$ depending on a single variable, the function $u(x,y) = \phi(P(x)Q^{-1}(y))$, denoted $\phi \circ g(x,y)$, has the derivatives

$$u'_x = \frac{P'(x)}{Q(y)}\phi'(P(x)Q^{-1}(y)),$$
$$u'_y = -Q'(y)\frac{P(x)}{Q^2(y)}\phi'(P(x)Q^{-1}(y)),$$

then u is solution of the differential equation

$$P(x)Q'(y)u'_x(x,y) + P'(x)Q(y)u'_y(x,y) = 0$$

for all x and y such that $Q(y)$ and $\phi'(P(x)Q^{-1}(y))$ are non null. This is extended to the ratio of two functions P and Q depending on (x,y), using their partial derivatives.

Proposition 6.3. *The functions $u = \phi(PQ^{-1})$ defined by functions $P(x,y)$ and $Q(x,y)$ are solutions of the differential equations*

$$(P'_y Q - PQ'_y)u'_x - (P'_x Q - PQ'_x)u'_y = 0$$

for all x and y such that $Q(x,y)$ and $\phi'(P(x,y)Q^{-1}(x,y))$ are non null.

The reverse argument may be used to solve explicitly all linear partial differential equations with integrable coefficients.

Theorem 6.3. *The differential equation*

$$A'_y u'_x - B'_x u'_y = 0$$

defined by functions $A(x,y)$ and $B(x,y)$ such that there exists a function $C(x,y)$ integrable with respect to the measures $A'_y(x,y)\,dy$ and $B'_x(x,y)\,dx$, has the solutions

$$u(x,y) = \int_{x_0}^{x} C(s,y)B'_x(s,y)\,ds + k_{1y} = \int_{y_0}^{y} C(x,t)A'_y(x,t)\,dt + k_{2x},$$

with functions k_{1y} depending only on y and k_{2x} depending only on x, fixed by the initial value of u at an arbitrary point (x_0, y_0).

Proof. Solutions of the differential equation have partial derivatives of the form $u'_x(x,y) = B'_x(x,y)C(x,y)$ and $u'_y = A'_y(x,y)C(x,y)$, then their primitive functions satisfy the equality of Proposition 6.3. Deriving the expression of this solution yields

$$u'_x(x,y) = B'_x(x,y)C(x,y) = \int_{y_0}^{y} \{A''_{xy}C + A'_y C'_x\}(x,t)\,dt + k'_{2x},$$

$$u'_y(x,y) = A'_y(x,y)C(x,y) = \int_{x_0}^{x} \{B''_{xy}C + B'_x C'_y\}(s,y)\,ds + k'_{1y}.$$

The second derivative u''_{xy} in both equations equals $B''_{xy}C + B'_x C'_y = A''_{xy}C + A'_y C'_x$, so that the equalities of the expressions for u'_x and u'_y, and therefore u are always true, with ad hoc functions k_1 and k_2. □

As a special case, the existence of a solution $u(x,y)$ of the differential equation $M\,dx + N\,dy = 0$ implies the equalities $u'_x = M$, $u'_y = N$ and $M'_y = N'_x$. A solution of the differential equation $du = C(M\,dx + N\,dy) = 0$ is

$$u(x,y) = \int_{x_0}^{x} M(s,y)\,ds + k_1(y) = \int_{y_0}^{y} N(x,t)\,dt + k_{2x} = C$$

with an arbitrary constant C. The functions k_j are $k_1(y) = u(x_0, y)$ and $k_{2x} = u(x, y_0)$. The partial derivatives of u are

$$u'_x(x,y) = \int_{y_0}^{y} N'_x(x,t)\,dt + k'_{2x} = M(x,y),$$
$$u'_y(x,y) = \int_{x_0}^{x} M'_y(s,y)\,ds + k'_1(y) = N(x,y)$$

therefore

$$k'_{2x} = u'_x(x,y_0) = M(x,y) - \int_{y_0}^{y} N'_x(x,t)\,dt,$$
$$k'_1(y) = u'_y(x_0,y) = N(x,y) - \int_{x_0}^{x} M'_y(s,y)\,ds.$$

Proposition 6.4. *The differential equation $yu'_x\,dx + xu'_y\,dy = 0$ has solutions $u(x,y)$ with partial derivatives $u'_x = xv(x,y)$ and $y'_x u'_y = -yv(x,y)$*

$$u(x,y) = \int_{x_0}^{x} sv(s,y)\,ds + u(x_0,y)$$
$$= -\int_{y_0}^{y} tv(x,t)\,dt + u(x,y_0),$$

with a function v of $C_1(\mathbb{R}^2)$ and $u(x,y_0) - u(x_0,y) = \int_{x_0}^{x} sv(s,y)\,ds + \int_{y_0}^{y} tv(x,t)\,dt$.

This is also a consequence of Poincaré's theorem. From Proposition 6.2, for every function ϕ of $C_1(\mathbb{R})$, the functions

$$u(x,y) = \phi(x^{-1}y), \quad u(x,y) = \phi(xy^{-1}),$$
$$u(x,y) = \phi(x^2 - y^2), \quad u(x,y) = \phi(y^2 - x^2)$$

are solutions of the differential equation $yu'_x + xu'_y = 0$. For instance, with the function $v(x,y) = 2(x-y)$, $u(x,y) = x^2 - y^2 + u(x_0,y) - u(x,y_0) = 0$ are solutions of $yu'_x + xu'_y = 0$.

Example 54. The differential equation $y\,dx + x\,dy = 0$ with initial conditions (x_0, y_0) at t_0 has the solution $u(x,y) = xy = u_0$, with a constant $u_0 = x_0 y_0$. It is obtained as solution of the equations

$$x'_t = x_t, \quad y'_t = -y_t$$

as $x_t = x_0 e^{t-t_0}$, $y_t = y_0 e^{-(t-t_0)}$, or from the equations

$$x'_t = -x_t, \quad y'_t = y_t$$

as $x_t = x_0 e^{-(t-t_0)}$, $y_t = y_0 e^{t-t_0}$.

Example 55. The differential equation $x^2\,dx - ay\,dy = 0$ has a solution $u(x,y)$ such that $u'_x(x,y) = x^2$ and $u'_y(x,y) = -ay$. The second equation implies

$$u(x,y) = u(x,y_0) - \frac{a}{2}(y^2 - y_0^2)$$

with an initial value u_0 at (x_0,y_0), therefore $u'_x(x,y) = u'_0(x) = x^2$ and $u_0(x) = \frac{1}{3}(x^3 - x_0^3) + u_0$.

The equation $y^2\,dx - ay\,dy = 0$ has the solution $y(x) = y_0 e^{\frac{x-x_0}{a}}$ and $y(x) = v(x)e^{\frac{x-x_0}{a}}$ is solution of $x^2 + y^2 - ayy'_x = 0$ with a function v such that

$$\frac{v'_x}{v_x} = \frac{x^2}{ay^2}$$

then

$$v(x) = v_0 \exp\Big(\frac{x^3 - x_0^3}{3ay^2}\Big).$$

The solutions of a first order partial differential equation with a non null control function $f(x,y)$ are sums of the solution of the differential equation equal to zero and a particular solution $h(x,y)$ of the differential equation with the non null function $f(x,y)$. The differential equation

$$M(x,y)\,dx + N(x,y)\,dy = f(x,y)\,dx$$

is equivalent to $M(x,y) + N(x,y)y'_x = f(x,y)$ and to

$$\begin{aligned}M(x,y)\,dx + N(x,y)\,dy = f(x,y)\,dx &= \frac{f}{y'_x}(x,y)\,dy\\ &= \frac{1}{2}\Big\{f(x,y)\,dx + \frac{f}{y'_x}(x,y)\,dy\Big\}.\end{aligned}$$

Solutions $u(x,y)$ have the form

$$\begin{aligned}u(x,y) &= \int_{x_0}^{x}\Big(M - \frac{f}{2}\Big)(s,y)\,ds + u(x_0,y)\\ &= \int_{y_0}^{y}\Big(N - \frac{f}{2y'_x}\Big)(x,t)\,dt + u(x,y_0)\end{aligned}$$

with partial derivatives

$$\begin{aligned}u'_x(x,y) &= \Big(M - \frac{f}{2}\Big)(x,y) = \frac{\partial}{\partial x}\int_{y_0}^{y}\Big(N - \frac{f}{2y'_x}\Big)(x,t)\,dt + u'(x,y_0),\\ u'_y(x,y) &= \Big(N - \frac{f}{2y'_x}\Big)(x,y) = \frac{\partial}{\partial y}\int_{x_0}^{x}\Big(M - \frac{f}{2}\Big)(s,y)\,ds + u'(x_0,y).\end{aligned}$$

The marginal functions of u and their difference are deduced by integration of their partial derivatives

$$u'(x, y_0) = M(x, y) - \int_{y_0}^{y} N'_x(x, t)\, dt - \frac{1}{2} f(x, y)$$
$$+ \frac{1}{2} \int_{y_0}^{y} \Big(\frac{f'_x}{y'_x} - f \frac{y''_{xx}}{y'^2_x} \Big)(x, t)\, dt,$$
$$u'(x_0, y) = N(x, y) - \int_{x_0}^{x} M'_y(s, y)\, ds + \frac{1}{2} \int_{x_0}^{x} f'_y(s, y)\, ds - \frac{f}{2y'_x}(x, y).$$

Example 56. A solution of the equation $xu'_x + yu'_y = f(x, y)$ is the sum of a solution $u_1(x, y) = \varphi(x^{-1}y)$, with φ in C_1, of $xu'_x + yy'_x u'_y = 0$ and a particular solution $u_2(x, y)$ of the equation with the second member f. Let $f(x, y) = xy$ and $u_2(x, y) = \frac{1}{2}xy$

$$u(x, y) = \frac{1}{2}xy + \varphi(x^{-1}y)$$

has the partial derivatives

$$u'_x = \frac{y}{2} - \frac{y}{x^2}\varphi'(x^{-1}y), \quad u'_y = \frac{x}{2} + \frac{1}{x}\varphi'(x^{-1}y).$$

Proposition 6.5. *Let f_1 and f_2 be solutions of the homogeneous partial differential equation $xu'_x\, dx + yu'_y\, dy = 0$. The differential equation*

$$xu'_x + yy'_x u'_y = xf_1 + yy'_x f_2$$

has the solution $u = xf_1 + yf_2$. The differential equation

$$xu'_x + yy'_x u'_y = x^{-1} f_1 + y^{-1} y'_x f_2$$

has the solution $u = -(x^{-1} f_1 + y^{-1} f_2)$. The differential equation

$$xu'_x + yy'_x u'_y = xf_1 + y^{-1} y'_x f_2$$

has the solution $u = xf_1 - y^{-1} f_2$ and the equation

$$xu'_x + yy'_x u'_y = x^{-1} f_1 + yy'_x f_2$$

has the solution $u = -x^{-1} f_1 + yf_2$.

Proof. The first order partial derivatives of $u(x, y) = xf_1(x, y) + yf_2(x, y)$ are $u'_x(x, y) = f_1(x, y) + xf'_{1x}(x, y) + yf'_{2x}(x, y)$ and $u'_y(x, y) = f_2(x, y) + xf'_{1y}(x, y) + yf'_{2y}(x, y)$, it follows that

$$x\{xf'_{1x}(x, y) + yf'_{2x}(x, y)\} + y\{xf'_{1y}(x, y) + yf'_{2y}(x, y)\} = 0$$

for all x and y, since f_1 and f_2 are solutions of the homogeneous equation. With the function $u = x^{-1}f_1 + y^{-1}f_2$, the differential equation is written as

$$\begin{aligned} xu'_x + yy'_x u'_y &= -x^{-1}f_1 - y^{-1}y'_x f_2 + x(x^{-1}f'_{1x} + y^{-1}f'_{2x}) \\ &+ yy'_x(x^{-1}f'_{1y} + y^{-1}f'_{2y}) \end{aligned}$$

and it equals $x^{-1}f_1 + y^{-1}y'_x f_2$. If $f = xf_1 + y^{-1}f_2$ or $f = x^{-1}f_1 + yf_2$, the calculus of the partial derivatives of $u = xf_1 - y^{-1}f_2$ or, respectively, $u = -x^{-1}f_1 + yf_2$ proves the result for these functions. □

Proposition 6.3 defined solutions of the partial differential equation $A'_y u'_x - B'_x u'_y(x, y) = f$ with $f = 0$. It is extended to the controlled equation with a non null function f.

Proposition 6.6. *The differential equation*

$$A'_y u'_x - B'_x u'_y(x, y) = f(x, y)$$

defined by functions $A(x, y)$, $B(x, y)$ and $f(x, y)$ of $C_1([x_0, x_1] \times [y_0, y_1])$ has the solutions

$$\begin{aligned} u(x, y) &= \int_{x_0}^{x} C(s, y)\Big\{B'_x(s, y) - \frac{f}{2}(s, y)\Big\}\, ds + u(x_0, y) \\ &= \int_{y_0}^{y} C(x, t)A'_y(x, t)\Big\{1 - \frac{f}{2B'_x}(x, t)\Big\}\, dt + u(x, y_0), \end{aligned}$$

for every function $C(x, y)$ such that the integrals are finite on $[x_0, x_1]$ and, respectively $[y_0, y_1]$.

Proof. The arguments of the proof are the same as for Proposition (6.3). The homogeneous equation implies $y'_x = A'^{-1}_y B'_x$ and the integral of the right-hand term with respect to y is the integral of $fy'^{-1}_x = B'^{-1}_x A'_y f$ with respect to y. □

In intervals where the function B'_x is different from zero, the solution of the equation normalized by B'_x is characterized as follows.

Proposition 6.7. *The differential equation $u'_x + a_x u'_y(x, y) = f(x, y)$ defined by functions $a(x, y)$ and $f(x, y)$ of $C_1([x_0, x_1] \times [y_0, y_1])$ has the solution*

$$\begin{aligned} u(x, y) &= v(x, y) + F(x, y) - \int_{x_0}^{x} aF'_y(s, y)\, ds \\ &= v(x, y) + F(x, y) - AF'_y(x, y) + \int_{x_0}^{x} AF''_{yy}(s, y)\, ds \end{aligned}$$

where $F(x,y)=\int_{x_0}^{x} f(s,y)\,ds$, $A(x,y)=\int_{x_0}^{x} a(s,y)\,ds$ *and the function* v *is solutions of the differential equation*

$$v'_x + a_x v'_y(x,y) = a\int_{x_0}^{x} aF'_y(s,y)\,ds - f(x,y).$$

The proof relies on integrations by parts including the second member. A normalization by B'_x leads to other expressions depending on a^{-1} instead of a for the solution of the same equation. Though the function $u(x,y)$ may be expressed in several different forms, they are equal and define a single solution under initial conditions.

These results proved for a function u in $\mathbb{R}^2$ are extended to higher dimensions. A function $u(x,y,z)$ solution of the differential equation

$$M(x,y,z)\,dx + N(x,y,z)\,dy + R(x,y,z)\,dz = 0$$

with continuous functions M, N and R of $C_1(\mathbb{R}^3)$ such that $M'_y = N'_x$, $M'_z = R'_x$ and $N'_z = R'_y$ is written as

$$\begin{aligned} u(x,y,z) &= \int_{x_0}^{x} M(s,y,z)\,ds + u(x_0,y,z) \\ &= \int_{y_0}^{y} N(x,t,z)\,dt + u(x,y_0,z) \\ &= \int_{z_0}^{z} R(x,y,\zeta)\,d\zeta + u(x,y,z_0). \end{aligned}$$

Its partial derivatives are

$$\begin{aligned} u'_x = M(x,y,z) &= \int_{y_0}^{y} N'_x(x,t,z)\,dt + u'_x(x,y_0,z) \\ &= \int_{z_0}^{z} R'_x(x,y,\zeta)\,d\zeta + u_x(x,y,z_0), \\ u'_x = M(x,y,z) &= \int_{y_0}^{y} N'_x(x,t,z)\,dt + u'_x(x,y_0,z) \\ &= \int_{z_0}^{z} R'_x(x,y,\zeta)\,d\zeta + u_x(x,y,z_0), \end{aligned}$$

$$\begin{aligned}
u'_y &= N(x, y, z) = \int_{x_0}^{x} M'_y(s, y, z)\, ds + u'_y(x_0, y, z) \\
&= \int_{z_0}^{z} R'_y(x, y, \zeta)\, d\zeta + u'_y(x, y, z_0), \\
u'_z &= R(x, y, z) = \int_{x_0}^{x} M'_z(s, y, z)\, ds + u'_z(x_0, y, z) \\
&= \frac{\partial}{\partial z} \int_{y_0}^{y} N(x, t, z)\, dt + u'_z(x, y_0, z).
\end{aligned}$$

The solutions of the simultaneous equations $M\,dx + cN\,dy = 0$ where z is considered as a constant and $(1-c)N\,dy + R\,dz = 0$ where x is considered as a constant provide particular solutions of a differential equation $M\,dx + N\,dy + R\,dz = 0$. The primitives of these equations depend on arbitrary functions of z and, respectively, x.

Example 57. A differential equation

$$A(y, z)\, dx + B(x, z)\, dy + C(x, y)\, dz = 0$$

has solutions $v(x, y, z) = xA(y, z) + yB(x, z) + zC(x, y) + k$, with a constant k. The expression of its partial derivatives $v'_x = A(y, z) + yB'_x(x, z) + zC'_x(x, y)$, $v'_y = xA'_y(y, z) + B(x, z) + zC'_y(x, y)$ and $v'_z = xA'_z(y, z) + yB'_z(x, z) + C(x, y)$ imply that the differential equation is equivalent to

$$\begin{aligned}
dv &= yB'_x(x, z)\, dx + zC'_x(x, y)\, dx + xA'_y(y, z)\, dy + zC'_y(x, y)\, dy \\
&\quad + xA'_z(y, z)\, dz + yB'_z(x, z)\, dz \\
&= x\, dA(y, z) + y\, dB(x, z) + z\, dC(x, y).
\end{aligned}$$

Example 58. A differential equation $b^{-1}(y, z)\, dx + a^{-1}(x)(y\, dy + z\, dz) = 0$ has a solution $u(x, y, z) = A(x) + B(y, z)$ constant if and only if A is a primitive of a, $B'_y = yb$ and $B'_z = zb$, which implies $yB'_z - zB'_y = 0$. From Theorem 6.3, a solution of the last equation is $B(y, z) = \frac{1}{2}(y^2 + z^2)$. Other solutions are defined by integrals of a function $C(y, z)$ as a constant function

$$\begin{aligned}
u(x, y, z) &= C(y, z)A(x) + \int_{y_0}^{y} zC(s, z)\, ds + B(y_0, z) \\
&= C(y, z)A(x) + \int_{z_0}^{z} zC(y, t)\, dt + B(y, z_0).
\end{aligned}$$

Proposition 6.8. *The differential equation*

$$A'_z u'_x\, dx\, dz + B'_x u'_y\, dx\, dy + C'_y u'_z\, dy\, dz = 0$$

defined by functions $A(x,y,z)$, $B(x,y,z)$ *and* $C(x,y,z)$ *of* $C_2(\mathbb{R}^3)$ *has a solution if there exist a function* v *of* C_2 *such that* $v''_{xz} + v''_{xy} + v''_{yz} = 0$ *and a continuous function* $h(x,y,z)$ *integrable with respect to* $A'_z u'_x$, $B'_x u'_y$ *and* $C'_y u'_z$, *such that*

$$\begin{aligned}
v(x,y,z) &= \int_{x_0}^{x}\int_{z_0}^{z} h(s,y,w)A'_z(s,y,w)u'_x(s,y,w)\,ds\,dw \\
&\quad + v(x_0,y,z_0) - v(x_0,y,z_0) - v(x,y,z_0) \\
&= \int_{x_0}^{x}\int_{y_0}^{y} h(s,t,z)B'_x(s,t,z)u'_y(s,t,z)\,ds\,dt \\
&\quad + v(x_0,y_0,z) - v(x_0,y,z) - v(x,y_0,z) \\
&= \int_{y_0}^{y}\int_{z_0}^{z} h(x,t,w)C'_y(x,t,w)u'_z(x,t,w)\,dt\,dw \\
&\quad + v(x,y_0,z_0) - v(x,y_0,z) - v(x,y,z_0).
\end{aligned}$$

Proof. Solutions $v(x,y,z)$ of the differential equation have partial derivatives of the form

$$\begin{aligned}
dv &= v'_x\,dx + v'_y\,dy + v'_z\,dz, \\
d^2v &= v''_{xx}\,dx^2 + v''_{yy}\,dy^2 + v''_{zz}\,dz^2 + 2v''_{xy}\,dx\,dy + 2v''_{xz}\,dx\,dz + 2v''_{yz}\,dy\,dz
\end{aligned}$$

the differential equation $A'_z u'_x + B'_x u'_y + C'_y u'_z = 0$ implies a linear relationship between the crossed derivatives, $v''_{xz} + v''_{xy} + v''_{yz} = 0$. This implies the existence of a function $h(x,y,z)$ such that

$$\begin{aligned}
v''_{xz} &= hA'_z u'_x = -h(B'_x u'_y + C'_y u'_z), \\
v''_{xy} &= hB'_x u'_y = -h(A'_z u'_x + C'_y u'_z), \\
v''_{yz} &= hC'_y u'_z = -h(A'_z u'_x + B'_x u'_y).
\end{aligned}$$

Then v'_x is obtained by integration of v''_{xy}, v''_{xz} or v'_y by integration of v''_{xy} or v''_{yz}, and v'_z by integration of v''_{xz} or v''_{yz}

$$\begin{aligned}
v'_x(x,y,z) &= \int_{z_0}^{z} h(x,y,w)A'_z(x,y,w)u'_x(x,y,w)\,dw + k_1(y,z) \\
v'_y(x,y,z) &= \int_{x_0}^{x} h(s,y,z)B'_x(s,y,z)u'_y(s,y,z)\,ds + k_2(y,z) \\
v'_z(x,y,z) &= \int_{x_0}^{x} h(s,y,z)A'_z(s,y,z)u'_x(s,y,z)\,ds + k_5(y,z)
\end{aligned}$$

and the expression of a solution follows from the integration of these derivatives. □

The differential equation

$$u'_x + au'_y + bu'_z = f$$

defined by functions a, b and f of $C_1([x_0, x_1] \times [y_0, y_1] \times [z_0, z_1])$ has the solution $u(x, y, z) = \int_{x_0}^{x} f(s, y, z)\, ds$ with y and z solutions of the implicit equations

$$y = A(x, y, z) = \int_{x_0}^{x} a(s, y, z)\, ds + A(x_0, y, z),$$
$$z = B(x, y, z) = \int_{x_0}^{x} b(s, y, z)\, ds + B(x_0, y, z)$$

under the necessary and sufficient conditions

$$\int_{x_0}^{x} a'_y(s, y, z)\, ds + A'_y(x_0, y, z), \quad \int_{x_0}^{x} b'_z(s, y, z)\, ds + B'_z(x_0, y, z) = 1.$$

6.3 Second order linear partial differential equations

A second order differential equation of the form

$$d^2 z = R(x, y)\, dx^2 + S(x, y)\, dx\, dy + T(x, y)\, dy^2 = 0, \tag{6.1}$$

is the differential of $dz = M(x, y)\, dx + N(x, y)\, dy$ constant if

$$R(x, y) = M'_x(x, y), \quad S(x, y) = M'_y(x, y) + N'_x(x, y), \quad T(x, y) = N'_y(x, y).$$

Necessary conditions for the integration of (6.1) is the existence of the primitive functions with arbitrary x_0 and y_0

$$M(x, y) = \int_{x_0}^{x} R(\xi, y)\, d\xi, \quad N(x, y) = \int_{y_0}^{y} T(x, \zeta)\, d\zeta$$

such that

$$S(x, y) = M'_y(x, y) + N'_x(x, y).$$

A solution $F(x, y) = 0$ exists if $M(x, y) = F'_x(x, y)$ and $N(x, y) = F'_y(x, y)$.

Let us consider (6.1) with homogeneous functions R, S and T of x and y, and y a function of x. By the homogeneity of the functions, one can denote $y = xu$ and its derivative is $dy = u\, dx + x\, du$. For every $x \neq 0$, (6.1) is equivalent to the equations

$$\begin{aligned} &R(1, u)\, dx^2 + 2S(1, u)\, dx\, (u\, dx + x\, du) \\ &\qquad + T(1, u)\, (u^2\, dx^2 + 2xu\, dx\, du + x^2\, du^2) = 0, \\ &\{R(1, u) + 2uS(1, u) + u^2 T(1, u)\}\, dx^2 + x^2 T(1, u)\, du^2 \\ &\qquad + 2x\{S(1, u) + uT(1, u)\}\, dx\, du = 0, \end{aligned}$$

this equation has the same form as (6.1) with a multiplicative functions $R(x,u) = r(u) = R(1,u) + 2uS(1,u) + u^2T(1,u)$, $S(x,u) = xh(u)$ with $h(u) = S(1,u) + uT(1,u)$ and $T(x,u) = x^2t(u) = x^2T(1,u)$. The first necessary condition for the existence of a solution is the existence of differentiable functions M and N such that

$$r(u) = M'_x(x,u), \quad xh(u) = M'_u(x,u) + N'_x(x,u), \quad x^2t(u) = N'_u(x,u).$$

The function $M(x,u)$ and $N(x,u)$ are not homogeneous.

Necessary conditions for the integration of (6.1) is the existence of a function F satisfying

$$\begin{aligned} F(x,y) &= \int_{x_0}^{x}\int_{x_0}^{t} R(s,y)\,ds\,dt + F(x_0,y) \\ &= \int_{y_0}^{y}\int_{x_0}^{t} S(s,t)\,ds\,dt + F(x_0,y) + F(x,y_0) + F(x_0,y_0) \\ &= \int_{y_0}^{y}\int_{y_0}^{t} T(x,s)\,ds\,dt + F(x,y_0), \end{aligned}$$

with partial derivatives

$$\begin{aligned} R(x,y) &= \int_{y_0}^{y} S'_x(s,y)\,ds + F''_{xx}(x,y_0) = \int_{y_0}^{y}\int_{y_0}^{t} T''_{xx}(x,s)\,ds\,dt + F''_{xx}(x,y_0), \\ S(x,y) &= \int_{x_0}^{x} R'_y(s,y)\,ds = \int_{y_0}^{y} T'_x(x,s)\,ds, \\ T(x,s) &= \int_{x_0}^{x}\int_{x_0}^{t} R''_{yy}(s,y)\,ds\,dt + F''_{yy}(x_0,y) = \int_{x_0}^{x} S'_y(s,y)\,ds + F''_{yy}(x_0,y). \end{aligned}$$

The resolution of second order partial differential equation is performed by identification of the partial derivatives. In particular, solutions of the differential equation $d^2u = f$ are the solutions of $\Delta u = f$ under the condition $u''_{xy} = 0$, they are therefore additive functions of x and y, solutions of $\Delta u = f$. Other solutions are solutions of $u''_{xy} = f$ under the condition $\Delta u = 0$, i.e.

$$u(x,y) = \int_{y_0}^{y}\int_{x_0}^{t} f(s,t)\,ds\,dt + u(x_0,y) + u(x,y_0) + u(x_0,y_0)$$

under the condition

$$\int_{y_0}^{y} f'_x(x,t)\,dt + u''_{xx}(x,y_0) + \int_{x_0}^{t} f'_y(s,y)\,ds + u''_{yy}(x_0,y) = 0$$

which is a condition for the boundary functions $u(x_0,y)$ and $u(x,y_0)$.

6.4 Multidimensional differential equations

Simultaneous differential equations of the same form for a variable set (x_t, y_t) may reduced to a single differential equation by linear combination with arbitrary constants. Its solutions constitute a family of functions depending on all constants and coefficients of this unique global differential equation. A simple method for their solution is to change them as two differential equation depending each on a single variable.

In an interval (t_0, t_1), let $Y_t = (y_{1t}, \ldots, y_{kt})^t$ be a k-dimensional variable with derivative Y_t', let A_t be a $k \times k$-dimensional continuous matrix and let F_t be a k-dimensional continuous variable. The first order linear differential equation $Y_t' = A_t Y_t + F_t$ with initial condition $Y_{t_0} = Y_0$ has an unique solution in (t_0, t_1).

If the matrix A can be transformed into a diagonal matrix $D = P^t AP$, then $A = PDP^t$ and the vector $Z = P^t Y$ satisfies the differential equation $Z_t' = D_t Z_t + P^t F_t$ which is solved components by components since the components of Z are independent due to the use of a diagonal matrix, then $Y = PZ$. If A can be transformed into a block-diagonal matrix, with m blocks, the equation with k components is split into m independent systems of smaller order.

A second order linear differential equation $Y_t'' = A_t Y_t' + B_t Y_t + F_t$ defined by continuous matrices A_t and B_t and a continuous vector F_t is written in a similar form denoting $U_t = Y_t'$ and $U_t' = Y_t''$. The vector $V_t = (U_t, U_t')$ has a first order derivative V_t' with components those of Y_t' and $A_t Y_t' + B_t Y_t + F_t$. It can be solved if an explicit expression of Y_t' is known. More generally, its solutions are written in similar forms as the ordinary second order differential equations. In particular, if A_t and B_t are constant, the k-dimensional differential equations are explicitly solved.

The solution is also obtained writing each component as an exponential function $y_{it} = e^{\rho_i t}$, the system of differential equations leads to a system of equations of order k with unknown variables the coefficients ρ_i. Each component y_{it} is a linear combination of k exponential variables having as coefficients the roots of this equation.

A system of first order and inhomogeneous linear differential equations

$$x_t' + px + qy = u,$$

$$y_t' + rx + sy = v$$

with p, q, r, s, u and v functions of a real variable t is written as a multidimentional differential equation for $X = (x, y)^t$, $X'_t + AX = F$ with

$$A = \begin{pmatrix} p & q \\ r & s \end{pmatrix}, \quad F = (u, v)^T.$$

As a linear first order differential equation with initial value $X(t_0) = X_0$, it has a unique solution $X_t = e^{\int_{t_0}^t A_s\,ds}\{X_0 + \int_{t_0}^t f_s e^{-\int_{t_0}^s A_s\,ds}\}$, with the matrix exponential defined as a series.

The homogeneous equations were solved in Section 1.4, now they reduce to independent systems of second order differential equations for x and, respectively, y

$$x'_t + px = u_1,$$
$$y'_t + sy = v_1,$$

with $qy = u - u_1$ and $rx = v - v_1$, and with functions u_1 and u_2 such that

$$x(t) = e^{-P_t}\Big(x_0 + \int_0^t u_{1s} e^{P_s}\,ds\Big) = \frac{v_t - v_{1t}}{r_t},$$
$$y(t) = e^{-S_t}\Big(y_0 + \int_0^t v_{1s} e^{S_s}\,ds\Big) = \frac{u_t - u_{1t}}{q_t},$$

where P and S are the primitives of p and, respectively s, with values zero at zero. Then introducing the function u_1 of the second equation in the first one defines $x(t)$ according to $y(t)$ and the functions u, v and v_1

$$\begin{aligned} x(t) &= e^{-P_t}\Big\{x_0 + \int_0^t (u - qy)e^{P_s}\,ds\Big\} \\ &= e^{-P_t}\Big\{x_0 + \int_0^t (u - qy_0 e^{P_s - S_s})\,ds - \int_0^t q e^{P_s - S_s}\int_0^s v_1 e^S\,ds\Big\}. \end{aligned}$$

The functions u_1 and v_1 are constrained by the equalities

$$\frac{v_t - v_{1t}}{r_t} = e^{-P_t}\Big\{x_0 + \int_0^t u_s e^{P_s}\,ds - \int_0^t q_s e^{P_s - S_s}\Big(y_0 + \int_0^s v_{1z} e^{S_z}\,dz\Big)\,ds\Big\},$$
$$\frac{u_t - u_{1t}}{q_t} = e^{-S_t}\Big\{y_0 + \int_0^t v_s e^{S_s}\,ds - \int_0^t r_s e^{S_s - P_s}\Big(x_0 + \int_0^s u_{1z} e^{P_z}\,dz\Big)\,ds\Big\},$$

they are solutions of second order ordinary differential equations depending only on the coefficients p_t, q_t, r_t, s_t and on (u_t, v_t)

$$\Big(\frac{v_t - v_{1t}}{r_t}\Big)'_t + p_t\frac{v_t - v_{1t}}{r_t} = u_t - q_t e^{-S_t}\Big(y_0 + \int_0^t v_{1z} e^{S_z}\,dz\Big)\,dt,$$

$$v_{1t} = \Big[\Big(s_t + \frac{1}{q_t}\Big)\Big\{u_t - \Big(\frac{v_t - v_{1t}}{r_t}\Big)'_t + p_t\frac{v_t - v_{1t}}{r_t}\Big\}\Big]'_t$$

and u_1 is solution of a similar differential equation.

Example 59. Deriving the second equation of the system

$$x'_t = y_t, \quad y'_t = -x_t$$

entails $y''_t = -x'_t = -y_t$ hence $x_t = a \sin t + b \cos t$ and $y_t = -b \sin t + a \cos t$. Then $x^2 + y^2 = a^2 + b^2$.

Example 60. The system of differential equations

$$x'_t = \lambda y_t - \mu x_t,$$
$$y'_t = -\mu y_t - \lambda k x_t$$

has solutions

$$x_t = b u_t - a v_t = -(a^2 + b^2) g_t e^{-\mu t},$$
$$y_t = a u_t + b v_t = (a^2 + b^2) f_t e^{-\mu t}$$

for all functions f_t and g_t such that

$$u_t = (a f_t - b g_t) e^{-\mu t},$$
$$v_t = (a g_t + b f_t) e^{-\mu t}.$$

Spiral. Let (x_t, y_t) satisfy the system of homogeneous differential equations

$$(a - t) x'_t = k y_t - x_t,$$
$$(a - t) y'_t = -y_t - k x_t, \tag{6.2}$$

for $t \neq a$, with initial conditions $x(0) = 0$, $y(0) = 0$. By a reparametrization of y as a function of x, the time-dependent differential equations (6.2) become

$$\frac{dy_t}{dx_t} = \frac{dy_t}{dt} \left(\frac{dx_t}{dt} \right)^{-1}$$
$$= -\frac{y_t + k x_t}{k y_t - x_t},$$
$$x_t y'_t - x'_t y_t = k (y_t y'_t + x_t x'_t)$$

and the solution (x_t, y_t) satisfies the equation

$$\frac{k}{2}(x_t^2 + y_t^2) = \int_0^t (x_s y'_s - x'_s y_s) \, ds.$$

The derivatives equations (6.2) are

$$x''_t = \frac{k}{a - t} y'_t + \frac{k}{(a - t)^2} y_t - \frac{1}{a - t} x'_t - \frac{1}{(a - t)^2} x_t,$$
$$y''_t = -\frac{1}{a - t} y'_t - \frac{1}{(a - t)^2} y_t - \frac{k}{a - t} x'_t - \frac{k}{(a - t)^2} x_t.$$

By the suppression of y_t and its derivatives (respectively x_t) in both equations for x_t (respectively y_t), Timmermans (1854, p.414) got two identical second order linear differential equations with separated variables

$$0 = (a-t)^2 x_t'' + (a-t)x_t' + (k^2+1)x_t,$$
$$0 = (a-t)^2 y_t'' + (a-t)y_t' + (k^2+1)y_t.$$

The variables x_t and y_t therefore have the same form and they vary independently. By the changes of variable $x(t) = (a-t)u(t)$, with derivatives such that $x_t' = -u + (a-t)u_t'$ and $x_t'' = -2u_t' + (a-t)u_t''$, or $y(t) = (a-t)u(t)$, the equations are equivalent to

$$(a-t)^2 u_t'' - (a-t)u_t' + k^2 u_t = 0.$$

Multiplying the equation by u_t' and integrating implies

$$(a-t)^2 u_t'^2 + k^2 u_t^2 = C.$$

The exact solutions are

$$x_t = \frac{r}{a}(a-t)\sin\log\Big(\frac{a}{a-t}\Big)^k,$$
$$y_t = \frac{r}{a}(a-t)\cos\log\Big(\frac{a}{a-t}\Big)^k$$

with the constant C such that $a^2C = k^2r^2$, therefore

$$y_t = \frac{(a-t)r}{a}\cos\arcsin\frac{ax_t}{(a-t)r}.$$

6.5 Epidemics

The spread of epidemics in a population with total size N is described by the vector of the rates of susceptible, infected and recovered individuals at time t, (S_t, I_t, R_t). They follow a nonlinear system of differential equations with constant coefficients

$$S_t' = -\beta S_t I_t - \mu S_t + \mu,$$
$$I_t' = \beta S_t I_t - (\lambda+\mu)I_t,$$
$$R_t' = \lambda I_t - \mu R_t.$$

with initial values $R_0 = 0$, S_0 and I_0 at $t = 0$. An unstable equilibrium holds as

$$S_t = \beta^{-1}(\lambda+\mu),$$
$$I_t = \{(\lambda+\mu)^{-1} - \beta^{-1}\}\mu,$$
$$\lambda I_t = \mu R_t.$$

The sum

$$X_t = S_t + I_t + R_t$$

has the derivative $X_t' = \mu(1-X_t)$ thus X_t is a decreasing function depending only on the recovery rate μ

$$X_t = 1 + (X_0 - 1)e^{-\mu t}.$$

Solving the last differential equation yields

$$R_t = \Big\{R_0 + \lambda \int_0^t I_s e^{\mu s}\, ds\Big\} e^{-\mu t},$$

and the sum of the two others $S_t + I_t$ has the derivative

$$S_t' + I_t' = \mu - \mu(S_t + I_t) - \lambda I_t,$$

its solution is

$$\begin{aligned} S_t + I_t &= e^{-\mu t}\Big\{S_0 + I_0 + e^{\mu t} - 1 - \lambda \int_0^t I_s e^{\mu s}\, ds\Big\} \\ &= 1 + (S_0 + I_0 - 1)e^{-\mu t} - \lambda \int_0^t I_s e^{-\mu(t-s)}\, ds. \end{aligned}$$

Solving separately the differential equations for the functions S_t and I_t provides implicit solutions depending on each other

$$\begin{aligned} I_t &= I_0 \exp\Big\{-(\lambda + \mu)t + \beta \int_0^t S_s\, ds\Big\}, \\ S_t &= S_0 \exp\Big\{-\mu t - \beta \int_0^t I_s\, ds\Big\} + \mu \int_0^t \exp\Big\{-\mu(t-s) - \beta \int_s^t I_u\, du\Big\} \end{aligned}$$

so I_t and R_t depend only on t and S_t. The sum $S_t + I_t$ is also written as a solution of the inhomogeneous equation

$$\begin{aligned} S_t' + I_t' &= -(\lambda + \mu)(S_t + I_t) + \lambda S_t + \mu, \\ S_t + I_t &= e^{-(\lambda+\mu)t}\Big\{S_0 + I_0 + \int_0^t (\lambda S_s + \mu)e^{(\lambda+\mu)s}\, ds\Big\} \\ &= \frac{\mu}{\lambda + \mu} + e^{-(\lambda+\mu)t}\Big\{S_0 + I_0 - \frac{\mu}{\lambda + \mu} + \lambda \int_0^t S_s e^{(\lambda+\mu)s}\, ds\Big\}, \end{aligned}$$

Plugging the expression of I_t in terms of S_t obtained from the first implicit equation yields a nonlinear equation for S_t independently of I_t and R_t

$$\begin{aligned} S_t = \frac{\mu}{\lambda + \mu} + e^{-(\lambda+\mu)t}\Big\{S_0 + I_0(1 - e^{\beta \int_0^t S_s\, ds}) \\ - \frac{\mu}{\lambda + \mu} + \lambda \int_0^t S_s e^{(\lambda+\mu)s}\, ds\Big\} \end{aligned}$$

where S_t is expressed as a function of t independent of the other variables. Using the second equation provides a differential equation for I_t independently of the other variables, R_t and S_t are deduced from its solution. The differential equations with variable coefficients are solved in the same way, replacing λt and μt by the primitives Λ_t and M_t of λ_t and μ_t.

The effect of vaccinations in the dynamics of epidemies with vertical transmission (rate ρ) was described by the system of differential equations

$$\begin{aligned}S_t' &= -\beta_t S_t I_t + \mu\rho I_t + \mu(S_t + R_t + V_t),\\ I_t' &= \beta_t S_t I_t - \mu\rho I_t - \lambda I_t,\\ R_t' &= \lambda I_t - \mu R_t\end{aligned}$$

with $S_t + I_t + R_t + V_t = 1$. They imply

$$\begin{aligned}V_t' &= -\mu(V_t + S_t),\\ V_t &= e^{-\mu t}\Big\{V_0 - \mu\int_0^t S_x e^{\mu x}\,dx\Big\},\\ R_t &= \lambda e^{-\mu t}\Big\{R_0 + \int_0^t I_x e^{\mu x}\,dx\Big\}\end{aligned}$$

and from the sum of the first two equations

$$S_t' + I_t' = -(\lambda+\mu)I_t + \mu.$$

The system has no equilibrium since all variables cannot be zero. Its integration yields

$$\begin{aligned}I_t &= \frac{\mu}{\lambda+\mu} + e^{-(\lambda+\mu)t}\Big\{I_0 - \frac{\mu}{\lambda+\mu} - \int_0^t e^{(\lambda+\mu)x}\,dS_x\Big\},\\ S_t + I_t &= \frac{\mu}{\lambda+\mu} + e^{-(\lambda+\mu)t}\Big\{S_0 + I_0 + (\lambda+\mu)\int_0^t S_x e^{\mu x}\,dx - \frac{\mu}{\lambda+\mu}\Big\},\\ S_t &= e^{-(\lambda+\mu)t}\Big\{S_0 + (\lambda+\mu)\int_0^t S_x e^{\mu x}\,dx + \int_0^t e^{(\lambda+\mu)x}\,dS_x\Big\}.\end{aligned}$$

Thus I_t, R_t and V_t vary as functions of t and S_t, whereas S_t is a function of t independent of the other functions. The implicit equation for S_t is approximated numerically on a grid $(t_i)_{i=0,\ldots,n}$ of the observation time interval in sub-intervals of equal length δ. Let $S_{i+1} = S_{t_{i+1}}$, it is approximated from the previous values $(S_j)_{j=0,\ldots,i}$ as

$$\begin{aligned}S_{i+1} &= e^{-(\lambda+\mu)t_{i+1}}\Big[S_0 + (\lambda+\mu)\delta\sum_{j=0}^{i}\{S_j e^{\mu t_j} + e^{(\lambda+\mu)t_j}(S_{j+1} - S_j)\}\Big]\\ &= \frac{1}{1-(\lambda+\mu)\delta e^{-(\lambda+\mu)\delta}}\Big\{e^{-(\lambda+\mu)\delta}S_i\\ &\qquad + (\lambda+\mu)\delta e^{-(\lambda+\mu)t_{i+1}}S_i e^{\mu t_i}(1 - e^{\lambda t_i})\Big\}.\end{aligned}$$

The solutions have the same form if the coefficients are functions of t.

The age a of an individual and the calendar time are linearly dependent with $a_0+t = a_t$, where a_0 is the birthdate and a varies in the interval $[0, A]$, therefore the first derivatives of a function $u(a,t)$ are $u'_t = u'_a$. A parameter concerning the individuals of age a at time t is a cumulative function on the birthdates upto a, $\rho_a = \rho(a_t) = \int_0^t \rho(a_s)\,ds = \int_0^a \rho(x)\,dx$. Introducing the age in the differential equations, they become

$$\begin{aligned}
S'_t(a,t) &= -\beta_a S(a,t)I(t) - \mu_a S(a,t) + \mu_a,\\
I'_t(a,t) &= \beta_a I(a,t)S(t) - (\lambda_a + \mu_a)I(a,t),\\
R'_t(a,t) &= \lambda_a I(a,t) - \mu_a R(a,t),
\end{aligned}$$

with the cumulative rates $I(t) = \int_0^A I(x,t)\,dx$ and $S(t) = \int_0^A S(x,t)\,dx$. Integrating a variable $\varphi(a)$ with respect to the time variable with an age in $[0,t]$ is equivalent to a convolution

$$\begin{aligned}
\int_0^t v(s)\varphi(a_s)\,ds &= \int_0^t \int_0^{t-s} v(s)\varphi(a_0+s)\,da_0\,ds\\
&= \int_0^t \int_0^t v(s)\varphi(x-s)\,dx\,ds
\end{aligned}$$

for every function v on $[0,t]$. The functions I and S are solutions of the implicit equations

$$\begin{aligned}
S(a,t) &= e^{-\int_0^t (\mu_{a_s}+\beta_{a_s} I_s)\,ds}\Big\{S_0(a) + \int_0^t \mu_{a_s} e^{\int_0^s (\mu_{a_x}+\beta_{a_x} I_x\,dx}\,ds\Big\},\\
R(a,t) &= e^{-\int_0^t \mu_{a_s}\,ds}\Big\{R_0(a) + \int_0^t I_s \lambda_{a_s} e^{\mu_{a_s}}\,ds\Big\},\\
I(a,t) &= I_0(a) e^{-\int_0^t (\lambda_{a_s}+\mu_{a_s}-\beta_{a_s} S_s)\,ds}.
\end{aligned}$$

An algorithm for the approximation of the solutions of these implicit equations is deduced.

Other models include the asymptomatic infectuousness which cannot be detected before symptoms of the disease appear, the migrations toward the population of individuals in every possible state and the loss of individuals due to all causes, they are analyzed as previously. An excess of mortality due to the epidemics is ratio of the death rate of the susceptible and contaminated sub-populations.

6.6 Lotka-Volterra equations

The simplest continuous model for the time variations of a population size $N(t)$ is the exponential model depending on the difference α of its rates of birth and death, where $N'(t) = \alpha N(t)$ hence $N(t) = N_0(t)e^{\alpha N(t)}$. The logistic model for a distribution F of population includes a control of the population growth as its size increases as $F'(t) = \alpha F(t)\{1 - kF(t)\}$, equivalently $F(t) = \frac{1}{k+e^{-\alpha t}}$.

The Lotka-Volterra equations for two the sizes of two competing populations is a system of time-dependent nonlinear differential equations

$$
\begin{aligned}
x'_t &= kx_t - ax_t y_t, \\
y'_t &= -ly_t + bx_t y_t,
\end{aligned} \tag{6.3}
$$

where the variable functions x_t and y_t represent the sizes of each population at t, with initial values x_0 and y_0 at $t = 0$. The function x_t is the size of a population of preys and y_t the size of a population of predators. The parameters are k, the reproduction rate of a prey population in expansion and $-l$, the death rate of a predator population in extinction, and the interaction parameters are a, the excess rate of mortality in the prey population due to the predators, and b, the excess of birth in the predator population in the presence of the other population. The system has a trivial stable state $x_t = y_t = 0$ and $x_t = b^{-1}l$, $y_t = a^{-1}k$.

The equations (6.3) may be solved numerically using series expansions of $C^\infty(\mathbb{R}_+)$ functions $x_t = \sum_{n=0}^\infty x_n t^n$ and $y_t = \sum_{n=0}^\infty y_n t^n$ with initial values x_0 and y_0 at $t = 0$, and such that

$$
\begin{aligned}
\sum_{n=1}^\infty n x_n t^{n-1} &= k \sum_{n=0}^\infty x_n t^n - a \sum_{n,m=0}^\infty x_n y_m t^n t^m, \\
\sum_{n=1}^\infty n y_n t^{n-1} t &= -l \sum_{n=0}^\infty y_n t^n y_t + b \sum_{n,m=0}^\infty x_n y_m t^n t^m,
\end{aligned}
$$

which imply

$$
\begin{aligned}
x_1 &= x_0(k - ay_0), \\
2x_2 &= x_1(k - ay_0) - ax_0 y_1, \\
nx_n &= kx_{n-1} - a \sum_{j=0}^n x_j y_{n-j},\ n \geq 2.
\end{aligned}
$$

The equations (6.3) also imply $bx'_t + ay'_t = bkx_t - aly_t$ and by their series expansions

$$ay_n = k^{-1}(cx_{n-1} - dy_{n-1}) - bx_n, \; n \geq 2.$$

The variables x_t and y_t are dependent and implicit solutions of (6.3) depend on the cumuluted densities

$$X_t = \int_{t_0}^{t} e^{l(s-t)} x_s \, ds, \quad Y_t = \int_{t_0}^{t} e^{-k(s-t)} y_s \, ds,$$

as

$$\begin{aligned} x_t &= x_0 e^{k(t-t_0)} - aY_t, \\ y_t &= x_0 e^{-l(t-t_0)} + bX_t. \end{aligned}$$

The determinant of the matrix A such that $\binom{x'_t}{y'_t} = A\binom{x_t}{y_t}$ is $abx_ty_t - kl$. If it is zero, y_t is determined by x_t and the system is reduced to a single equation $x'_t = kx_t - b^{-1}kl$, then

$$x_t = e^{k(t-t_0)}\{x_0 - b^{-1}l(e^{-kt} - e^{-kt_0})\}.$$

If $abx_ty_t - kl$ is not zero, the differential equation

$$bx'_t + ay'_t = kbx - lay$$

has the implicit solutions

$$\begin{aligned} bx_t + ay_t &= e^{k(t-t_0)}\Big\{c_0 - a(k+l)\int_{t_0}^{t} y_s e^{-k(s-t_0)} \, ds\Big\}, \\ &= e^{-l(t-t_0)}\Big\{c_0 + b(k+l)\int_{t_0}^{t} x_s e^{l(s-t_0)} \, ds\Big\}, \end{aligned}$$

where $c_0 = bx_0 + ay_0$. By a reparametrization of y as a function of x, the time-dependent differential equations (6.3) become

$$\begin{aligned} \frac{dy}{dx} &= \frac{dy}{dt}\Big(\frac{dx}{dt}\Big)^{-1} \\ &= \frac{y(bx-l)}{x(k-ay)}, \\ \frac{dy}{y\,dt}(k-ay) &= \frac{dx}{x\,dt}(bx-l) \end{aligned}$$

and both terms are equal to a constant c since they do not depend on the same variable. Then the solutions of (6.3) satisfy the marginal differential equations

$$\frac{dx}{x} = \frac{c\,dt}{bx_t - l}, \quad \frac{dy}{y} = \frac{c\,dt}{k - ay_t},$$

so x_t and y_t are solutions of the separated implicit equation

$$x_t = x_0 \exp\Big\{c \int_{t_0}^t (bx_s - l)^{-1}\, ds\Big)\Big\},$$

$$y_t = y_0 \exp\Big\{c \int_{t_0}^t (k - ay_s)-1\, ds\Big)\Big\}.$$

The system of equations (6.3) implies

$$xy'(k - ay) + x'y(l - bx) = 0,$$

it has solutions (x_t, y_t) satisfying the implicit equation $F(x_t, y_t) = c$, with a constant c and for every t in a finite interval $[t_0, t_1]$ if

$$\frac{\partial F}{\partial t}(x_t, y_t) = x_t' \frac{\partial F}{\partial x}(x_t, y_t) + y_t' \frac{\partial F}{\partial y}(x_t, y_t) = 0.$$

Identifying both equations leads to $F(x, y) = F_{l,b}(x) + F_{k,a}(y)$ with derivatives

$$F_{l,b}'(x) = \frac{l - bx}{x}, \quad F_{k,a}'(y) = \frac{k - ay}{y}$$

so $F_{l,b}(x) = \log(x^l) - bx$ and $F_{k,a}(y) = \log(y^k) - ay$.

The constant parameters of the Lotka-Volterra equation (6.3) may be modified according to the time-varying environmental conditions and the equations are written as

$$\begin{aligned} dx_t &= k_t x_t - a_t x_t y_t, \\ dy_t &= -l_t y_t + b_t x_t y_t, \end{aligned} \tag{6.4}$$

with the initial condition (x_0, y_0).

Equation (6.4) is the differential of a function $F(x_t, y_t) = c$, with a constant c and for every t in a finite interval $[t_0, t_1]$ if

$$F(x_t, y_t) = F_{l_t, b_t}(x_t) + F_{k_t, a_t}(y_t)$$

with $F_{l_t, b_t}(x_t) = \log(x_t^{l_t}) - b_t x_t$ and $F_{k_t, a_t}(y_t) = \log(y_t^{k_t}) - a_t y_t$.

Time variable implicit solutions are now

$$x_t = x_0 \exp\Big\{\int_{t_0}^t (k_s - a_s y_s)\, ds\Big\},$$

$$y_t = y_0 \exp\Big\{\int_{t_0}^t (b_s x_s - l_s)\, ds\Big\}.$$

These solutions are approximated over a grid of n small sub-intervals $(]t_{i-1}, t_i])_{i \leq n}$ of equal length δ of $[t_0, t_1]$. If t belongs to $[t_{j-1}, t_j[$, they are approximated $x_j = x_{t_j}$ and $y_j = y_{t_j}$ calculated from the initial values in the form

$$x_j = x_0 \exp\Big\{\delta \sum_{i=0}^{j-1} (k_i - a_i y_i)\Big\},$$
$$y_j = y_0 \exp\Big\{\delta \sum_{i=0}^{j-1} (b_{t_i} x_{t_i} - l_{t_i})\Big\},$$

they imply recursive formulas

$$x_{j+1} = x_j e^{\delta(k_j - a_j y_j)},$$
$$y_{j+1} = y_j e^{\delta(b_{t_i} x_{t_i} - l_{t_i})}.$$

The Lotka-Volterra equations have been extended by adding functions $f(x, y)$ and, respectively, $g(x, y)$ to the right-hand side as control functions

$$\begin{aligned} x'_t &= kx_t - ax_t y_t + f(x_t, y_t), \\ y'_t &= -ly_t + bx_t y_t + g(x_t, y_t). \end{aligned} \tag{6.5}$$

They are equivalent to the implicit equations

$$\begin{aligned} x_t = \Big[x_0 + \int_{t_0}^t f(x_s, y_s) \exp\Big\{-\int_{t_0}^s (k_z - a_z y_z)\, dz\Big\}\, ds\Big] \\ \cdot \exp\Big\{\int_{t_0}^t (k_s - a_s y_s)\, ds\Big\}, \\ y_t = \Big[y_0 + \int_{t_0}^t g(x_s, y_s) \exp\Big\{-\int_{t_0}^s (b_z x_z - l_z)\, dz\Big\}\, ds\Big] \\ \cdot \exp\Big\{\int_{t_0}^t (b_s x_s - l_s)\, ds\Big\}. \end{aligned}$$

Another extension of the equations modifies the interaction between the populations as

$$\begin{aligned} x'_t &= kx_t - b\frac{x_t y_t}{1 + cx_t}, \\ y'_t &= -ly_t + b\frac{x_t y_t}{1 + cx_t}, \end{aligned} \tag{6.6}$$

with $c > 0$, it follows that

$$\begin{aligned} x'_t + y'_t &= kx_t - ly_t, \\ \frac{x'_t + y'_t}{x_t + y_t} &= k - (k + l)\frac{y_t}{x_t + y_t}, \\ x_t + y_t &= (x_0 + y_0) \exp\Big\{kt - (k + l) \int_{t_0}^t \frac{y_s}{x_s + y_s}\, ds\Big\}. \end{aligned}$$

These solutions are approximated over a grid as previously. The convergence of x_t to zero as t increases implies that the ratio y_s over $x_s + y_s$ tend to one, moreover y_t vanishes with the exponential e^{-lt}. If y_t tends to zero as t increases, $x_t + y_t$ becomes equivalent to e^{kt} and x_t tends to infinity.

In systems with n interacting populations, the differential equations for the proportion of species i in the whole population becomes

$$x'_{i,t} = x_i\Big(k_i - \sum_{j\neq i, j=1}^{n} a_{ij}x_j\Big), \; i = 1, \dots, n,$$

with $\sum_{j=1}^{n} x_j = 1$. At the equilibrium, $x_i = 0$ or $\sum_{j\neq i,j=1}^{n} a_{ij}x_j = k_i$, for every $i = 1, \dots, n$. For every $n \geq 2$, there exists a linear combination of the variables such that

$$\sum_{i=1}^{n} \lambda_i x'_{i,t} = \sum_{i=1}^{n} \lambda_i k_i x_i$$

and $\sum_{i=1}^{n} \lambda_i x_i \sum_{j\neq i,j=1}^{n} a_{ij}x_j = 0$. Solving the linear equation provides n implicit equations

$$\sum_{i=1}^{n} \lambda_i x_{i,t} = e^{k_j t}\Big\{c_0 + \sum_{i=1}^{n}(k_i - k_j)\int_{t_0}^{t} x_{i,s}e^{k_j s}\,ds\Big\}, \; j = 1, \dots, n.$$

The constants of the Lotka-Volterra equations may be replaced by functions of the population sizes, in the form

$$\begin{aligned} x'_t &= f(x_t) - g(x_t)y_t, \\ y'_t &= ag(x_t)y_t - by_t, \end{aligned}$$

with specified functions f and g, where b is the death rate of the predators. A solution for the population size of predators depends simply on x as

$$y_t = y_0 \exp\Big[\int_0^t \{ag(x_s) - b\}\,ds\Big].$$

More generally the functions f and g may depend on x and y

$$\begin{aligned} x'_t &= f(x_t, y_t)x_t - g(x_t, y_t)y_t, \\ y'_t &= ag(x_t, y_t)y_t - by_t \end{aligned}$$

and implicit solutions are deduced.

6.7 Birth-and-death differential equations

The birth-and-death system of two time-dependent linear differential equations (1.14) may be written in several equivalent forms which imply several expressions for their solutions. Let U_t be the vector with components x_t and y_t, and let M be the matrix

$$M = \begin{pmatrix} a_{11} & -a_{12} \\ -a_{21} & a_{22} \end{pmatrix},$$

then (1.14) is equivalent to the linear differential equation $U_t' = MU_t$ with solution $U_t = U_0 e^{Mt}$ defined as the exponential of the matrix Mt.

From the linear combinaisons

$$\begin{aligned} a_{21}x_t' + a_{11}y_t' &= (a_{11}a_{22} - a_{12}a_{21})y_t \equiv Ay_t, \\ a_{22}x_t' + a_{12}y_t' &= (a_{11}a_{22} - a_{12}a_{21})x_t \equiv Ax_t, \end{aligned} \tag{6.7}$$

of the birth-and-death differential equations with constant coefficients, we get $x_t(a_{21}x_t' + a_{11}y_t') = y_t(a_{22}x_t' + a_{12}y_t')$. Equations (6.7) are simultaneous first order linear partial differential equations. They have implicit solutions

$$\begin{aligned} x_t &= \Big[x_0 + \exp\Big\{-\frac{a_{12}}{a_{22}} \int_{t_0}^t e^{-a_{22}^{-1}A(s-t_0)} y_s' \, ds\Big\}\Big] \exp\Big\{a_{22}^{-1}A(t-t_0)\Big\}, \\ y_t &= \Big[y_0 + \exp\Big\{-\frac{a_{21}}{a_{11}} \int_{t_0}^t e^{-a_{11}^{-1}A(s-t_0)} x_s' \, ds\Big\}\Big] \exp\Big\{a_{11}^{-1}A(t-t_0)\Big\}. \end{aligned}$$

They are approximation on a grid $(x_i, y_i, t_i)_{i=1,\dots,n}$ with time increments $t_{i+1} - t_i = \delta$. If t belongs to $]t_{i+1}, t_{i+2}]$, x_t and y_t are approximated by $x_{i+1} = x(t_{i+1})$ and $y_{i+1} = y(t_{i+1})$ such that

$$\begin{aligned} x_{i+1} &= \Big[x_0 + \exp\Big\{-\frac{a_{12}}{A} \sum_{j=0}^i e^{-a_{22}^{-1}At_j} y_j'(1 - e^{-a_{22}^{-1}A\delta})\Big\}\Big] \\ &\qquad \cdot \exp\Big\{a_{22}^{-1}A(t_{i+1} - t_0)\Big\} \\ y_{i+1} &= \Big[y_0 + \exp\Big\{-\frac{a_{21}}{A} \sum_{j=0}^i e^{-a_{11}^{-1}At_j} x_j'(1 - e^{-a_{11}^{-1}A\delta})\Big\}\Big] \\ &\qquad \cdot \exp\Big\{a_{11}^{-1}A(t_{i+1} - t_0)\Big\} \end{aligned}$$

where $x_{i+1}' = \delta^{-1}(x_{i+1} - x_i)$ and $y_{i+1}' = \delta^{-1}(y_{i+1} - y_i)$. An interpolation provides a better approximation

$$\begin{aligned} \widetilde{x}_t &= x_{i+1} + x_{i+1} + \delta^{-1}(t - t_{i+1})(x_{i+2} - x_{i+1}), \\ \widetilde{y}_t &= y_{i+1} + y_{i+1} + \delta^{-1}(t - t_{i+1})(y_{i+2} - y_{i+1}), \end{aligned}$$

and

$$\begin{aligned}\widetilde{x}'_t &= x'_{i+1} + \delta^{-1}(x_{i+1} - x_i)(x'_{i+2} - x'_{i+1})\\ &= \delta^{-1}(x_{i+1} - x_i) + \delta^{-2}(x_{i+2} - 2x_{i+1} + x_i),\end{aligned}$$

the approximation of y'_t is similar.

A second approach relies on the second derivatives of the birth-and-death rates. From (6.7) we get two independent second order differential equations for the birth and death rates

$$\begin{aligned}0 &= a_{11}^{-1}Ax''_t + a_{11}^{-1}a_{12}y'_t - Ax'_t\\ &= a_{11}^{-1}Ax''_t - a_{11}^{-1}(a_{22} + a_{11})Ax'_t + a_{11}^{-1}A^2x_t,\\ 0 &= a_{22}^{-1}Ay''_t + a_{22}^{-1}a_{21}x'_t - Ay'_t\\ &= a_{22}^{-1}Ay''_t - a_{22}^{-1}(a_{11} + a_{22})Ay'_t + a_{22}^{-1}A^2y_t,\end{aligned}$$

they satisfy the same differential equation

$$u''_t - (a_{22} + a_{11})u'_t + Au_t = 0. \tag{6.8}$$

The characteristic polynomial of the second order differential equation (6.8) is $P = r^2 - (a_{11} + a_{22})r + A = 0$, its discriminant $(a_{11} - a_{11})^2 + 4a_{12}a_{21}$ is strictly positive since $a_{12}a_{21} > 0$ and there exist two distinct real roots. The independent solutions x_t and y_t of these equations are respectively defined by Proposition 4.12 as linear combinations of real exponential functions, they are the product of the exponential function $e^{\frac{1}{2}(a_{11}+a_{22})(t-t_0)}$ and the same trigonometric sine and cosine functions with parameter $\{(a_{11}-a_{22})^2+4a_{12}a_{21}\}^{\frac{1}{2}}$, they differ only by their initial values. They are periodic and remain bounded as t tends to infinity. If x_t and y_t have close values at time t_1, they remain close later. If $x_t + y_t$ is constant, it remains constant for all constant coefficients and the behaviour of x_t and y_t are opposite.

If the coefficients a_{ij} are functions of t, the solutions of the differential equation (1.14) have the same form $x_t = v_te^{B_{11,t}}$ and $y_t = w_te^{B_{22,t}}$ with the primitives $B_{ii,t} = \int_{t_0}^t a_{ii}^{-1}(s)A(s)\,ds$, for $i = 1, 2$, and with functions

$$v_t = x_0 + \exp\Big\{-\int_{t_0}^t \frac{a_{12,s}}{a_{22,s}}e^{-B_{11,s}}y'_s\,ds\Big\},$$

$$w_t = y_0 + \exp\Big\{-\int_{t_0}^t \frac{a_{21,s}}{a_{11,s}}e^{-B_{22,t}}x'_s\,ds\Big\}$$

These implicit equations are approximated over a grid of small sub-intervals $(]t_i, t_{i+1}])_{i\leq n}$ of equal length δ of the observation interval. If t belongs to $]t_i, t_{i+1}]$, v_t is approximated by $\widetilde{v}_t$ in the same form as $\widetilde{x}_t$ with

$$v_{i+1} = x_0 + \prod_{j=0}^{i} \exp\Big\{-\delta y_j' \frac{a_{12,j}}{a_{22,j}} e^{-B_{11,j}}\Big\}$$
$$v_{i+1} - x_0 = (v_i - x_0) \exp\Big\{-\delta y_i' \frac{a_{12,i}}{a_{22,i}} e^{-B_{11,i}}\Big\},$$

where the coefficient are index by j for t_j. The function w_t has a similar approximation. The second derivatives of the differential equations (1.14) with variable coefficients cannot be solved is with constants.

The migrations between populations modify the birth-and-death differential equations, let z_t be the global immigration rate, the birth-and-death equations become

$$\begin{aligned} x_t' &= a_{11}x_t - a_{12}y_t + b_1 z_t, \\ y_t' &= -a_{21}x_t + a_{22}y_t + b_2 z_t, \\ z_t' &= f_t, \end{aligned} \tag{6.9}$$

with a positive or negative function f_t possibly dependent of x_t and y_t or their difference and with coefficients $b_i \geq 0$ if $f_t \geq 0$ and $b_i \geq 0$ if $f_t \geq 0$, $b_i \geq 0$ if $f_t \geq 0$, for $i = 1, 2$. The system of equations (6.9) has an equilibrium in the homogeneous equation without migrations with $A = 0$ and $y_t = a_{12}^{-1} a_{11} x_t$.

Let $z_t = F_t$ a primitive of f_t, the differential equations (6.9) are inhomogeneous and their solutions are written as the sum of the solutions of the homogeneous equations (1.14) and particular solutions of (6.9). With constant coefficients, they have the implicit solutions

$$x_t = \Big[x_0 + \exp\Big\{\int_{t_0}^{t} e^{-a_{11}(s-t_0)}(b_1 f_s - a_{12}y_s)\, ds\Big\}\Big] \exp\Big\{a_{11}(t - t_0)\Big\},$$
$$y_t = \Big[y_0 + \exp\Big\{\int_{t_0}^{t} e^{-a_{22}(s-t_0)}(b_2 f_s - a_{21}x_s)\, ds\Big\}\Big] \exp\Big\{a_{22}(t - t_0)\Big\}$$

and they are approximated on a grid as previously.

The linear combinaisons (6.7) become

$$\begin{aligned} a_{21}x_t' + a_{11}y_t' &= \equiv Ay_t + B_1 F_t, \\ a_{22}x_t' + a_{12}y_t' &= Ax_t + B_2 F_t, \end{aligned} \tag{6.10}$$

where $B_1 = a_{21}b_1 + a_{11}b_2$ and $B_2 = a_{22}b_1 + a_{12}b_2$.

The unique homogeneous equation (6.8) is replaced by

$$\begin{aligned}(a_{11}B_2 - B_1)f_t &= Ax_t'' + a_{12}y_t' - a_{11}Ax_t'\\ (a_{11}B_2 - B_1)f_t + AB_2F_t &= Ax_t'' - (a_{22} + a_{11})Ax_t' + A^2x_t,\\ (a_{22}B_1 - B_2)f_t &= a_{22}^{-1}Ay_t'' + a_{22}^{-1}a_{21}x_t' - Ay_t'\\ (a_{22}B_1 - B_2)f_t + AB_1F_t &= Ay_t'' - (a_{11} + a_{22})Ay_t' + A^2y_t.\end{aligned}$$

Let $f_{1t} = (a_{11}B_2 - B_1)f_t + AB_2F_t$ and $f_{2t} = (a_{22}B_1 - B_2)f_t + AB_1F_t$, and let F_{1t} and F_{2t} be their primitives. Denoting x_{Ht} and y_{Ht} the exponential-sinusoïdal solutions of the homogeneous equation (6.8), particular solutions of the inhomogeneous equations are obtained by variation of the constant as $x_t = h_t x_{Ht}$ and $y_t = k_t y_{Ht}$ with functions h and k solutions of the differential equations

$$\begin{aligned}A^{-1}f_{1t} &= h_t'' + h_t'(2x_{H,t}' - a_{11} - a_{22}),\\ A^{-1}f_{2t} &= k_t'' + k_t'(2y_{H,t}' - a_{11} - a_{22}),\end{aligned}$$

therefore

$$\begin{aligned}h_t' &= A^{-1}e^{-2x_{H,t}+(a_{11}+a_{22})(t-t_0)}\int_{t_0}^t f_{1s}e^{-2x_{H,s}+(a_{11}+a_{22})(s-t_0)}\,ds,\\ k_t' &= A^{-1}e^{-2y_{H,t}+(a_{11}+a_{22})(t-t_0)}\int_{t_0}^t f_{2s}e^{-2y_{H,s}+(a_{11}+a_{22})(s-t_0)}\,ds\end{aligned}$$

and their primitives give the solutions.

With functional coefficients, the inhomogeneous equations (6.9) have the implicit solutions

$$\begin{aligned}x_t &= \Big[x_0 + \exp\Big\{\int_{t_0}^t e^{-\int_{t_0}^s a_{11,\varsigma}\,d\varsigma}(b_{1,s}f_s - a_{12,s}y_s)\,ds\Big\}\Big]\exp\Big\{\int_{t_0}^t a_{11,s}\,ds\Big\},\\ y_t &= \Big[y_0 + \exp\Big\{\int_{t_0}^t e^{-\int_{t_0}^s a_{22,\varsigma}\,d\varsigma}(b_{2,s}f_s - a_{21,s}x_s)\,ds\Big\}\Big]\exp\Big\{\int_{t_0}^t a_{22,s}\,ds\Big\}.\end{aligned}$$

6.8 Differential equations for multi-states dynamic models

The dynamics of populations concern the variations of the population size and the individual behaviors between random states. The motions towards the activities is described by the following graph.

A system of differential equations describes the variations in time of the probabilities of being in each possible state of the activities according to the entrance of individuals in the process or their departure. The model corresponding to Figure (6.1) with four transient states, has 4 probabilities

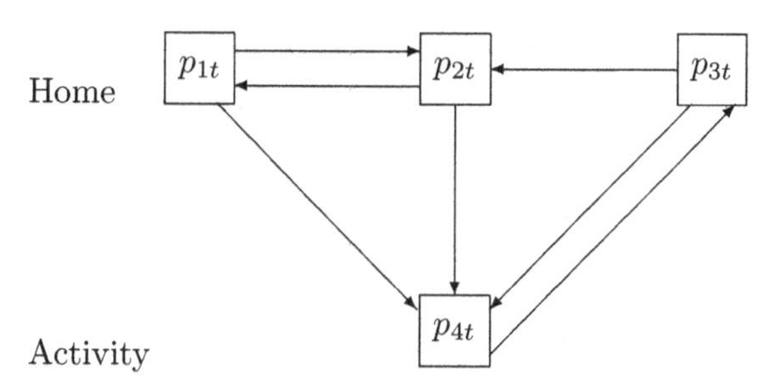

Fig. 6.1 Graph of the dynamic in a population

of sojourn in states p_{1t}, p_{2t}, p_{3t} and p_{4t}, with a total sum $p_{1t} + p_{2t} + p_{3t} + p_{4t} \leq 1$.

At the initial time $t_0 = 0$ of the daily activity, most individuals are staying in S_0, then some of them look for an activity and enter in S_1 where they collect informations from the more experienced individuals of the population. The departures from S_1 towards S_4 entail the beginning of the activity of a proportion α of individuals from S_1. When individuals come back from S_4, they enter in S_3 where they can choose between the three other states and move between them according to the rate of activity in each of them.

In a simplified model, transition rates are denoted as follows: $1-\alpha$ is the constant probability of exit of the system from S_3, the individual leaving S_3 in the system are directed to several possible states with the rates β of entrance in S_1, γ_2 of entrance in S_4 and θ of entrance in S_4, with a sum $\beta + \gamma_2 + \theta = 1$. With $\alpha < 1$, the sum of the sojourn probabilities is strictly smaller than one. The parameter γ_1 is the transition rate from S_4 to S_2, η is the exit rate from S_2 to S_3, h_t is the transition function from S_1 to S_2. Their variations follow the differential equations

$$\begin{aligned}
\frac{dp_{1t}}{dt} &= \alpha\beta p_{3t} - h_t p_{1t}, \\
\frac{dp_{2t}}{dt} &= h_t p_{1t} + \gamma_1 p_{4t} + \alpha\gamma_2 p_{3t} - \eta p_{2t}, \\
\frac{dp_{3t}}{dt} &= \eta p_{2t} - \alpha p_{3t}, \\
\frac{dp_{4t}}{dt} &= \alpha\theta p_{3t} - \gamma_2 p_{4t}.
\end{aligned}$$

The system of equations has only an equilibrium at the extinction where all individuals become inactive and have gone out of all states, hence the four probabilities are zero. Let $H_t = \int_0^t h_s\,ds$ be the primitive of the transition function. The solution of each equation defines implicit equations depending on the solutions of the other equations

$$\begin{aligned}
p_{1t} &= \alpha\beta \int_0^t p_{3s} e^{-(H_t - H_s)}\,ds,\\
p_{2t} &= \int_0^t (h_s p_{1s} + \gamma_1 p_{4s} + \alpha\gamma_2 p_{3s}) e^{-\eta(t-s)}\,ds,\\
p_{3t} &= \int_0^t \eta p_{2s} e^{-\alpha(t-s)}\,ds,\\
p_{4t} &= \int_0^t \alpha\theta\, p_{3s} e^{-\gamma_2(t-s)}\,ds.
\end{aligned} \tag{6.11}$$

The proportion of individuals in activity in S_2 is deduced from the probabilities of staying in states S_3 and S_4, according to the equation

$$p_{2t} = \int_0^t \{h_s\alpha\beta p_{3s} e^{-(H_t-H_s)} + \alpha\gamma_2 p_{3s} + \gamma_1 p_{4s}\} e^{-\eta(t-s)}\,ds, \tag{6.12}$$

where p_{3t} and p_{4t} are expressed as integrals of p_{2t} in (6.11) and in the equation

$$p_{4t} = \int_0^t \alpha\theta\eta p_{2s} e^{-(\alpha+\gamma_2)(t-s)}\,ds.$$

The probability p_{2t} is therefore written as solution of an implicit equation depending only on the function p_{2t}

$$p_{2t} = \alpha \int_0^t \{h_s\beta e^{-(H_t-H_s)} + \theta\gamma_1 e^{-\gamma_2(t-s)} + \gamma_2\}\eta p_{2s} e^{-(\alpha+\eta)(t-s)}\,ds. \tag{6.13}$$

The probability p_{3t} of is also written as solution of an implicit equation with a single probability function, due to (6.11) and the equality

$$p_{2t} = \alpha \int_0^t (\beta h_s e^{-(H_t-H_s)} + \theta\gamma_1 e^{-\gamma_2(t-s)} + \gamma_2) e^{-\eta(t-s)} p_{3s}\,ds. \tag{6.14}$$

An approximation of the integrals is performed over a set of n_t small subintervals $(I_i)_{i\le n_t}$ of equal length δ of $[0,t]$, where the integrands are considered as constant. Approximate solutions of the system of implicit equations (6.11) are deduced recursively as sums starting from the initial probability of sojourn in each state at t_0, $p_{j,0}$. The initial probabilities are estimated by the proportion of the observed numbers of individuals in each state at t_0. For every t in I_i, let $h_t = h_i$, $H_t = H_i$ and $p_{j,i} = p_{jt}$, for

$j = 1, \dots, 4$. As when δ tends to zero, the system of equations (6.11) is approximated by the following one

$$
\begin{aligned}
p_{1t} &= \alpha\beta \sum_{i=0}^{n_t-1} p_{3,i} \int_{I_i} e^{-(H_t - H_s)}\, ds, \\
p_{2t} &= \frac{1}{\eta} \sum_{i=0}^{n_t-1} (h_i p_{1,i} + \gamma_1 p_{4,i} + \alpha\gamma_2 p_{3,i})\, e^{-\eta(t-t_i)}(1 - e^{-\eta\delta}) \\
&\sim \delta \sum_{i=0}^{n_t-1} (h_i p_{1,i} + \gamma_1 p_{4,i} + \alpha\gamma_2 p_{3,i})\, e^{-\eta(t-t_i)}, \\
p_{3t} &= \frac{\eta}{\alpha} \sum_{i=0}^{n_t-1} p_{2,i}\, e^{-\alpha(t-t_i)}(1 - e^{-\alpha\delta}) \\
&\sim \eta\delta \sum_{i=0}^{n_t-1} p_{2,i}\, e^{-\alpha(t-t_i)}, \\
p_{4t} &= \alpha\frac{\theta}{\gamma_2} \sum_{i=0}^{n_t-1} p_{3,i}\, e^{-\gamma_2(t-t_i)}(1 - e^{-\alpha\delta}) \\
&\sim \alpha\theta\delta \sum_{i=0}^{n_t-1} p_{3,i}\, e^{-\gamma_2(t-t_i)}.
\end{aligned}
$$

An equivalent approximation of p_{2t} is deduced from (6.13). As δ tends to zero, p_{2t} is equivalent to

$$
\alpha\eta\delta \sum_{i=0}^{n_t-1} p_{2i}\{\beta h_i e^{-(H_t - H_i)} + \theta\gamma_1 e^{-\gamma_2(t-t_i)} + \gamma_2\} e^{-(\alpha+\eta)(t-t_i)}.
$$

An approximation of p_{3t} is deduced from (6.13).

The model is extended by allowing motions towards several distinct areas. Let A and B be two subsystems, the probability functions p_{jt} are replaced by functions specific to A or B, $p_{j,At}$ and $p_{j,Bt}$, for $j = 1, \dots, 4$, and they follow similar equations with transition probabilities indexed by A or B. The functions from the first state to the second ones $S_{2,A}$ and $S_{2,B}$ are denoted h_{At} for A and h_{Bt} for B. With a single state S_1, the differential equation for p_{1t} becomes

$$
\frac{dp_{1t}}{dt} = \alpha(\beta_A p_{3,At} + \beta_B p_{3,Bt}) - (h_{At} + h_{Bt})p_{1t}. \tag{6.15}
$$

When the model is symmetric, the transition probabilities of the areas A and B are identical, their solutions are therefore defined by the same equations. The equation for p_{2t} is identical to the equation for $p_{2,At} + p_{2,Bt}$, depending on p_{1t} and on the sums $p_{3,At} + p_{3,Bt}$ and $p_{4,At} + p_{4,Bt}$, etc. The implicit solutions of the differential equations are therefore identical to the previous ones. This property is generalized to k equivalent subsystems. In a model of k locations with unequal transition and sojourn probabilities, the dimension of the system of differential equations becomes $4k$ in the previous model and they are solved and approximated as previously. In Equation (6.15) for p_{1t}, the transition function h_t is the sum of the transition functions of the subsystems and

$$p_{1t} = \alpha \int_0^t (\beta_A p_{3,As} + \beta_B p_{3,Bs}) e^{-(H_t - H_s)} \, ds$$

where $p_{i,Bt}$ are solutions of implicit equations similar to (6.11).

Finally, the constant transition probabilities can be replaced by functional transition probabilities and the exponential with constant parameters are replaced by exponential functions of the primitives of the transition functions.

6.9 Poincaré-Lorenz differential system

Poincaré-Lorenz's system of nonlinear differential equations describes the variations of a spatial fluid in a domain Ω of $\mathbb{R}^3$ as

$$\begin{aligned} x_t' &= \sigma(y_t - x_t), \\ y_t' &= \rho x_t - y_t - x_t z_t, \\ z_t' &= x_t y_t - \beta z_t \end{aligned}$$

has two instable equilibrium points with coordinates $x = y = \pm(\beta(\rho - 1)^{\frac{1}{2}}$ and $z = \rho - 1$, on the opposite sides of the set Ω. The coordinates of the system are solutions of the implicit equations

$$\begin{aligned} x_t &= e^{-\sigma t}\Big(x_0 e^{\sigma t_0} + \sigma \int_{t_0}^t e^{\sigma s} y_s \, ds\Big), \\ y_t &= (y_0 e^{t_0} + u_t) e^{-t}, \\ z_t &= e^{-\beta t}\Big(z_0 e^{\beta t_0} + \int_{t_0}^t e^{\beta s} x_s y_s \, ds\Big) \\ u_t &= \rho \int_{t_0}^t e^s x_s \, ds - \int_{t_0}^t z_s x_s e^s \, ds. \end{aligned} \tag{6.16}$$

Using the expressions of x_t and z_t, the function u can be written with integrals depending only on y_t, they yield an implicit function $y_t = u(y_t, t)e^{-t}$ that does not depend on x_t and z_t.

The functions x_t, y_t and z_t are approximated over a grid of n small subintervals $(]t_{i-1}, t_i])_{i \leq n}$ of equal length δ of the domain $[t_0, t_1]$ of the variable t. Approximated solutions of the system of implicit equations (6.16) are calculated recursively as sums starting from the initial values at t_0. For $i \geq 1$

$$
\begin{aligned}
x_{i+1} &= x_{t_{i+1}} \approx e^{-\sigma t_{i+1}} \Big(x_0 e^{\sigma t_0} + \sigma\delta \sum_{k=0}^{i} e^{\sigma t_k} y_k \Big) \\
&= x_i + e^{-\sigma\delta}(x_i + \sigma\delta y_i), \\
y_{i+1} &= y_{t_{i+1}} \approx e^{-t_{i+1}} \Big\{ y_0 e^{t_0} + \delta \sum_{k=0}^{i} e^{t_k}(\rho - z_k) x_k \Big\} \\
&= y_i + e^{-\delta}\{y_i + \delta(\rho - z_{t_i}) x_i\}, \\
z_{i+1} &= z_{t_{i+1}} \approx e^{-\beta t_{i+1}} \Big(z_0 e^{\beta t_0} + \delta \sum_{k=0}^{i} e^{\beta t_k} x_k y_k \Big) \\
&= z_i + e^{-\beta\delta}(z_i + \delta x_i y_i).
\end{aligned}
$$

The variation $x_{i+1} - x_i$ is bounded by $(e^{-\sigma\delta} - 1)x_i + e^{-\sigma\delta}\sigma\delta y_i)$. If the coordinates are bounded by constants M_k, $k = 1, 2, 3$

$$
|x_{i+1} - x_i| \leq (e^{-\sigma\delta} - 1)M_1 + e^{-\sigma\delta}\sigma\delta M_2.
$$

The approximation error of x_t by x_i is smaller than $|x_{i+1} - x_i|$ for every t in $[t_i, t_{i+1}[$. The variations $y_{i+1} - y_i$ and $z_{i+1} - z_i$ and the approximation errors of y_t and z_t have similar bounds. For every i, the variations between t_i and t_{i+1} converge to zero as δ tends to zero.

Roessler's system of differential equations is defined as

$$
\begin{aligned}
x'_t &= -(y_t + z_t), \\
y'_t &= x_t - a y_t, \\
z'_t &= b - c z_t + x_t z_t
\end{aligned}
$$

it has two instable equilibrium points with coordinates $x, y, z)$ such that $y = -z = a^{-1}x$ with $az^2 + cz - b = 0$. The components y_t and z_t of the system are expressed as integrals of exponential functions depending on the primitive $X_t = X_0 + \int_{t_0}^t x_s \, ds$ of x_t

$$x_t = x_0 - \int_{t_0}^t (y_s + z_s) \, ds, \tag{6.17}$$

$$y_t = e^{-at}\Big(y_0 e^{at_0} + \int_{t_0}^t e^{as} \, dX_s\Big), \tag{6.18}$$

$$z_t = e^{X_t - ct}\Big(z_0 e^{ct_0 - X_0} + b\int_{t_0}^t e^{cs - X_s} \, ds\Big),$$

they are written as equations depending on the primitives $x_t = x_0 + Y_0 - Y_t + Z_0 - Z_t$, y_t and z_t depending on X_t. The primitives of y_t and z_t are

$$\begin{aligned} Y_0 - Y_t &= \frac{1}{a}\Big(y_0 e^{-a(t-t_0)} + e^{-at}\int_{t_0}^t e^{as} \, dX_s + X_0 - X_t\Big) \\ &= \frac{1}{a}(y_t + X_0 - X_t), \\ Z_t - Z_0 &= z_0 e^{ct_0 - X_0}\int_{t_0}^t e^{X_s - cs} \, ds + b\int_{t_0}^t e^{-(cw - X_w)}\int_{t_0}^w e^{cs - X_s} \, ds \, dw, \\ x_t &= x_0 + \frac{1}{a}(y_t + X_0 - X_t) + Z_0 - Z_t. \end{aligned}$$

The coordinate x_t satisfies the differential equation $x_t'' +' (y_t' + z_t') = 0$ which leads to an integro-differential equation for x

$$x_t'' + x + ay(x, t) + (x - c)z(x, t) + b = 0.$$

A numerical solution of the differential equations is calculated iteratively from the initial value $x_0, y_0, z_0)$ at t_0 on a spatial grid (x_i, y_i, z_i) of the components of the system at a time sequence $(t_i)_{i=0,\dots,n}$ of equal length $t_{i+1} - t_i = \delta$. From the implicit equations (6.18), the values at t_{i+1} are approximated by

$$\begin{aligned} x_{i+1} &= x_0 + \frac{1}{a}\Big(y_{i+1} - \delta\sum_{k=0}^i x_k\Big) - \delta\sum_{k=0}^i z_k, \\ y_{i+1} &= e^{-at_{i+1}}\Big\{y_0 e^{at_0} + \sum_{k=0}^i e^{at_k}(X_{k+1} - X_k)\Big\}, \\ z_{i+1} &= e^{X_{i+1} - ct_{i+1}}\Big(z_0 e^{ct_0 - X_0} + b\delta\sum_{k=0}^i e^{ct_k - X_k}\Big). \end{aligned}$$

The variations of the coordinates between t_i and t_{i+1} lead to a sequential algorithm

$$x_{i+1} = \Big(1 - \frac{\delta}{a}\Big)x_i + \frac{1}{a}(y_{i+1} - y_i) - \delta z_i,$$
$$y_{i+1} = y_i + e^{-a\delta}\{y_i + \delta x_i\},$$
$$z_{i+1} = z_i + e^{(x_i - c)\delta}(z_i + b\delta).$$

If the coordinates are bounded by constants M_k, $k = 1, 2, 3$

$$|y_{i+1} - y_i| \le y_0 e^{-a(t_i - t_0)}(e^{-a\delta} - 1) + \delta M_1$$

it is a $O(\delta)$ hence $|x_{i+1} - x_i| = O(\delta)$. Similarly, $|z_{i+1} - z_i| = O(\delta)$ and they converge to zero with δ.

6.10 Exercises

Solve the partial differential equations

6.8.1. $u'_y + axu = 0.$

6.8.2. $u''_{xy} = 2ax + 2by.$

6.8.3. $xu'_y + yu'_x = 0.$

6.8.5. $xu'_y - yu'_x = 0.$

6.8.6. $au'_y + bu'_x = 0.$

6.8.7. $au'_y + bu'_x = f(ax - by).$

6.8.8. $u'_x + bu'_y + u'_z = a(x + z).$

6.8.9. $u'_x - u'_y = u\frac{x-y}{(x+y)^2}.$

6.8.10. $xu'_y + yu'_x = u.$

6.8.11. $x^2u'_x + y^2u'_y = \frac{x^4}{y}.$

6.8.12. $xyu'^2_x + ayu'_y + azu'_z = 0.$

6.8.13. $x(y-1)+y(x-1)y'_x=(y-1)f_x$.

6.8.14. $xy^{-1}y'_x+xz^{-1}z'_x+2z=0$.

6.8.15. $xu'_x-yu'_y=\frac{x^2}{y}$.

6.8.16. $xu'_x+yu'_y=\frac{xy}{u}$.

6.8.17. $y^2u'_y-xyu'_x=axu$.

6.8.18. $x^2u''_{xx}+y^2u''_{yy}=xu'_x+yu'_y-4x^{-4}u$.

6.8.19. $u''_{xx}+\frac{2u'_x}{x}+\frac{2yu'_y}{x}-u=\frac{2e^x}{x}$.

6.8.20. $u''_{xx}+u''_{yy}-\left(\frac{1}{x}+\frac{1}{y}\right)u=A(y)\left(\frac{1}{y}-\frac{1}{x}\right)$, where A is the Airy function.

Find solutions of the simultaneous differential equations

6.8.21. $x'=y+f(x),\quad y'=x$.

6.8.22. $x'_t=y_t^2,\quad y'_t=ax_t$.

6.8.23. $x'_t=y_t^3,\quad y'_t=\cos t$, with $x_0=0=y_0$.

6.8.24. $x'_t=-y+ax(x^2+y^2)$ and $y'_t=x+ay(x^2+y^2)$.

6.8.25. $x'_t=a_t(l-y_t),\quad y'_t=a_tx_t$.

6.8.26. $x'_t=ar_t\frac{\partial R}{\partial x},\quad y'_t=br_t\frac{\partial R}{\partial y}$ for a function R of $r=(x^2+y^2)^{\frac{1}{2}}$.

6.8.27. $x'_t=a_1y+b_1x+c_1,\quad y'_t=a_2y+b_2x+c_2$, with constant coefficients.

6.8.28. $x'_t=ay_tz_t,\quad y'_t=bx_tz_t,\quad z'_t=cx_ty_t$.

Chapter 7

Special functions

Several Eulerian integral functions have been defined by Legendre, Riemann, Jacobi and other mathematicians, they are related to the function Gamma and their recursive properties rely on integrations. Some of them are detailed in the first section which originates from Legendre (1825), he defined the functions I and L and calculated tables of the numerical values of the function Beta. Here we present them with simple proofs and different methods to get their numerical values. Expansions of other functions solutions of second order differential equations are explicited, in particular the Airy, Bessel, Hermite and Laguerre functions. Byerly (1893) proved result about Bessel's functions and the results about the other equations are new.

7.1 Eulerian functions

Several integral functions with two real exponents generalize the function Gamma, they satisfy invariance properties under the permutation of their exponents. The function Beta is defined for strictly positive integers p and q as

$$B_{p,q} = \int_0^1 x^{p-1}(1-x)^{q-1}\,dx. \tag{7.1}$$

Denoting $x = \sin^2\theta$ or $x = \cos^2\theta$, it is equivalent to

$$\begin{aligned} B_{p,q} &= 2\int_0^1 \sin^{2p-1}\theta\cos^{2q-1}\theta\,d\theta \\ &= 2\int_0^1 \cos^{2p-1}\alpha\sin^{2q-1}\alpha\,d\alpha = B_{q,p}. \end{aligned}$$

Denoting $1-x=(1+y)^{-1}$, it is also written as

$$B_{p,q}=\int_0^\infty \frac{x^{p-1}}{(1+x)^{p+q}}\,dx. \tag{7.2}$$

Integrating by part the first and the last equalities provides recurrence formulas, for all $p\geq 1$ and $q\geq 1$

$$\begin{aligned}
B_{p+1,q} &= \frac{p}{q}B_{p,q+1},\\
B_{p,q} &= \frac{p+q}{p}B_{p+1,q}=\frac{p+q}{q}B_{p,q+1},
\end{aligned}$$

with the initial values $B_{1,q}=\int_0^1(1-x)^{q-1}\,dx=\frac{1}{q}$ and $B_{p,1}=\frac{1}{p}$. These properties imply

$$\begin{aligned}
B_{p,q} &= \frac{p}{p+q}B_{p-1,q}=\frac{p(p-1)}{(p+q)(p+q-1)}B_{p-2,q}=\cdots=\frac{p!q!}{(p+q)!}\\
B_{p,q} &= \frac{\Gamma_p\Gamma_q}{\Gamma_{p+q}}
\end{aligned}$$

and

$$B_{p+1,q+1}=\frac{(p+1)(q+1)}{(p+q+1)(p+q+2)}B_{p,q}.$$

The real function Beta

$$B_{x,y}=\frac{\Gamma_x\Gamma_y}{\Gamma_{x+y}},$$

with $B_{1,y}=\frac{1}{y}$ and $B_{x,1}=\frac{1}{x}$ for all real $x\geq 1$ and $y\geq 1$, is equivalently defined by the integral (7.1). It has the same recurrence properties as the integer function. The first partial derivatives B'_x and B'_y are similar

$$\begin{aligned}
B'_x(x,y) &= \Big(\frac{\Gamma'_x}{\Gamma_x}-\frac{\Gamma'_{x+y}}{\Gamma_{x+y}}\Big)B_{x,y},\\
B'_x-B'_y &= \Big(\frac{\Gamma'_x}{\Gamma_x}-\frac{\Gamma'_y}{\Gamma_y}\Big)B_{x,y}.
\end{aligned}$$

From the equality $\Gamma_{1+x}=x\Gamma_x$ we deduce

$$B_{x,y}=\frac{(x+y)(x+y+1)}{xy}B_{x+1,y+1}$$

for all $x>0$ and $y>0$, in particular $B_{x,x}=2B_{x,x+1}$ and the addition of variables induces a change of scale of the function.

The properties established for the integers generalize to all strictly positive real x and y. According to Euler, the function Beta is characterized by the additive-multiplicative property

$$\begin{aligned} B_{x,y}B_{x+y,z} &= \frac{\Gamma_x\Gamma_y\Gamma_z}{\Gamma_{x+y+z}} = B_{x,z}B_{x+z,y} \\ &= B_{y,z}B_{y+z,x}, \end{aligned} \tag{7.3}$$

for all strictly positive real x, y, z. For all integers $p > 0$ and $q > 0$

$$p\Gamma_{\frac{p}{q}} = q\Gamma_{\frac{p+q}{q}}, \tag{7.4}$$

for instance

$$\Gamma_{2+\frac{1}{2}} = \frac{3}{2}\Gamma_{1+\frac{1}{2}} = \frac{3}{4}\Gamma_{\frac{1}{2}}. \tag{7.5}$$

For every a in $]0,1[$, $\Gamma_a = k^{1-a}m_{ak-1;k}$ is calculated from (3.10) where $m_{ak-1;k}$ is the $(ak-1)$th moment of a variable with density proportional to $e^{-\frac{k^2}{2}}$ on $\mathbb{R}_+$, for every integer k such that $ak \geq 1$. From this expression, we get

$$\begin{aligned} B_{1+\frac{1}{p},\frac{1}{p}} &= \frac{1}{2}B_{\frac{1}{p},\frac{1}{p}}, \\ B_{1+\frac{1}{p},1+\frac{1}{p}} &= \frac{1}{2(2+p)}B_{\frac{1}{p},\frac{1}{p}}, \\ B_{1+\frac{1}{p},\frac{1}{q}} &= \frac{p+q}{q}B_{\frac{1}{p},\frac{1}{q}}, \\ B_{1+\frac{1}{p},1+\frac{1}{q}} &= \frac{pq}{(p+q)(pq+p+q)}B_{\frac{1}{p},\frac{1}{p}}. \end{aligned}$$

Values of the function Gamma with fractional indices are calculated from the moments of a variable X with density proportional to $e^{-\frac{x^k}{k}}$, by (3.10). Under the normal distribution, with $k = 2$ and denoting $2x = y^2$, we have $dx = \sqrt{2x}\,dy$ and

$$\begin{aligned} \Gamma_{\frac{1}{2}} &= \int_0^\infty x^{-\frac{1}{2}}e^{-x}\,dx \\ &= \sqrt{2}\int_0^\infty e^{-\frac{y^2}{2}}\,dy = \sqrt{\pi}, \\ \Gamma_{k+\frac{1}{2}} &= \frac{\sqrt{\pi}}{2^k}m_{2k} \end{aligned}$$

where $m_{2k} = E(X^{2k})$. Replacing m_{2k} by $2^{1-k}\Gamma_k^{-1}\Gamma_{2k}$ for $k \geq 0$ entails

$$\begin{aligned} \Gamma_k\Gamma_{k+\frac{1}{2}} &= \Gamma_{\frac{1}{2}}\Gamma_{2k}2^{1-2k}, \\ B_{k,\frac{1}{2}}^{-1}\,B_{k,k} &= 2^{1-2k}. \end{aligned} \tag{7.6}$$

The recurence formula provides another way to compute tables of $\Gamma_k\Gamma_{k+\frac{1}{2}}\Gamma_{2k}^{-1}$ and the values of $B_{p,\frac{q}{2}}$ and $B_{\frac{p}{2},\frac{q}{2}}$ are deduced, for all integers $p \geq 1$ and $q \geq 1$. For example, from (7.6) or (7.5)

$$\Gamma_{\frac{3}{2}} = \frac{\sqrt{\pi}}{2}, \quad \Gamma_{\frac{5}{2}} = 3\frac{\sqrt{\pi}}{4}$$

then

$$B_{\frac{1}{2},\frac{1}{2}} = \pi, \quad B_{\frac{3}{2},\frac{1}{2}} = \frac{\pi}{4},$$
$$B_{\frac{1}{2},\frac{5}{2}} = \frac{\pi}{8}, \quad B_{\frac{3}{2},\frac{5}{2}} = \frac{\pi}{2^6}.$$

Replacing $\frac{1}{2}$ by $\frac{1}{3}$, $\Gamma_{k+\frac{2}{3}} = 2^{-k+\frac{1}{3}+\frac{1}{2}}\sqrt{\pi}m_{2k+\frac{1}{3}}$ where the constant $m_{2k+\frac{1}{3}}$ can be computed numerically, replacing the even moments of a normal variable X by $m_a = E(|X|^a)$ for every real $a > 0$. The property (7.3) allows us to calculate values of the functions Gamma and Beta for all rational variables. The values of the function Beta calculated between consecutive integers are used to calculate its values on the other real intervals. The functions Beta and Gamma cannot be extended to negative variables from their definitions since the integrations by parts are not definite on $\mathbb{R}_-$. The extension to $\mathbb{R}_-$ of the recurrence formula for the function Gamma gives $\Gamma_{-\frac{1}{2}} = -2\sqrt{\pi}$ and $\Gamma_{-\frac{3}{2}} = \frac{4}{3}\sqrt{\pi}$ hence $B_{\frac{3}{2},-\frac{1}{2}} = -\pi$ and $B_{\frac{5}{2},-\frac{3}{2}} = \pi$.

The function $B_{x,1-x} = \Gamma_x\Gamma_{1-x}$ is defined on $]0,1[$ and

$$B_{x,1-x} = \frac{\Gamma_x\Gamma_{2-x}}{(1-x)} = \cdots = \frac{\Gamma_x\Gamma_{k-x}}{(1-x)\cdots(k-1-x)}.$$

Its value at $\frac{1}{2}$ is π and it is symmetric at $\frac{1}{2}$ due to its invariance under a permutation of its variables. It tends to infinity at zero and one (Table 1).

Table 1. Values of the function $B_{x,1-x}$ on $]0,.45[$.

.05	.1	.15	.2	.25	.3	.35	.4	.45
20.791	9.707	6.920	5.345	4.443	3.883	3.526	3.303	3.181

For every real x in $]1,2[$, denoting $s(1-z) = z$, i.e. $z = (1+s)^{-1}s$, we have $z^{-1}\,dz = s^{-1}(1+s)^{-1}\,ds$, $dz = (1+s)^{-2}\,ds$ and

$$B_{x,1-x} = \int_0^1 z^x(1-z)^{-x}z^{-1}\,dz$$
$$= \int_0^\infty \frac{s^{x-1}}{1+s}\,ds = 2\int_0^{\frac{\pi}{2}} \tan^{2x-1} s\,ds, \tag{7.7}$$

integrating by parts the integral of $s^{x-1}(1+s)^{-1}$ yields

$$B_{x,1-x} = \frac{1}{x}\int_0^\infty \frac{s^x\,ds}{(1+s)^2} = \frac{1}{x}\int_0^1 z^x(1-z)^{-x}\,dz$$

therefore

$$xB_{x,1-x} = B_{1+x,1-x}.$$

By a change of variable, it is also written as

$$B_{x,1-x} = 4\Gamma_{\frac{3}{2}}\int_0^1 y^{1-x}(1-y)^x\,dy.$$

The derivatives of $B_{1+x,1-x}$ in (7.7) are

$$B_{x,1-x}^{(k)} = \int_0^\infty \frac{s^{x-1}\log^k s}{1+s}\,ds,\ k \geq 1.$$

They extend to $x < 0$ and $x > 1$ from the definition of Γ by recurrence to $\mathbb{R}_-$. Legendre reported Euler's calculus $B_{a,1-a} = \pi\sin^{-1}(a\pi)$ on $]0,1[$.

The function Beta is computed on $\mathbb{R}_+^2$ using the recurrence formulas for an extrapolation of its values on intervals $(k, k+1)$ to every real interval. Equation (7.6) extends to every $x > 0$

$$\Gamma_x\Gamma_{x+\frac{1}{2}} = \Gamma_{\frac{1}{2}}\Gamma_{2x}2^{1-2x}$$

equivalently, as proved by Legendre

$$B_{x,\frac{1}{2}} = 2^{2x-1}B_{x,x}. \tag{7.8}$$

As n tends to infinity, the approximation

$$\Gamma_{x+n} \sim \Gamma_n n^x,\ x > 1 \tag{7.9}$$

proved in Exercise (7.8.7) implies

$$\begin{aligned} B_{n,x} &\sim n^{-x}\Gamma_x, \\ B_{n+x,x} &\sim B_{n,x}, \\ B_{n,n+x} &\sim 2^{-x}B_{n,n}. \end{aligned}$$

Eulerian integrals with fractional variables are defined with three indices, at fixed n

$$I_{p,q} = \int_0^1 \frac{x^{p-1}}{(1-x^n)^{1-\frac{q}{n}}}\,dx$$

for all integers p and q. By the change of variable $x^n = 1 - y^n$ on $[0,1]$, the integrals $I_{p,q}$ and $I_{q,p}$ are identical. From (7.2) and with the change of variable $x^n = y$, for all integers $n > 1$, p and q

$$I_{p,q} = \frac{1}{n} B_{\frac{p}{n},\frac{q}{n}}. \tag{7.10}$$

The question to find out all integers p and q such that $I_{p,q}$ have the same value at fixed n is then transposed to the function Beta. If $p > n$, the function I satisfies the recurrence formula

$$I_{p,q} = \frac{p-n}{p+q-n} I_{p-n,q},$$

and for every real $a > 0$

$$I_{a,n+1} = I_{n+1,a} = \frac{1}{a+1} I_{1,a}$$

this is equivalent to

$$I_{a,n+1} = \frac{1}{a+1} \Gamma_{\frac{a}{n}} \Gamma_{\frac{1}{n}} \Gamma^{-1}_{\frac{a+1}{n}}.$$

If p or q are larger than n and by the recurrence formula, the exponents of the integral are reduced to integers p and q smaller than or equal to n. Let $p = kn + p'$ with $p' > 0$

$$I_{p,q} = \prod_{j=1}^{k} \frac{p-nj}{p+q-nj} I_{p',q}.$$

In the same way, let $q = ln + q'$ with $q' > 0$

$$I_{q,p} = \prod_{j=1}^{l} \frac{q-nj}{p+q-nj} I_{q',p}$$

and by the permutation of the indices

$$I_{p,q} = \prod_{j=1}^{k} \frac{p-nj}{p+q-nj} \prod_{j=1}^{l} \frac{q-nj}{p+q-nj} I_{p',q'}.$$

In particular, from (7.10)

$$\begin{aligned} I_{nq+q',np+p'} &= \frac{1}{n} B_{p,q} && \text{if } p' = 0, q' = 0, \\ &= \frac{B_{p,p+q-k} B_{q,p+q-l}}{B_{p-k,p+q} B_{q-l,p+q}} I_{q',p'} && \text{if } 0 < p' < n, 0 < q' < n. \end{aligned}$$

The recurrence formulas written for the function Beta on $]0,1[$ can be used for I with p and $q < n$.

The function I is related to the functions B and Γ according to the value of n, for all strictly positive real x and y

$$\begin{aligned} I_{x,ny} &= B_{x,y}, \qquad \text{if } n = 1, \\ &= \frac{1}{n} B_{\frac{x}{n}-1,y}, \quad \text{if } n > 1. \end{aligned}$$

Replacing ny by y determines $I_{x,y}$ from the function $B_{x,\frac{y}{n}}$. By a change of variable, for all real x and y and for every integer $n \geq 1$

$$I_{x,x} = 2^{1-2n^{-1}x} B_{x,n},$$

the properties of the function I are still true replacing the integer n by a real $a > 0$. Equation (7.10) generalizes as

$$I_{ax,ay} = a^{-1} B_{x-1,y+1},$$

by a permutation of the indices, it implies

$$B_{x,1-x} = aI_{a(x+1),-ax} = aI_{a(x-1),a(2-x)},$$

for all real $x > 0$ and $a > 0$, the function I is therefore calculated on $\mathbb{R}_+^2$ from the function $B_{x,1-x}$ by a change of variable.

The function I generalizes to $\mathbb{R}^2$ like Beta and from (7.8)

$$\begin{aligned} I_{ax,ax} &= \frac{1}{a} B_{x-1,x+1} = \frac{x}{a(x-1)} B_{x,x} \\ &= \frac{x}{a2^{2x-1}(x-1)} B_{x,\frac{1}{2}} \\ &= \frac{x}{2^{2x-1}(x-1)} I_{a(x+1),-\frac{a}{2}} \end{aligned}$$

therefore

$$I_{ax,ax} = 2^{1-2x} \frac{x}{x-1} I_{\frac{3a}{2},a(x-1)}.$$

All properties of the functions Beta apply to the function $I_{ax,ay}$.

The integral function

$$\begin{aligned} L_{a,b} &= \int_0^1 \log^b\Big(\frac{1}{x}\Big) x^{a-1}\, dx \\ &= \int_0^\infty y^b e^{-ay}\, dy \end{aligned}$$

is a transformed function Gamma defined for every real $a > 0$, it has the property

$$L_{a,b} = \frac{1}{a^{b+1}} \Gamma_{b+1}.$$

For every real $a \neq 0$, $L_{a,0} = a^{-1}$ and for every real b

$$L_{1,b} = \int_0^1 \log^b\Big(\frac{1}{x}\Big)\,dx = \int_0^\infty x^b e^{-x}\,dx = \Gamma_{b+1}.$$

Its first derivatives are

$$L'_a(a,b) = -L(a,b+1),$$
$$L'_b(a,b) = bL(a,b-1)$$

and its higher derivatives are deduced.

7.2 Airy function

The Airy function A solution of the differential equation (4.16) in $\mathbb{R}^+$ is also defined as the unique solution of the implicit equation

$$A(x) = \int_0^x (s^{-1}x - 1)A(s)\,ds,$$

it has the value zero at zero. Another implicit equation of this function is obtained writing the first term of the differential equation $x^2u''_x - xu_x = 0$ as the derivative of a function since $x^2u''_x = (x^2u'_x)' - 2xu'_x$. The differential equation becomes $(x^2u'_x)' - 2xu'_x - xu_x = 0$ and its primitive is

$$x^2u'_x - 2\int_0^x s\,du_s - \int_0^x su_s\,ds = a$$

with a constant a. Its value at zero is $a = 0$ and, integrating by parts, this is equivalent to the implicit equation

$$A(x) = -2x^{-1}\int_0^x s\,dA(s) + \int_0^x \Big(x^{-1}s - 1\Big)A(s)\,ds,$$

hence

$$A(x) = \int_0^x \Big(1 + 2x^{-1} - x^{-1}s\Big)A(s)\,ds.$$

Assuming the existence of a series $A(x) = \sum_{n=0}^\infty a_n x^n$, Airy's differential equation (4.16) is equivalent to equations for the series

$$0 = \sum_{n=1}^\infty \{n(n-1)a_n x^{n-1} - a_n x^n\},$$

$$0 = \sum_{k=1}^\infty \{k(k+1)a_{k+1} - a_k\}x^k,$$

$$a_{k+1} = \frac{a_k}{k(k+1)},\quad k \geq 1,$$

with $a_0 = 0$ an arbitrary constant a_1. The series is convergent for every real x and the coefficients are defined from an arbitrary constant a_1 as

$$a_{k+1} = a_1\Big\{1 + \frac{1}{2} + \frac{1}{2^2.3} + \frac{1}{2^2.3^2.4} + \cdots + \frac{1}{2^2.3^2.4^2\cdots k^2(k+1)}\Big\}. \quad (7.11)$$

From (7.11), the generating function of A is solution of the first order differential equation

$$xu'_x + u = a_1\Big\{1 + x + \frac{x^2}{2} + \cdots + \frac{x^k}{2^2.3^2.4^2\cdots k^2} + \cdots\Big\}.$$

A parametric Airy function A_λ with parameter λ is solution of the differential equation

$$xu''_x - \lambda^2 u_x = 0$$

and it has the same behaviour. By the change of variable $v(x) = x^n A(x)$ with an integer n, the function v is solution of the differential equation

$$x^2 v''_x - 2nxv'_x + \{n(n+1) - \lambda^2 x)v_x = 0$$

and its limiting behaviour is similar as x increases. With an integer $n > 1$ and $A(x) = x^n v(x)$, the function v is solution of the differential equation

$$x^2 v''_x + 2nxv'_x + \{n(n-1) - \lambda^2 x\}v_x = 0.$$

By an expansion $A_\lambda(x) = \sum_{n=0}^\infty a_n x^n$, the differential equation for A_λ is equivalent to equations for the series

$$0 = \sum_{n=1}^\infty \{n(n-1)a_n x^{n-1} - \lambda^2 a_n x^n\},$$

$$a_{k+1} = \frac{\lambda^2 a_k}{k(k+1)},\ k \geq 1,$$

with $a_0 = 0$.

The series is convergent for every real x if $\lambda < 1$ and the coefficients are defined from an arbitrary constant a_1 as

$$a_{k+1} = a_1\Big\{1 + \frac{\lambda^2}{2} + \frac{\lambda^4}{2^2.3} + \cdots + \frac{\lambda^{2k}}{2^2.3^2.4^2\cdots k^2(k+1)}\Big\}.$$

7.3 Bessel's function

Bessel's equations have the form

$$x^2 u''_x + xu'_x + (x^2 - \alpha^2)u_x = 0,\ x \in \mathbb{R}_+.$$

Writing $x^2u''_x + xu'_x = x(xu'_x)'$, Bessel's equation is equivalent to

$$(xu'_x)' + (x - x^{-1}\alpha^2)u_x = 0.$$

Its integration gives an implicit equation

$$\begin{aligned} x_1u'_1 &= xu'_x + \int_x^{x_1} (s - s^{-1}\alpha^2)u_s\,ds \\ u_x &= u_0 + x_1u'_1 \log\frac{x}{x_0} - \int_{x_0}^{x} t^{-1} \int_t^{x_1} (s - s^{-1}\alpha^2)u_s\,ds\,dt \\ &= u_0 + x_1u'_1 \log\frac{x}{x_0} - \int_{x_0}^{x} \log\Big(\frac{s}{x_0}\Big)\,(s - s^{-1}\alpha^2)u_s\,ds \\ &\quad - \log\Big(\frac{x}{x_0}\Big) \int_x^{x_1} (s - s^{-1}\alpha^2)u_s\,ds. \end{aligned}$$

With $\alpha = 0$, Bessel's equation reduces to Fourier's equation

$$xu''_x + u'_x + xu_x = 0$$

and its solutions are Bessel's functions

$$\begin{aligned} J_0(x) &= 1 - \frac{x^2}{2^2} + \frac{x^4}{2^24^2} - \frac{x^6}{2^24^26^2} + \dots, \\ K_0(x) &= J_0(x)\log x + \frac{x^2}{2^2} - \frac{x^4}{2^24^2}\Big(1 + \frac{1}{2}\Big) + \frac{x^6}{2^24^26^2}\Big(1 + \frac{1}{2} + \frac{1}{3}\Big) + \dots \end{aligned}$$

and their linear combinations. Higher orders Bessel's functions have the same form with coefficients depending on the constant.

Denoting $u(x) = x^\alpha v(x)$, Byerly (1893) proved that Bessel's equation is equivalent to

$$xv''_x + (2\alpha + 1)v'_x + xv = 0$$

and he obtained an integral solution of Bessel's equation

$$J_\alpha(x) = \frac{x^\alpha}{2^\alpha\sqrt{\pi}\,\Gamma_{\alpha+\frac{1}{2}}} \int_0^\pi \sin^{2\alpha}\phi \cos(x\cos\phi)\,d\phi.$$

With integer constants, the functions J_n satisfy the differential equation $\{x^nJ_n(x)\}'_x = x^nJ_{n-1}(x)$, $n \geq 1$, and the recurrence formula

$$J_n(x) = \frac{x}{2n}\{J_{n-1}(x) + J_{n-1}(x)\},$$

with $J'_0(x) = -J_1(x)$. A function K_α is also solution of the differential equation with parameters α.

7.4 Boyd's function

Boyd's differential equation on $\mathbb{R}_*$

$$xu''_x + (1 - \lambda x)u_x = 0, \; 0 < \lambda < 1,$$

differs from the Airy equation by the coefficient of u depending on x. Assuming that it has a solution $B(x) = \sum_{n=0}^{\infty} b_n x^n$, Boyd's equation is equivalent to the algebraic equations

$$0 = \sum_{n=1}^{\infty} \{n(n-1)b_n x^{n-1} + (1 - \lambda x)b_n x^n\},$$

$$0 = \sum_{k=1}^{\infty} \{k(k+1)b_{k+1} + b_k - \lambda b_{k-1}\}x^k, k \geq 2,$$

with $b_0 = 0$ and b_1 arbitrary. They have an unique solution

$$b_{k+1} = \frac{\lambda b_{k-1} - b_k}{k(k+1)}$$

so $b_2 = -\frac{1}{2}b_1$, $b_3 = \frac{1}{6}b_1(\frac{1}{2} + 2\lambda)$, and so on. The coefficients are alternatively positive and negative, the sign b_k being $(-1)^{k+1}\text{sign}(b_1)$, and $|b_k|$ is decreasing, the series is convergent for every real $x > 0$. As k tends to infinity

$$b_{k+1} \sim \frac{b_{k-1}}{k(k+1)} \sim \frac{b_{k-1}}{k^2},$$

the coefficients are therefore equivalent to the coefficients of the Airy function. The solution $B(x)$ of the differential equation converges to zero at zero and it goes to infinity at infinity, with a sign depending on the ratio of the series $B_1(x) = \sum_{k=0}^{\infty} b_{2k+1}x^{2k+1}$ and $B_2(x) = \sum_{k=1}^{\infty} b_{2k}x^{2k}$. By the asymptotic equivalence of the coefficients b_{k+1}, the expansion of $B(x) = \lim_{k\to\infty} B_k(x)$ is such that $B - B_k$ is asymptotically equivalent the sum of the corresponding terms in the expansion of Bessel's function J_0.

7.5 Hermite's function

Hermite's equations have the form

$$u''_x - 2xu'_x + 2\alpha^2 u_x = 0, \; x \in \mathbb{R}. \tag{7.12}$$

An implicit equation equivalent to (7.12) is obtained by integration by parts of $u'_x = u'_0 + 2\int_0^x s\,du_s - 2\alpha^2 \int_0^x u_s\,ds$

$$\begin{aligned} u_x &= u_0 + xu'_0 + 2\int_0^x \int_0^t s\,du_s\,dt - 2\alpha^2 \int_0^x \int_0^t u_s\,ds\,dt \\ &= u_0 + xu'_0 + 2\int_0^x (x-s)s\,du_s \\ &\quad - \alpha^2\Big\{x^2 u_x - \int_0^x t^2\,du_t - 2\int_0^x (x-s)s\,du_s\Big\} \end{aligned}$$

therefore the solution of Hermite's differential equation is also solution of an integral equation defining u_x from u'_x under the initial conditions

$$u_x = \frac{u_0 + xu'_0}{1+\alpha^2 x^2} + 2\frac{1+\alpha^2}{1+\alpha^2 x^2} \int_0^x (x-s)s\,du_s$$

for every real x. The implicit equations are not unique. Using the reparametrization technique (4.15) with $x_t = e^{-t}$, Hermite's differential equation is equivalent to

$$y''_t + (1+2e^{-2t})y'_t + 2\alpha^2 e^{-2t} y_t = 0.$$

Denoting $y_t = e^{h_t}$, it becomes

$$h''_t + h'^2_t + (1+2e^{-2t})h'_t + 2\alpha^2 e^{-2t} = 0. \tag{7.13}$$

The solution of the homogeneous equation $h''_t + h'^2_t + (1+2e^{-2t})h'_t = 0$ satisfies

$$\begin{aligned} h'_t e^{h_t} &= k_1 \exp\{e^{-2t} - t - 1\}, \\ e^{h_t} &= k_2 \int_0^t \exp\{e^{-2s} - s\}\,ds \end{aligned}$$

with constants k_j determined by boundary conditions. The whole equation (7.13) has a solution h_t such that $y_t = e^{h_t} = u_t e^{h_{1t}}$ with h_1 the solution of the homogeneous equation and u satisfying the nonlinear differential equation

$$u''_t + 2u'_t + b_t u_t = 0$$

where $b_t = 2\alpha^2 e^{-2t} e^{-h_{1t}}$ is a known function.

Assuming that $u(x) = \sum_{n=0}^{\infty} a_n x^n$, the solution of Hermite's differential equation is equivalent to equations for the series

$$\begin{aligned} 0 &= \sum_{n=1}^{\infty} \{n(n-1)a_n x^{n-2} - 2na_n x^n + 2\alpha^2 a_n x^n\}, \\ 0 &= \sum_{k=1}^{\infty} \{(k+1)(k+2)a_{k+2} - 2(k-\alpha^2)a_k\}x^k, \end{aligned}$$

therefore

$$a_{k+2} = a_k \frac{2(k-\alpha^2)}{(k+1)(k+2)}$$

for every $k \geq 0$, with arbitrary constants a_0 and a_1. The coefficients define two series, S_0 for the sum of the even coefficients and S_1 for the sum of the odd coefficients

$$\begin{aligned} S_0(x) &= 1 - \alpha^2 x^2 \Big\{ 1 + x^2 \frac{2-\alpha^2}{6} + x^4 \frac{(2-\alpha^2)(4-\alpha^2)}{6.15} \\ &\quad + x^6 \frac{(2-\alpha^2)(4-\alpha^2)(6-\alpha^2)}{6.15.28} \\ &\quad + x^8 \frac{(2-\alpha^2)(4-\alpha^2)(6-\alpha^2)(8-\alpha^2)}{6.15.28} + \cdots \Big\}, \\ S_1(x) &= x \Big\{ 1 + x^2 \frac{1-\alpha^2}{3} + x^4 \frac{(1-\alpha^2)(3-\alpha^2)}{3.10} \\ &\quad + x^6 \frac{(1-\alpha^2)(3-\alpha^2)(5-\alpha^2)}{3.10.21} + \cdots \Big\}. \end{aligned}$$

The series $S_0(x)$ and $S_1(x)$ are convergent for every x in $[-1,1]$ and the solutions of Hermite's differential equation are the linear combinations $u(x) = a_0 S_0(x) + a_1 S_1(x)$ with arbitrary constants a_0 and a_1.

7.6 Laguerre's functions

Laguerre's differential equation

$$x u_x'' + (1-x) u_x' + \lambda u_x = 0$$

is equivalent to $(x u_x')' - x u_x' + \lambda u_x = 0$ with $x \geq 0$. It is integrated as

$$x u_x' - \int_0^x s \, du_s + \lambda \int_0^x u_s \, ds = 0$$

and it is equivalent to the implicit equation

$$u_x = u_0 + \int_0^x t^{-1} \int_0^t s \, du_s \, dt - \lambda \int_0^x t^{-1} \int_0^t u_s \, ds \, dt.$$

By an expansion of its solution as $u(x) = \sum_{n=0}^\infty a_n x^n$, Laguerre's differential equation is equivalent to

$$\begin{aligned} 0 &= \sum_{n=0}^\infty \{ n(n-1) a_n x^{n-1} + n a_n (1-x) x^{n-1} + \lambda a_n x^n \}, \\ 0 &= \sum_{k=0}^\infty \{ (k+1)^2 a_{k+1} - (k-\lambda) a_k \} x^k, \end{aligned}$$

therefore

$$a_{k+1} = a_k \frac{k-\lambda}{(k+1)^2}$$

for every $k \geq 0$, with arbitrary constants a_0. As k tends to infinity, a_{k+1} is equivalent to $k^{-1}a_k$ and the sum $a_k + a_{k+1} + \cdots + a_{k+n}$ is equivalent to $S_{k,n} = a_k(k-1)^{-1}(k-k^{-n})$. Since $S_k = \sum_{j=k}^{\infty} a_j = a_k(k-1)^{-1}k \sim a_k$ as k tends to infinity, they define a convergent series for every x in $]-1,1[$. This proves the existence of a solution and provides an algorithm for computing it. All coefficients a_k such that $k > \lambda$ have the same sign and the signs of a_{k+1} and a_k are the opposite if $k < \lambda$, moreover all coefficients are zero after an integer $\lambda = k$. Then u_x tends to infinity as n tends to infinity, and the sign of u at the infinity is $(-1)^{k+1}$ if $a_0 > 0$ and $k < \lambda \leq k+1$.

The differential equation

$$\frac{d}{dx}(x^{\alpha+1}e^{-x}u'_x) + \lambda x^{\alpha}e^{-x}u = 0,$$

defined another Laguerre's function with real coefficients $\alpha > 0$ and $\lambda > 0$. It is equivalent to

$$xu''_x + (\alpha+1-x)u'_x + \lambda u_x = 0$$

and a series expansion of its solution as $u(x) = \sum_{n=0}^{\infty} a_n x^n$ yields a convergent series for every x in $]-1,1[$, with coefficients

$$a_{n+1} = \frac{a_n}{n+1} \cdot \frac{n-\lambda}{n+\alpha+1}, \quad n \geq 2,$$

$$a_1 = a_0 \frac{\lambda}{\alpha+1}$$

and its asymptotic behaviour is similar to the previous case.

7.7 Hydrogen atom equation and others

The hydrogen atom equation on $\mathbb{R}_+$ is

$$u''_x - \Big(\frac{k}{x} + \frac{h}{x^2} - \lambda\Big)u_x = 0$$

with independent real parameters k and $h > -\frac{1}{4}$ and λ arbitrary (Everitt, 2005). Let a, $b > 0$ and $c > 0$ be real constants, the function $u_x = ax^{-b}e^{-cx}$ is solution of the equation

$$u''_x - \Big(\frac{2bc}{x} + \frac{b(b+1)}{x^2} + c^2\Big)u_x = 0$$

with fixed parameters depending only on c and b such that $k = 2bc$ and $h = b(b+1)$ satisfy the constraint, it tends to zero as x tends to infinity. Other solutions of the same equation are $u_x = a_2 x^{-b} e^{cx}$ and linear combinations of the hyperbolic functions divided by x^b. This function is solution of the equation with parameter λ determined by the other parameters.

Another solution of this equation with is obtained with an expansion of u as $u_x = \sum_{n \geq 0} a_n x^n$, for every real x. The coefficients are solutions of the equation

$$\sum_{n \geq 2} \{n(n-1)a_n x^n - nka_n x^{n+1} - ha_n x^{n+2} + \lambda a_n x^n = 0$$

with $a_0 = a_1 = 0$. Under the condition $h = 2$, this equation has a non trivial solution depending on an arbitrary constant a_2

$$a_n = \frac{ka_{n-1} + \lambda a_{n-2}}{n^2 - n + h}, \quad n \geq 2.$$

These coefficients define a convergent series which differs from the hyperbolic solution. As n tends to infinity, the coefficients are asymptotically free from the parameter h and they depend linearly on the two previous coefficients.

7.8 Exercises

7.8.1. Prove $B_{\frac{1}{2},x} = 2^{2x-1} B_{x,x}$ by a change of variable.

7.8.2. Prove (7.10).

7.8.3. Prove that $B_{x,k-x}$ converges to zero as k tends to infinity, for every real x in $]0,1[$.

7.8.4. Calculate $B^{(k)}_{x,1-x}$, for $k \geq 1$.

7.8.5. Let $\varphi = \log \Gamma$, prove that for all $x > 0$ and y in $]0,1[$

$$\frac{\varphi(x+y) - \varphi(x)}{y} \leq \varphi(x+1) - \varphi(x)$$

and $\Gamma'_x \leq \log(x+1)\Gamma_x$.

7.8.6. Let $U_n(x) = x\log n + \log B_{n,x}$ and $g_n(x) = \sum_{k=2}^{n} U_k(x) - \log x$, prove

$$g_n(x) - x\log\frac{n}{n-1} \leq \log\Gamma_x \leq g_n(x)$$

for every x in $]0,1[$ and the convergence of $g_n(x)$ to $\log\Gamma_x$. Deduce the approximation (7.9) as n tends to infinity.

7.8.7. Prove the expansion $\Gamma^{-1}(x) = xe^{\gamma x}\prod_{n=1}^{\infty}\left\{\left(1+\frac{x}{n}\right)e^{-\frac{x}{n}}\right\}$, with Euler's constant $\gamma = \lim_{n\to\infty}\gamma_n = \sum_{k=1}^{n} k^{-1} - \log n$, and the convergence of the expansion.

7.8.8. Calculate the derivative of $\log\Gamma$ from (7.8.7).

7.8.9. Write recurrences for Euler's function $L_{x,y}$.

7.8.10. Prove

$$I_1 = \int_0^1 \frac{y^{a-1} - y^{n-a-1}}{1-y^n}\,dy = \frac{\pi}{n}\cot\frac{a\pi}{n}$$

aand deduce the values of

$$I_2 = \int_0^1 \frac{y^{a-1} + y^{n-a-1}}{1-y^n}\log\frac{1}{y}\,dy, \quad I_3 = \int_0^1 \frac{y^{a-1} + y^{n-a-1}}{1-y^n}\log^2\frac{1}{y}\,dy.$$

7.8.11. Solve Littlewood-Mclead's equation $u''_x - (x\sin x - \lambda)u_x = 0$.

7.8.12. Solve $(x+a)y''_x + \{2 - (x+a)^{\frac{1}{2}}\}y'_x = 0$, for all strictly positive and distinct real x and a.

7.8.13. Solve the differential equation $(1-x^2)y''_x - xy'_x + a^2y_x = 0$ on $[0,1]$ using a Fourier series for y.

Chapter 8

Solutions

In this chapter the title of the sections are those of the chapters where the exercises were given and the references of the exercises are the same.

8.1 Integral and differential calculus

1.6.1. The function is homogeneous in x and y, with

$$u(x, y) = x^n \Big\{ \varphi\Big(\frac{y}{x}\Big) + \Big(\frac{y}{x}\Big)^n \psi\Big(\frac{y}{x}\Big) \Big\}$$

and its partial derivatives are

$$\begin{aligned} u'_x(x, y) &= n x^{-1} u - y x^{n-2} \Big\{ \varphi'\Big(\frac{y}{x}\Big) + \Big(\frac{y}{x}\Big)^n \psi'\Big(\frac{y}{x}\Big) \Big\} \\ &\quad - n y^n x^{-1} \psi\Big(\frac{y}{x}\Big), \\ u'_y(x, y) &= x^{n-1} \Big\{ \varphi'\Big(\frac{y}{x}\Big) + \Big(\frac{y}{x}\Big)^n \psi'\Big(\frac{y}{x}\Big) \Big\} + n y^{n-1} \psi\Big(\frac{y}{x}\Big), \end{aligned}$$

which implies $x u'_x + y u'_y = n u$.

1.6.2. The partial derivatives of u are

$$\begin{aligned} u'_x(x, y, z) &= \Big(\frac{1}{y} - \frac{z}{x^2}\Big) \varphi'\Big(\frac{x}{y}, \frac{y}{z}, \frac{z}{x}\Big), \\ u'_y(x, y, z) &= \Big(\frac{1}{z} - \frac{x}{y^2}\Big) \varphi'\Big(\frac{x}{y}, \frac{y}{z}, \frac{z}{x}\Big), \\ u'_z(x, y, z) &= \Big(\frac{1}{x} - \frac{y}{z^2}\Big) \varphi'\Big(\frac{x}{y}, \frac{y}{z}, \frac{z}{x}\Big), \end{aligned}$$

therefore $x u'_x + y u'_y + z u'_z = 0$.

1.6.3. From the partial derivative

$$u'_y(x,y) = a\{\varphi'(ay+bx)\psi(ay-bx) + \varphi(ay+bx)\psi'(ay-bx)\},$$

it follows $u(x,y) = u_0 \exp(\int_0^y u^{-1}u'_y)$.

1.6.4. The partial derivatives of r are

$$u'_x = a\varphi'(ax+cz) = a\psi'(ax-by),$$
$$u'_y = -b\psi'(ax-by),$$
$$u'_z = c\varphi'(ax+cz),$$

which imply $2a^{-1}u'_x + b^{-1}u'_y - c^{-1}u'_z = 0$.

1.6.5. The partial derivatives of u^2 are

$$2uu'_x = y - \frac{y}{x^2}\varphi'\Big(\frac{y}{x}\Big), \quad 2uu'_y = x + \frac{1}{x}\varphi'\Big(\frac{y}{x}\Big),$$

they satisfy the differential equation $(xu'_x + yu'_y)u = xy$.

1.6.6. A solution $u(x,y,z)$ constant has the derivatives $u'_x = x+y$, $u'_y = x+z$, $u'_z = x+y$ then $u(x,y,z) = xy+xz+yz$ is a solution.

1.6.7. Since $\{u(1-u)\}^{-1} = u^{-1} + (1-u)^{-1}$, the differential equation is equivalent to

$$\frac{y'}{y} + \frac{y'}{1-y} = 1,$$
$$\frac{y(1-y_0)}{y_0(1-y)} = e^{t-t_0}$$

and y is the logistic function

$$y_x = \frac{k}{k+e^{-(t-t_0)}}, \qquad k = \frac{y_0}{1-y_0}.$$

1.6.8. $y_x = \tan(ax)$, $y_x = \tan(ax) + (2a)^{-1}b$, $y(x) = x^{-1}\tan x$.

1.6.9. The solution must be a linear combination of $\cos x$ and $\sin x$, therefore $y(x) = \frac{1}{2}(\cos x + \sin x)$.

1.6.10. The derivative of $u(x) = (1-x^2)y^{\frac{1}{2}}(x)$ satisfies

$$y'_x - 4\frac{xy}{1-x^2} = \frac{2y^{\frac{1}{2}}}{1-x^2}u'(x),$$

$$y'_x + \frac{xy}{1-x^2} = \frac{2y^{\frac{1}{2}}}{1-x^2}u'(x) + 5\frac{xy}{1-x^2} = xy^{\frac{1}{2}},$$

$$u'(x) = \frac{1}{2}x(1-x^2) - \frac{5}{2}xy^{\frac{1}{2}} = \frac{1}{2}x(1-x^2) - \frac{5x}{2(1-x^2)}u(x).$$

The solution of the equation $2(1-x^2)u'_x + 5xu = 0$ is the real or complex root $z(x) = c(x^2-1)^{\frac{5}{4}}$ and $u = zv$ satisfies $2z(x)v'(x) = x(1-x^2)$ i.e. $2cv'(x) = -x(x^2-1)^{-\frac{1}{4}}$ hence $2cv(x) = -\frac{2}{3}(x^2-1)^{\frac{3}{4}}$ and

$$u = -\frac{1}{3}(x^2-1)^2 = (1-x^2)y^{\frac{1}{2}}(x), \qquad y(x) = \frac{1}{9}(x^2-1)^2.$$

1.6.11. By subtraction of the differential equations at a and b

$$\begin{aligned}
0 &= \{(x+a)-(x+b)\}y''_x - \{(x+a)^{\frac{1}{2}} - (x+b)^{\frac{1}{2}}\}y'_x,\\
0 &= \{(x+a)^{\frac{1}{2}} + (x+b)^{\frac{1}{2}}\}y''_x - y'_x.
\end{aligned}$$

Let $y'_x = u\{(x+a)^{\frac{1}{2}} - (x+b)^{\frac{1}{2}}\}$, then

$$\begin{aligned}
y''_x &= \frac{u}{2}\{(x+a)^{-\frac{1}{2}} - (x+b)^{-\frac{1}{2}}\} + \frac{u'_x}{u}y'_x,\\
0 &= \frac{u}{2}\left\{\frac{(x+b)^{\frac{1}{2}}}{(x+a)^{\frac{1}{2}}} - \frac{(x+a)^{\frac{1}{2}}}{(x+b)^{\frac{1}{2}}}\right\} + \{(x+a)-(x+b)\}u'_x\\
&\quad - \{(x+a)^{\frac{1}{2}} - (x+b)^{\frac{1}{2}}\}u\\
0 &= \{(x+a)^{\frac{1}{2}} + (x+b)^{\frac{1}{2}}\}u'_x - u - \frac{u}{2}\{(x+a)^{-\frac{1}{2}} + (x+b)^{-\frac{1}{2}}\}.
\end{aligned}$$

Since the above equalities are valid for all a and b, the last equality is also true for b and a third constant c, and by subtraction

$$\begin{aligned}
0 &= \{(x+a)^{\frac{1}{2}} - (x+c)^{\frac{1}{2}}\}u'_x - \frac{u}{2}\{(x+a)^{-\frac{1}{2}} - (x+c)^{-\frac{1}{2}}\},\\
0 &= 2\{(x+a)^{\frac{1}{2}}(x+c)^{\frac{1}{2}}\}u'_x - u,\\
u_x &= (x+a)^{\frac{1}{2}} - (x+c)^{\frac{1}{2}}.
\end{aligned}$$

Choosing $b = c$

$$\begin{aligned}
y'_x &= (x+a) + (x+b) - 2(x+a)^{\frac{1}{2}}(x+b)^{\frac{1}{2}},\\
y_x &= \frac{1}{2}\{(x+a)^2 + (x+b)^2\} - 2I,\\
I &= \int_0^x (s+a)^{\frac{1}{2}}(s+b)^{\frac{1}{2}}\,ds = \int_0^x \left\{\left(s+\frac{a+b}{2}\right)^2 - k^2\right\}^{\frac{1}{2}} ds,\\
k &= \frac{a-b}{2} < s + \frac{a+b}{2},\\
I &= k\int_0^{y_x} (y^2-1)^{\frac{1}{2}}\,dy = k\int_0^{u_x} \sinh^2 u\,du = \frac{k}{2}\int_0^{u_x} \{\cosh(2u) - 1\}\,du\\
&= \frac{k}{4}\{\sinh(2\operatorname{arg\,cosh} y_x) - 2\operatorname{arg\,cosh} y_x\},
\end{aligned}$$

where $y_x = k^{-1}\{x + \frac{1}{2}(a+b)\} = \cosh u_x$.

1.6.12. Deriving $y = u(a+bv)^{-1}$ entails

$$y'_x = \frac{u'_x}{a+bv} - \frac{buv'_x}{(a+bv)^2} = \frac{u'_x(a+bv) - buv'_x}{(a+bv)^2} = \frac{ay}{ax+bx^{n+1}}.$$

The assumption $u'_x v - uv'_x = 0$ implies $v = cu$ and $y = u(a+bcu)^{-1}$ with an arbitrary constant c

$$y'_x = \frac{au'_x}{(a+bcu)^2} = \frac{au'_x y^2}{u^2} = \frac{ay}{ax+bx^{n+1}}$$

since it does not depend on c, one can choose $c = 1$. This equality is true if y is replaced by y^n and with $u = x^n$, then the first equality becomes $y^{-(n+1)}y'_x = ax^{-(n+1)}$. Its primitive is

$$y_x = \left(\frac{x^n}{a+bx^n}\right)^{\frac{1}{n}}.$$

1.6.13. $x^2 + x^{-2} - x^{-1}y^2 = k$, constant.

1.6.14. $F(x) = x\arcsin(ax) + a^{-2}(1-a^2x^2)^{\frac{1}{2}}$.

1.6.15. $F(x) = x\arctan(ax) - \log(1+a^2x^2)^{\frac{1}{2a^2}}$.

1.6.16. Denoting $u_x = \sin x$, the primitive of f is

$$F(x) = \int_0^{\sin x} \frac{du}{a^2+u^2(b^2-a^2)} = \frac{1}{a^2}\int_0^{\sin x} \frac{du}{1+u^2a^{-2}(b^2-a^2)}.$$

If $b^2 > a^2$, $F(x) = a^{-2}\arctan\{a^{-1}\sqrt{b^2-a^2}\sin x\}$. If $b < a$, F is the primitive of the function $(1-c^2u^2)^{-1} = (1+cu)^{-1} + uc(1-cu)^{-1}$ where $c^2 = a^{-2}(a^2-b^2)$, and

$$F(x) = \frac{1}{a\sqrt{a^2-b^2}}\log\frac{a+\sqrt{a^2-b^2}\sin x}{\{a^2-(a^2-b^2)\sin^2 x\}^{\frac{1}{2}}}.$$

1.6.17. The primitive of $e^{ax}\cos(nx)$, $a > 0$, is

$$F(x) = \frac{e^{ax}}{a^2+n^2}\{a\cos(nx) + n\sin(nx)\}.$$

1.6.18. The primitive of $e^{ax}\sin(nx)$, $a > 0$, is

$$F(x) = \frac{e^{ax}}{a^2+n^2}\{a\sin(nx) - n\cos(nx)\}.$$

1.6.19. Writing $t_x^2 = 1 \pm 2xz + x^2 = (x \pm z)^2 + 1 - z^2$ and $u^2 = 1 - z^2$, at fixed z, it follows that $t\,dt = (x \pm z)\,dx = (t_x^2 - u^2)^{\frac{1}{2}}\,dx$ and

$$F(x) = \int_0^x \frac{ds}{t_s} = \int_0^{t_x} \frac{dt}{(t^2-u^2)^{\frac{1}{2}}} = \frac{1}{u}\int_0^{t_x} \frac{dt}{\{(u^{-1}t)^2-1\}^{\frac{1}{2}}},$$

$$= \int_1^{u^{-1}t_x} \frac{dy}{(y^2-1)^{\frac{1}{2}}},$$

with the notation $y = u^{-1}t > 1$. Since the primitive of $(s^2 - 1)^{-\frac{1}{2}}$ is $\arg\cosh s$

$$F_z(x) = \arg\cosh(u^{-1}t_x) = \arg\cosh\Big(\frac{1 \pm 2xz + x^2}{1 - z^2}\Big)^{\frac{1}{2}}.$$

1.6.20. Since $x^2 + 6x + 10 = (x+3)^2 + 1$, the primitive is

$$F(x) = \int_0^x \frac{dt}{\{(t+3)^2 + 1\}^{\frac{1}{2}}} = \arg\sinh(x+3),$$

using the equality $\cosh^2 y - \sinh^2 y = 1$.

1.6.21. The first derivatives of (1.15) are $x'_t = -k \sin t \ \cos^{k-1} t$ and $y'_t = k cost \ \sin^{k-1} t$. The second derivatives are

$$\begin{aligned} x''_t &= k(k-1)\sin^2 t \ \cos^{k-2} t - k\cos^k t, \\ y''_t &= k(k-1)\cos^2 t \ \sin^{k-2} t - ksin \sin^k t \end{aligned}$$

and Figure (8.1) is their graph with $k = 3$.

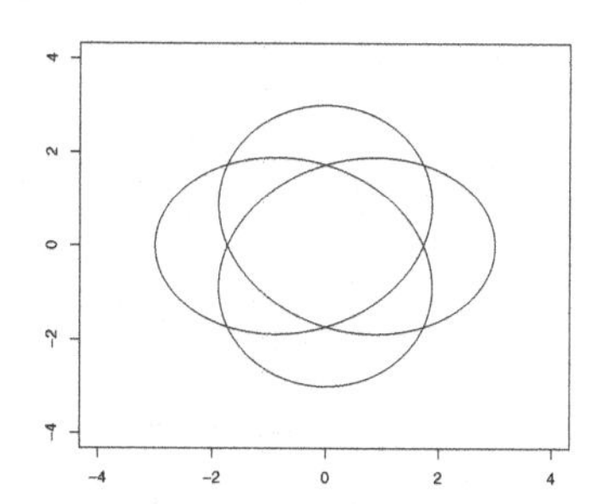

Fig. 8.1 Graph of a leaf-like hypocycloïd.

1.6.22. Let $v(x) = (a^2 + x^2)^{\frac{1}{2}}$ and $z_x^n = x + v(x)$, then

$$\begin{aligned} x &= \frac{z^{2n} - a^2}{2z^n}, \\ dx &= n\frac{z^{2n} + a^2}{2z^{n+1}}\, dz, \\ v(x) &= z^n - x = \frac{z^{2n} + a^2}{2z^n}, \end{aligned} \tag{8.1}$$

and $I = \int_0^x v^{-1}(x)\, dx = n\int_{z_0}^{z_x} z^{-1}\, dz = \log z_x^n - \log a$.

1.6.23. With the same notations and $x_0 > 0$

$$f(x) = \frac{z^n}{x} - 1 = \frac{z^{2n} + a^2}{z^{2n} - a^2},$$

$$F(x) - F(x_0) = n \int_{z_0}^{z_x} \frac{(z^{2n} + a^2)^2}{2z^{n+1}(z^{2n} - a^2)}\, dz,$$

$$= \frac{n}{2} \int_{z_0}^{z_x} \left\{ z^{n-1} + \frac{3a^2}{z^{n+1}} + \frac{4a^4}{z^{n+1}(z^{2n} - a^2)} \right\} dz,$$

$$F(x) = \frac{z_x^n}{2} - \frac{3a^2}{2z_x^n} + a^3 I_x,$$

$$I_x - I_{x_0} = 2an \int_{z_0}^{z_x} \left\{ \frac{z^{n-1}}{z^{2n} - a^2} - \frac{a}{z^{n+1}} \right\} dz,$$

$$I_x = \frac{2}{az_x^n} + \log \frac{z_x^n - a}{z_x^n + a},$$

finally, with $z_x^n = x + (a^2 + x^2)^{\frac{1}{2}}$

$$F(x) = \frac{1}{2} \left\{ z_x^n + \frac{a^2(2a^2 - 3)}{z_x^n} + a^4 \log(2xz_x^n) \right\}.$$

1.6.24. Let $t = \sin u$, so

$$\int_0^x (1 - t^2)^{\frac{1}{2}}\, dt = \int_0^x (1 - \sin u^2)^{\frac{1}{2}} \cos u\, du$$

$$= \int_0^x \cos^2 u\, du,$$

then integrating by parts with $f(x) = \sin t$ and $g(x) = \cos t$ entails

$$\int_0^x \cos^2 t\, dt = \sin x \cos x + \int_0^x \sin^2 t\, dt$$

$$= x - \int_0^x \cos^2 t\, dt + \sin x \cos x$$

$$\int_0^x \cos^2 t\, dt = \frac{1}{2}\{x + \sin x \cos x\}$$

therefore $\int_0^x (1 - t^2)^{\frac{1}{2}}\, dt = \frac{1}{2}\{\arccos x + x(1 - x^2)^{\frac{1}{2}}\}$.

1.6.25. By the same arguments as for Ex.1.6.22, we obtain

$$\int_0^x (1 + t^2)^{\frac{1}{2}}\, dt = \frac{1}{2}\{x(1 + x^2)^{\frac{1}{2}} + \operatorname{arc\,sinh} x\}.$$

1.6.26. Using (8.1) and with $z^{2n} = x + (a^2 + x^2)^{\frac{1}{2}}$

$$f(x)\,dx = \frac{n}{4}\Big(z^{2n-1} - \frac{2a^2}{z} + \frac{a^4}{z^{2n-1}}\Big)\,dz,$$
$$F(x) = \frac{1}{8}\Big(z_x^{2n} - \frac{a^4}{z_x^{2n}} - 2a^2 \log z^{2n}\Big).$$

1.6.27. Let $F(x) = P_3(x)(1-x^2)^{-\frac{3}{2}} + P_2(x)(1-x^2)^{-\frac{1}{2}}$ where P_3 is a third degree polynomial and P_1 is a first degree polynomial

$$f(x) = \frac{2P_3'(1-x^2) + 3xP_3}{2(1-x^2)^{\frac{5}{2}}} + \frac{2P_1'(1-x^2) + xP_1}{2(1-x^2)^{\frac{3}{2}}},$$
$$F(x) = -\frac{6 - 3x - 6x^2 + x^3}{6(1-x^2)^{\frac{3}{2}}} + \frac{2+x}{2(1-x^2)^{\frac{1}{2}}}.$$

1.6.28. Let the primitive be the ratio of a fifth degree polynomial and $(1+x^2)^{-\frac{5}{2}}$ and derivativing it, we get

$$F_{\frac{7}{2}}(x) = \frac{x(15 + 20x^2 + 8x^4)}{15(1+x^2)^{\frac{5}{2}}}.$$

1.6.29. $I = \frac{\pi}{2} \log 2$.

1.6.30. The differential equation is equivalent to

$$y_x'(1 + \tan^2 y) = a(1 + \tan^2 x)$$

hence $\tan y = a \tan x$ and $y(x) = \arctan(a \tan x)$.

1.6.31. Writing $\gamma_n = \sum_{k=1}^n k^{-1} - \log n = \sum_{k=1}^n \{k^{-1} - \log(1 + k^{-1})\} + \log(1 + n^{-1}) = \sum_{k=1}^n 0(k^{-2}) + \log(1 + n^{-1})$ proves its convergence.

1.6.32. By an expansion of $\log(1+x)$

$$\int_0^1 \log(1+s)\,ds = \sum_{k=1}^\infty k^{-1} \int_0^1 (-s)^k\,ds = \sum_{k=1}^\infty \frac{(-1)^k}{k(k+1)}.$$

Moreover the primitive of $\log(1+x)$ is $(1+x)\log(1+x) - x$ therefore $\sum_{k=0}^\infty (-1)^k \{k(k+1)\}^{-1} = 2\log 2$. Similarly

$$\sum_{k=1}^\infty (-1)^k k^{-2} = -\sum_{k=1}^\infty k^{-1} \int_0^1 (-s)^{k-1}\,ds$$
$$= \int_0^1 s^{-1} \log(1+s)\,ds = \frac{\pi^2}{12}.$$

1.6.33. As previously, we have

$$\begin{aligned}\zeta(2) &= \sum_{k=1}^{\infty} k^{-1} \int_0^1 s^{k-1}\, ds = \int_0^1 s^{-1} \log(1-s)\, ds \\ &= \int_0^{\infty} \frac{y\, dy}{e^y - 1} = \frac{\pi^2}{6}, \\ \zeta(3) &= \int_0^1 x^{-1} \int_0^x \sum_{k=1}^{\infty} k^{-1} s^{k-1}\, ds\, dx \\ &= \int_0^1 x^{-1} \int_0^x s^{-1} \log(1-s)\, ds\, dx = \int_0^1 s^{-1} \log(1-s) \log s\, ds \\ &= \int_0^{\infty} s \log(1 - e^{-s})\, ds\end{aligned}$$

and the function ζ expressed in the same form by successive integrations $\zeta(k) = \int_0^{\infty} s^{k-2} \log(1-e^{-s})\, ds$ for every $k \geq 3$. This integral generalizes replacing the integer k by a to real or complex variable.

8.2 Orthogonal polynomials

2.9.1. By a change of variable, for all real y and integer $n \geq 1$

$$EL_n(X-y) = \int_y^{\infty} L_n(x-y) e^{-x}\, dx = e^{-y} \int_0^{\infty} L_n(x) e^{-x}\, dx$$

and $\int_0^{\infty} L_n(x) e^{-x}\, dx = 0$ since L_0 and L_n are orthogonal.

2.9.2. The function $u(x) = x^m$ is a particular solution, a general solution is a linear combination of u and uv such that

$$\begin{aligned}0 &= \frac{v''_x}{v'_x} + \frac{(2m-1)u'_x}{u}, \\ v'_x &= c_1 u^{-(2m-1)}, \\ v_x &= v c_2 - \frac{c_1}{2(m-1)x^{2m-2}}, \\ y(x) &= \frac{1}{x}\Big\{c_2 x^{m+1} - c_1 \frac{1}{x^{2m-1}}\Big\},\end{aligned}$$

for every $x > 0$ and with arbitrary constants.

2.9.3. From the derivatives

$$\begin{aligned}u'_x &= 2Amx(x^2-1)^{m-1} - 2Bmx(x^2-1)^{-(m+1)}, \\ u''_x &= 4Am(m-1)x^2(x^2-1)^{m-2} + 2Bm(m+1)x^2(x^2-1)^{-(m+1)} \\ &\quad + 2Am(x^2-1)^{m-1} - 2Bm(x^2-1)^{-(m+1)}\end{aligned}$$

and, denoting $a_m = 2Am(m-1)$ and $b_m = 4Bm(m+1)$, we get

$$\begin{aligned}(x^2-1)u''_x &= 2a_m(x^2-1)^m + 2b_m(x^2-1)^{-m} + 2a_m(x^2-1)^{m-1} \\ &\quad + 2b_m(x^2-1)^{-(m+1)} + 2Am(x^2-1)^m - 2Bm(x^2-1)^{-m} \\ &= 2m(m+1)u - 4Am^2(x^2-1)^m + \frac{4m^2}{x^2-1}u \\ &\quad - \frac{2}{x}u'_x + \frac{2(x^2-1)}{x}u'_x - 4Am^2(x^2-1)^m,\end{aligned}$$

$$(1-x^2)u''_x + 2m\Big\{(m+1) + \frac{2m}{x^2-1}\Big\}u + 2\Big\{x - \frac{1}{x}\Big\}u'_x = 4Am^2(x^2-1)^m.$$

2.9.4. From (2.10), the derivatives satisfy

$$\begin{aligned}0 &= (1-x^2)u''_x - 4xu'_x + (m^2+m-2)u, \\ 0 &= (1-x^2)u_x^{(k+2)} - 2(k+1)xu_x^{(k+1)} + \{m(m+1) - k(k+1)\}u^{(k)},\ k \ge 1.\end{aligned}$$

2.9.5. From (2.10)

$$\begin{aligned}0 &= \frac{d^2}{d\theta^2}P_m(\cos\theta) - 2\cos\theta\frac{d}{d\theta}P_m(\cos\theta) + m(m+1)P_m(\cos\theta), \\ 0 &= \sin^2\theta P''_m(\cos\theta) + (2\cos\theta\sin\theta - \cos\theta)P'_m(\cos\theta) + m(m+1)P_m(\cos\theta), \\ 0 &= \sin^2\theta P''_m(\cos\theta) + \{\sin(2\theta) - \cos\theta\}P'_m(\cos\theta) + m(m+1)P_m(\cos\theta).\end{aligned}$$

2.9.6. The trigonometric formulas imply $T_{n+1}(x) = 2xT_n(x) - T_{n-1}(x)$, for every $n \ge 2$. Their generating function is

$$\begin{aligned}G(z;\cos\theta) &= \sum_{n\ge 0}\mathrm{Re}\{(ze^{i\theta})^n\} = \lim_{n\to\infty}\mathrm{Re}\frac{1-(ze^{i\theta})^n}{1 - z\cos\theta - iz\sin\theta} \\ &= \frac{(1-z\cos\theta)^2 + z^2\sin^2\theta}{(1-z\cos\theta)^2 - z^2\sin^2\theta}.\end{aligned}$$

Let $t = \cos\theta$

$$G(z;t) = \frac{1-zt)^2 + z^2(1-t^2)}{(1-zt)^2 - z^2(1-t^2)} = \frac{1-2zt+z^2}{(1-zt)^2 - z^2(1-t^2)}.$$

From the derivative

$$\frac{d}{d\theta}T_n(\cos\theta) = -\sin\theta\, T'_n(\cos\theta) = -n\sin(n\theta),$$

with $x'_\theta = -\sin\theta$, with deduce

$$T'_n(x) = n\frac{\sin(n\theta)}{\sin\theta} = n\frac{\{1-T_n^2(x)\}^{\frac{1}{2}}}{(1-x^2)^{\frac{1}{2}}},$$

$$\frac{d}{dx}\Big\{(1-x^2)^{\frac{1}{2}}T'_n(x)\Big\} = -nT'_n(x)T_n(x)\{1-T_n^2(x)\}^{-\frac{1}{2}},$$

therefore

$$(1-x^2)^{\frac{1}{2}}\frac{d}{dx}\Big\{(1-x^2)^{\frac{1}{2}}T_n'(x)\Big\}+n^2T_n(x)=0,$$
$$(1-x^2)T_n''(x)-xT_n'(x)+n^2T_n(x)=0.$$

2.9.7. Let $y_n = u_n v_n$ with Chebyshev's polynomial $u_n = T_n$ and v_n solution of the differential equation

$$(1-x^2)v_{n,x}''+2(1-x^2)u_n^{-1}u_{n,x}'v_{n,x}'-xv_{n,x}'=0$$

gives $v_n(x)=k(1-x^2)^{-\frac{1}{2}}T_n^{-2}(x)$ and $y_n(x)=T_n(x)+k(1-x^2)^{-\frac{1}{2}}T_n^{-1}(x)$, with an arbitrary constant k.

2.9.8. Applying Proposition 2.13

$$\begin{aligned}Eg(X-w) &= \sum_{k\geq 0} E\{g(X)H_k(X)\}\frac{w^k}{k!},\\ &= \sum_{k\neq j\geq 0} E\{g(X)H_k(X)H_j(X)e^{\frac{-X^2}{2}}\}\frac{w^{j+k}}{(j+k)!}\\ &= \sqrt{2\pi}\sum_{k\neq j\geq 0} E\{g(X)f_N^{(k)}(X)f_N^{(j)}(X)\}\frac{w^{j+k}}{(j+k)!}.\end{aligned}$$

2.9.9. From Proposition 2.13, the Fourier transform of H_k has the expansion

$$\widehat{H}_k(t)=\sum_{j\geq k+1}\frac{(-it)^{j-k}}{(j-k)!}e^{\frac{t^2}{2}}\int_{\mathbb{R}} H_j(x+it)\,dx,\ k\geq 1.$$

2.9.10. The constants are $\alpha_0^{-2}=\int_I f(x)\,dx$ and $\alpha_j^{-2}=\int_I Q_j^2(x)f(x)\,dx$, for every $k\geq 1$. They are written as $Q_0(x)=\alpha_0^{-1}$

$$Q_1(x)=\frac{1}{\alpha_1}\Big\{\frac{f'(x)}{f(x)}G(x)+G'(x)\Big\},$$
$$Q_2(x)=\frac{1}{\alpha_2}\Big\{\frac{f''(x)}{f(x)}G^2(x)+4\frac{f'(x)}{f(x)}G(x)G'(x)+2G(x)G''(x)+2G'^2(x)\Big\},$$
$$Q_k(x)=\frac{1}{\alpha_k}\Big\{\sum_{j=0}^{k}\frac{f^{(j)}(x)}{f(x)}\frac{d^{n-j}}{dx^{n-j}}G^n(x)\Big\}.$$

2.9.11. The first derivatives of $X_{mn}(x)=x^m(x-1)^n$ are solutions of the equations

$$\begin{aligned}0&=x(1-x)y_x'-\{(m+n)x-m\}y,\\ 0&=x(1-x)y_x''+\{m-1-(m+n-2)x\}y_x'-(m+n)y,\\ 0&=x(1-x)y_x''+\{m-2-(m+n+4)x\}y_x'-2(m+n+1)y,\\ 0&=x(1-x)y_x''+\{m-k-1-(m+n+2(k+1))x\}y_x'\\ &\quad-\{(k+1)(m+n)+k(k+1)\}y\end{aligned}$$

for the kth derivative then

$$x(x-1)y''_x + \{1 + (3m+n+2)x\}y'_x - (m+1)(2m+n)y = 0$$

for the mth derivative.

2.9.12. Let $A_{m,n}$ we have $\varphi'(z) = A_{m,n}\varphi(z)$ and $\psi'(z) = -A_{n,m}\psi(z)$, and this relation is generalized.

8.3 Calculus and optimization

3.7.1. The extrema of an ellipse in the plane are solutions of the differential equations

$$\begin{aligned} 0 &= \frac{x-x_0}{a^2} + y'_x \frac{y-y_0}{b^2}, \\ 0 &= y'_x, \\ 0 &= \frac{1}{a^2} + \frac{1}{b^2}(y'^2_x + (y-y_0)y''_x), \end{aligned}$$

the first two equations imply $x = x_0$, therefore $y = y_0 \pm b$, with $b > 0$. The sign of the second derivative

$$y''_x = -\frac{1}{y-y_0}\left(y'^2_x + \frac{b^2}{a^2}\right)$$

is positive if $y - y_0$ is negative so the minimum of the ellipse is reached at $(x_0, -b)$ and its maximum is reached at (x_0, b). The extrema along the ab-sisse axis are obtained from the derivatives with respect to y, the maximum is reached at (a, y_0) and the minimum at $(-a, y_0)$.

3.7.2. Let $u(x,y) = y^2 + 2x^2y + 6x - 3 = 0$, it has the derivative $u'_x(x,y) = 6 + 4xy + 2y'_x(y + x^2)$, and $y'_x = 0$ if $xy = \frac{3}{2}$ where $x^2u = \frac{9}{4} + 9x^3 - 3 = 0$, $x^3 = \frac{1}{12}$ and $y = \frac{3}{2(12)^{\frac{1}{3}}}$. The second derivative at this value is $u''_x(x,y) = 4y + 2y''_x(y + x^2) = 0$ therefore $y''_x < 0$ and y is maximum at this point.

3.7.3. Let $a < 0$, $u'_x(x,y) = (3y^2 + ax)y'_x + 6x^2 + ay = 0$ and $y'_x = 0$ implies either $x = 0$ and $y = 0$, or $6x^2 + ay = 0$. In the second case, $x = -\frac{a}{3.2^{\frac{1}{3}}}$ and $y = -\frac{2^{\frac{1}{3}}a}{3}$. The second derivative is $u''_x(x,y) = (3y^2 + ax)y''_x + 12x = 0$, at the second point $y''_x = -\frac{12x}{3y^2+ax} = \frac{12}{2^{\frac{1}{3}}a} < 0$ and y is maximum.

3.7.4. The partial derivative $u'_x = 2y'_x y + 2x - 2ay - 2axy'_x = 0$ gives $y'_x = (ax-y)^{-1}(x-ay)$ and $y''_x = (ax-y)^{-2}(xy'_x - y)(1-a^2)$. Then

$$\begin{aligned}\frac{1}{2}u''_x &= y''_x y + y'^2_x + 1 - ay'_x - axy''_x \\ &= -\frac{u(1-a^2)}{(ax-y)^2} + \frac{(x-ay)^2}{(ax-y)^2} + 1 - ay'_x \\ &= \frac{a^2(y^2+x^2+axy) - 3axy}{(ax-y)^2}.\end{aligned}$$

If $a = 1$, $u''_x > 0$ and there is a minimum. Let $xy > 0$ and $a > 1$, then $(a^3 - 3a)xy > -2a^2xy$ and $u''_x > 0$, if $a < 1$ then $u''_x < 0$. If $xy > 0$, the conclusions are the opposite.

3.7.5. Denoting $u = 2cx+b$ the derivative of $t_x^2 = a+bx+cx^2$, $x = u-\frac{1}{2}bc$ and $t_x^2 = c(u^2+\alpha) = c\alpha(v^2+1)$ with the real or complex notations

$$\alpha = \frac{a}{c} - \frac{b^2}{4c}, \quad v = \frac{u}{\alpha^{\frac{1}{2}}},$$

$$dv = \frac{du}{\alpha^{\frac{1}{2}}} = 2c\frac{dx}{\alpha^{\frac{1}{2}}}$$

it follows that

$$F(x) = \int_0^x \frac{ds}{t_s} = \frac{1}{2c^{\frac{3}{2}}}\int_0^{k_x} \frac{dv}{(v^2+1)^{\frac{1}{2}}},$$

$$k_x = \alpha^{-\frac{1}{2}}(2cx+b)$$

and the primitive of f is obtained from the primitive $\arg\sinh x$ of $(x^2+1)^{-\frac{1}{2}}$

$$F(x) = \frac{1}{2c^{\frac{3}{2}}} \arg\sinh \alpha^{-\frac{1}{2}}(2cx+b).$$

3.7.6. $\frac{1}{2}(1+\sin x)(1-\sin x) = \arcsin x$.

3.7.7. Integrating by parts

$$\begin{aligned}I &= \int_a^b x\log(x)(1-x^2)^{-\frac{1}{2}}\,dx = \log\frac{x}{1-x^2}\Big|_a^b + \frac{1}{2}\int_a^b (1-x^2)\log x\,dx \\ &= \log\frac{b(1-a^2)}{a(1-b^2)} + \frac{1}{2}\Big\{\log\frac{b^b}{a^a} - b - a\Big\} - \frac{1}{6}\{x^3(\log x - \frac{1}{3})\Big|_a^b.\end{aligned}$$

3.7.8. The sequence of integrals are calculated from $u_0 = \frac{\pi}{2}$ and $u_1 = 1$ by integration of the expansions

$$\sin^{2n} x = 2^{2n-1}\sum_{m=0}^{n}(-1)^{m+n}a_m\cos\{2(n-m)x\},$$

$$\sin^{2n+1} x = 2^{2n}\sum_{m=0}^{n}(-1)^{m+n}a_m\sin\{2n-2m+1)x\},$$

with the binomial constants a_m. They are also calculated iteratively from u_0 and u_1 with integrations by parts

$$u_n = \int_0^{\frac{\pi}{2}} \sin^{n-1} x \, d(-\cos x) = (n-1) \int_0^{\frac{\pi}{2}} \sin^{n-2} x \cos^2 x \, dx$$
$$= (n-1)\Big\{\int_0^{\frac{\pi}{2}} \sin^{n-2} x \, dx - \sin^n x\Big\} \, dx$$

so that for $n \geq 1$, $u_n = \frac{n-1}{n} u_{n-2}$

$$u_{2n} = \frac{\pi}{2} \prod_{m=1}^{n-1} \frac{2m-1}{2m} = \frac{\pi}{2} \cdot \frac{1}{3} \cdot \frac{3}{4} \cdot \frac{5}{6} \cdots \frac{2n-1}{2n},$$

$$u_{2n+1} = \prod_{m=1}^{n-1} \frac{2m}{2m+1} = \frac{2}{3} \cdot \frac{4}{5} \cdot \frac{6}{7} \cdots \frac{2n}{2n+1}.$$

They lead to Wallis's expansion of π

$$\frac{\pi}{2} = \frac{2}{3} \cdot \frac{3}{1} \cdot \frac{4}{3} \cdot \frac{4}{5} \cdot \frac{6}{5} \cdot \frac{6}{7} \cdots$$

3.7.9. Integrating by parts

$$y_{m,n} = \frac{n-1}{m+1} \int_0^{\frac{\pi}{2}} \sin^{m+1} x \cos^{n-2} x \, dx$$
$$= \frac{n-1}{m+1} y_{m+1,n-2} = \frac{n-1}{m+1} \cdot \frac{n-3}{m+2} y_{m+2,n-4}$$
$$= y_{m,1} \prod_{k=0}^{\frac{n}{2}-1} \frac{n-2k-1}{m+k+1}, \text{ if } n \text{ is even},$$
$$= y_{m,0} \prod_{k=0}^{\frac{n-1}{2}} \frac{n-2k-1}{m+k+1}, \text{ if } n \text{ is odd},$$

with $y_{m,1} = \frac{1}{m+1}$. Let $a > 0$ be real, the derivative of $y_{a,0}$ with respect to a is $I_a = a \int_0^{\frac{\pi}{2}} \sin^{a-1} x \cos x \, dx = 1$ and its primitive is $y_a = a + y_0 = a + \frac{\pi}{2}$, in particular $y_{m,0} = m + \frac{\pi}{2}$.

3.7.10. The function $u_{m,n}(x) = \sin^m x \cos^{-n} x = \sin^m x (1 - \sin^2 x)^{-\frac{n}{2}}$ has a Taylor expansion

$$u_{m,n}(x) = \sin^m x \sum_{k \geq 0} \frac{n(n+2)\cdots(n+2k-2)}{2^k} \frac{\sin^{2k}}{k!} (\sin^2 x)^{(k)},$$

$$(\sin^2 x)' = 2\sin(x)\cos(x),$$
$$(\sin^2 x)'' = 2(\cos^2 x - \sin^2 x) = 2(1 - 2\sin^2 x),$$
$$(\sin^2 x)^{(3)} = -8\sin(x)\cos(x)$$

and the following derivatives of $\sin^2 x$ are deduced. If k is odd, the integrals of the functions $\sin^m x \sin^{2k}(\sin^2 x)^{(k)}$ are proportional to the integrals $(m+2k+2)^{-1}\sin^{m+2k+2} x$ of $\sin^{m+2k+1} x \cos x$. If k is even, the functions $\sin^m x \sin^{2k}(\sin^2 x)^{(k)}$ depend only on $\sin^m x \sin^{2k}$ and $\sin^m x \sin^{2k+2}$, their integrals are calculated like u_n in the previous exercise.

3.7.11. Let $x = \cos\theta$, $y_{m,n} = \int_0^1 x^{\frac{n}{2}}(1-x^2)^{\frac{m}{2}-1}\,d\theta = y_{n,m}$, replacing x^2 by $1-x^2$. The proof is similar for $u_{n,m}$.

3.7.12. Writing $\cos(bx)$ and $\sin(bx)$ as exponentials and integrating

$$A = \frac{a}{a^2+b^2}, \qquad B = \frac{b}{a^2+b^2}.$$

3.7.13. Derivating with respect to a

$$I'_a(y) = I'_a(y_0) - \int_{y_0}^{y} \sin(ax)\,dx = I'_a(y_0) + a^{-1}\{\cos(ay) - \cos(ay_0)\}.$$

Its primitive at $a > a_0 > 0$ is

$$\begin{aligned}\int_{a_0}^{a} I'_s(y)\,ds &= \int_{a_0}^{a} I'_s(y_0)\,ds + \int_{a_0}^{a} s^{-1}\{\cos(sy) - \cos(ay_0)\}\,ds\\ &= \int_{a_0}^{a} I'_s(y_0)\,ds + I_y(a) - I_y(a_0) - I_{y_0}(a) + I_{y_0}(a_0),\end{aligned}$$

$$I_a(y) - I_{a_0}(y) = I_a(y_0) + I_y(a) - I_y(a_0) - I_{y_0}(a).$$

At $y_0 = \frac{\pi}{2a}$ and $y_0 = 0$, we have

$$\begin{aligned}I_a(y) - I_a\Big(\frac{\pi}{2a}\Big) &= I_{a_0}(y) - I_{a_0}\Big(\frac{\pi}{2a}\Big) + I_y(a) - I_y(a_0),\\ I_a(y) - I_a(0) &= I_{a_0}(y) - I_{a_0}(0) + I_y(a) - I_y(a_0) + \log\frac{a}{a_0},\\ I_{a_0}(0) - I_{a_0}\Big(\frac{\pi}{2a}\Big) &= I_a(0) - I_a\Big(\frac{\pi}{2a}\Big) + \log\frac{a}{a_0}.\end{aligned}$$

3.7.14. The integral of a series expansion of $(1-e^{i\theta})^{-1}$ yields 2π.

3.7.15. It has the value $I(\frac{1}{2}) = (\frac{\pi}{2})^{\frac{1}{2}}$ and the derivative

$$I'(a) = -\frac{1}{2a}I(a)$$

it follows that $\sqrt{a}\,I(a) = \frac{\sqrt{\pi}}{2}$.

3.7.16. By a change of variable

$$I_p = \int_0^1 e^{-y} y^{p-1}\,dy - p\int_0^1 (1-y)^{p-1}\int_0^y e^{-s}s^{p-1}\,ds\,dy$$
$$= \int_0^1 e^{-y} y^{p-1}\,dy - \int_0^1 e^{-y} y^{p-1}(1-y)^p\,dy$$
$$A_p = \int_0^1 e^{-y} y^{p-1}\,dy = (p-1)!\Big\{1+\frac{1}{2!}+\ldots+\frac{1}{(p-1)!}\Big\}.$$

we get $I_p = \frac{1}{2}A_p$.

3.7.17. By a change of variable, $I_k = \Gamma_k$.

3.7.18. $y(x) = a\tanh x + b\tanh^{-1} x$.

3.7.19. $a = b = c = (4\pi)^{-\frac{1}{3}}(3V)^{\frac{1}{3}}$.

3.7.20. Integrating by parts, $I_{m,n} = \int_0^1 x^{m-1}\log^{n-1}\frac{1}{x}\,dx = \frac{n}{m}I_{m,n-1}$ and $I_{m,1} = \Gamma_n$, then $I_{m,n} = m^{-(n+1)}n!$.

3.7.21. By an expnsion of the exponential, the primitive of $f_1(x) = x^{-1}e^{ax}$ is

$$F_1(x) = \log x + \sum_{k\geq 1\frac{a^k x^k}{k\,k!}}$$

and for $n>1$, integrations by parts yield

$$F_n(x) = e^{ax}\sum_{k=1}^{n-1}\frac{(n-k-1)!\,a^{k-1}}{x^{n-k}\,n!} + \frac{a^{n-2}}{(n-1)!}F_1(x).$$

3.7.22. The limit is the integral of $x^{-1}\sin x$ on $\mathbb{R}_+$, which is $\frac{\pi}{2}$.

3.7.23. Since $\sin^2 x - \cos^2 x = -\cos 2x$, this is true for $k=1$ and their integral is $\frac{\pi}{4}$, the same argument applies iteratively for every integer and their integral is

$$\frac{1.3.7\cdots(2k-1)}{2^{k+2}\pi}.$$

3.7.24. The convolution of functions f and g on $\mathbb{R}_+$ is $\int_0^\infty f(x-y)g(y)\,dy = \int_0^\infty f(-t)g(t+x)\,dt$ therefore

$$n\int_0^\infty \sin^{n-1}(x-y)\cos(x-y)e^{-y}\,dy = (-1)^n\int_0^\infty \sin^n(t)e^{-x-t}\,dt = 0$$

and its derivative at $x = 0$ gives the result. Deriving twice this equation gives

$$\frac{n+(-1)^n}{n(n-1)} \int_0^\infty \sin^n(t)e^{-t}\,dt = (n-2)(n-3)\int_0^\infty \sin^{n-4}(y)\cos^4(y)e^{-y}\,dy$$
$$-(5n-8)\int_0^\infty \sin^{n-2}(y)\cos^2(y)e^{-y}\,dy,$$

the $2k$th derivatives of the equation of convolutions provide a relationship between the integrals $\int_0^\infty \sin^{n-2k}(y)\cos^{2k}(y)e^{-y}\,dy$ and $\int_0^\infty \sin^{n-2(k-1)}(y)\cos^{2(k-1)}(y)e^{-y}\,dy$.

3.7.25. Writing the convolution of $\sin^n$ and $\cos^m$ like in (3.7.23)

$$0 = n\int_0^{\frac{\pi}{2}} \sin^{n-1}(x-y)\cos^m y\,dy + (-1)^n \sin^n x,$$

$$0 = n(n-1)\int_0^{\frac{\pi}{2}} \sin^{n-2}(x-y)\cos^m y\,dy - n\int_0^{\frac{\pi}{2}} \sin^n(x-y)\cos^m y\,dy$$
$$-(-1)^n n \sin^{n-1} x\cos x$$

and the result is obtained with $x = 0$.

3.7.26. Let $x = \tan u$,

$$I = \int_0^{\frac{\pi}{2}} \frac{du}{1+\tan^2 u} = \int_0^{\frac{\pi}{2}} \cos^2 u\,du$$
$$= \frac{1}{2}\int_0^{\frac{\pi}{2}} (1+\cos 2u)\,du = \frac{\pi}{4}.$$

3.7.27. Let $2a < 1$, $\Gamma_a\Gamma_{1-a}$ is written as a product of integrals A_1 and A_2 using the change of variable $x = r^2\sin^2\theta$ and $y = r^2\cos^2\theta$

$$A_1 = 2\int_0^\infty re^{-r^2}\,dr = \int_0^\infty e^{-x}\,dx = 1,$$

$$A_2 = -2\int_0^{\frac{\pi}{2}} \cos^{1-2a}\theta\sin^{2a-1}\theta\,d\theta$$
$$= \int_0^1 x^{-a}(1-x)^{a-1}\,dx.$$

3.7.28. The first Euler-Lagrange condition is $f+\lambda^2 f'' = 0$ hence $f(x) = a\sin(\lambda^{-1}x) + b\cos(\lambda^{-1}x)$ and the second condition is satisfied (Theorems 3.1-3.2). The minimum achieved by this function is $4ab\sin^2\lambda^{-1}$.

3.7.29. The first Euler-Lagrange condition is $f + af'' + bf^{(4)} = 0$. Assuming that $a^2 \geq 4b$, there exist $\omega_1 > 0$ and $\omega_2 > 0$ such that the equation has solutions $f(x) = c_1 e^{-\omega_1 x} + c_2 e^{-\omega_2 x}$ satisfying the integrability conditions. The integral is strictly positive if $c_1^2\omega_1^{-1} + c_2^2\omega_2^{-1} + 4c_1c_2\omega_1^{-1}\omega_2^{-1} - a(c_1^2\omega_1 + c_2^2\omega_2 + 4c_1c_2) + b(c_1^2\omega_1^3 + c_2^2\omega_2^3 + 4c_1c_2\omega_1\omega_2) > 0$.

8.4 Linear and nonlinear differential equations

4.7.1. A particular solution is $y_1(x) = x^{-1}\sin x$, let $y_2 = uv$ be another solution then $\frac{v_x''}{v_x'} + \frac{2\cos x}{\sin x} = 0$ and there exists a constant k such that

$$\begin{aligned}
k &= v_x' \sin^2 x, \\
v_x &= v_0 + k\int_{x_0}^{x} \sin^{-2} s\, ds, \\
I_x &= \int_{x_0}^{x} \sin^{-2} s\, ds = -\int_{\sin^{-1} x_0}^{\sin^{-1} x} \frac{dy}{(1-y^2)^{\frac{1}{2}}} \\
&= \arcsin(\sin^{-1} x_0) - \arcsin(\sin^{-1} x)
\end{aligned}$$

and a solution is the sum of y_1 and y_2.

4.7.2. A solution is a linear combination of $y_1(x) = \exp(m\cos ax)$ and $y_2 = uv$ such that $v_x'' = mav_x' \sin(ax)$, by integration $v_x' = ke^{-m\cos(ax)}$ and

$$v_x = -\frac{k}{a}\int_{\cos(ax_0)}^{\cos(ax)} \frac{e^{-mz}}{\sqrt{1-z^2}}\, dz$$

where the last integral is calculated with an integration by parts.

4.7.3. $y(x) = (c_1 x^{-1} + x^{-2})e^{-x}$.

4.7.4. A solution is given by Proposition 4.12, with $\Delta = (a-b)^2$, $\alpha = -\frac{1}{2}(a+b)$ and $\beta = \frac{1}{2}|a-b|$.

4.7.5. $y(x) = (1+x)e^{-x} + x(\log x - 1)e^{-x}$.

4.7.6. A solution is $y(x) = \cos(2x)$.

4.7.7. Let $y = e^{-ax^2}u$ with derivatives $y'(x) = e^{-ax^2}(u' - 2axu)$ and $y''(x) = e^{-ax^2}(u'' - 4axu' - 2au + 4a^2x^2u)$ such that

$$(y'' + 4axy' - 4ax^2y)e^{ax^2} = u'' - 2au = e^{ax^2} f(x).$$

A solution $u(x) = e^{bx}v(x)$ is such that $b = \pm\sqrt{2a}$ and

$$e^{-bx}(u'' - 2au) = (v'' + 2bv') = e^{ax^2 - bx} f(x),$$

its solutions with initial value $y(0) = y_0$ follow from Theorem 4.1

$$v(x) = u_0 e^{-2bx} + e^{-2bx} \int_{x_0}^{x} e^{at^2 + bt} f_t \, dt.$$

4.7.8. By the change of variable $x = e^t$ and $u(t) = y(x)$

$$\begin{aligned}
u_t^{(1)} &= xy'_x, \\
u_t^{(2)} &= x^2 y_x^{(2)} + xy'_x, \\
u_t^{(3)} &= x^3 y_x^{(3)} + 3x^2 y_x^{(2)} + xy'_x, \\
x^2 y_x^{(2)} &= u_t^{(2)} - u_t^{(1)}, \\
x^3 y_x^{(3)} &= u_t^{(3)} - 3u_t^{(2)} + 2u_t^{(1)},
\end{aligned}$$

the differential equation is equivalent to the differential equation with constant coefficients $u^{(3)} + (a-3)u^{(2)} + (b-a+2)u^{(1)} + cu = 0$. Its characteristic polynomial $p(r) = r^3 + (a-3)r^2 + (b-a+2)r + c = 0$ has three real or complex roots that define three exponential solutions $u_1(t), \ldots, u_3(t)$, then $y(x) = c_1 y_1(x) + c_2 y_2(x) + c_3 y_3(x)$ is the solution of the equation.

4.7.9. A solution in the form $y(x) = \exp\{\int_0^x (b + cs)(a - s)^{-2} \, ds\}$ with derivatives

$$y'(x) = \frac{b + cx}{(a - x)^2} y, \quad y''(x) = y\Big\{\frac{(b + cx)^2}{(a - x)^4} + \frac{c}{(a - x)^2} + \frac{2(b + cx)}{(a - x)^3}\Big\}.$$

It satisfies the equation with the polynomials $P_2(x) = (a - x)^2$ and $P_1(x) = (b + cx)^2 + 2(a - x)(b + cx) + c(a - x)^2$.

4.7.10. $y(x) = \sin \frac{x^2}{2}$.

4.7.11. $y(x) = a \cos x + b \sin b - \frac{a}{3} \cos(3x)$ is solution with $a = 1 + \frac{1}{8}$.

4.7.12. A solution is $y(x) = u(x)e^{-ax}$ with derivatives

$$\begin{aligned}
y'(x) &= u'(x)e^{-ax} - ay(x), \\
y''(x) &= u''(x)e^{-ax} - 2au'(x)e^{-ax} + a^2 y(x)
\end{aligned}$$

then $e^{ax}(y'' + 2ay' + a^2 y) = u'' = h(x)$ and u is deduced by integration.

4.7.13. $y_1(x) == x(\log x - 1)$ and

$$y_2(x) = \frac{x^{n+1}}{n+1}\Big(\log x - \frac{1}{n+1}\Big).$$

4.7.14. Let $y(x) = (1-x^2)u^{\frac{1}{2}}(x)$, its first derivative is

$$y'(x) = \frac{1}{2}u'(x)u^{-\frac{1}{2}}(x)(1-x^2) - 2xu^{\frac{1}{2}}(x)$$

it satisfies

$$\begin{aligned} 2y'_x + 2\frac{xy}{1-x^2} &= (1-x^2)u'(x)u^{-\frac{1}{2}}(x) - 2xu^{\frac{1}{2}}(x) \\ &= 2x(1-x^2)^{\frac{1}{2}}u^{\frac{1}{4}}(x). \end{aligned}$$

The solution of the homogeneous equation $(1-x^2)u'(x) = 2xu(x)$ is $u(x) = u_0(1-x^2)^{-1}$, let $y = uv$ be solution of the inhomogeneous equation such that $uv' = xu^{\frac{1}{2}}v^{\frac{1}{2}}$ i.e. $v'v^{-\frac{1}{2}} = u^{-\frac{1}{2}}x(1-x^2)^{\frac{1}{2}} = u_0^{-\frac{1}{2}}x(1-x^2)$ then

$$v^{\frac{1}{2}}(x) = v_0^{\frac{1}{2}} + u_0^{-\frac{1}{2}}x^2(1-\frac{1}{2}x^2)$$

and the initial values are $v(0) = u_0 = y_0^{\frac{1}{2}}$.

4.7.15. The homogneous differential equation has solutions y such that $xy(x)$ is constant. Assuming that $xy(x) = h(x)$ implies $h'(x)h^2(x) = ax$ and

$$h(x) = \Big(\frac{3a}{2}\Big)^{\frac{1}{3}}x^{\frac{2}{3}}, \quad y(x) = \Big(\frac{3a}{2}\Big)^{\frac{1}{3}}x^{-\frac{1}{3}}.$$

4.7.16. The homogneous differential equation has the solutions $y(x) = y_0e^{\frac{a}{x}}$. Let $y = uv$ such that u is solution of the homogneous equation then $vv' = bx^{-3}e^{-\frac{2a}{x}}$ and

$$v^2(x) = y_0^2 - 4a^2b\int_{2ax^{-1}}^{\infty} e^{-t}t\,dt.$$

4.7.17. The homogneous differential equation has the solution $y(x) = y_0\exp\{\frac{a}{2x^2}\}$. Let $y = uv$ such that u is solution of the homogneous equation then the differential equation is equivalent to $xu^2vv' = f(x)$ and

$$v^2(x) = y_0^2 + 2\int_0^x s^{-1}f(s)\exp\{-\frac{a}{s^2}\}\,ds.$$

4.7.18. Denoting $y = xu$ for $x \neq 0$, the equation is equivalent to $xu'_x + u^3 - u^2 = 0$ and

$$\frac{1}{x} = \frac{u'_x}{u^2-u^3} = \frac{u'_x}{1-u} + \frac{u'_x}{u} + \frac{u'_x}{u^2}.$$

The solution for u different from 0 and 1 is such that

$$x = c\frac{1-u}{u^3}, \quad c = \frac{x_0 u_0^3}{1-u_0}.$$

4.7.19. $y = (y_0 + x^2)e^{-ax}$, such that $y(0) = y_0$.

4.7.20. $y(x) = \sin x \cos x$.

4.7.21. The equation is equivalent to

$$\frac{a}{x} + b\frac{y'_x}{y} + mx^{m-1}y^n + nx^m y^{n-1}y'_x = 0,$$

hence $x^a y^b = C\exp\{-x^m y^n\}$.

4.7.22. The equation $u'_x - (1-a)x^{-1}u - u = 0$ has the solution

$$u(x) = u_0\left(\frac{x}{x_0}\right)^{1-a} e^{x-x_0},$$

let $z(x) = v(x)(\frac{x}{x_0})^{1-a}e^{x-x_0}$ be solution of $z'_x - (1-a)x^{-1}z - z + 1 = 0$ with initial value $z_0 = z(x_0) = v(x_0)$, then

$$v'_x = e^{-(x-x_0)}(\frac{x}{x_0})^{a-1}$$

and $v(x) = z_0 + \int_{x_0}^{x} e^{-(t-x_0)}(\frac{t}{x_0})^{a-1}\,dt$.

4.7.23. $y(x) = y_0 x^a u(x)$ with $u''(x) = h$.

4.7.24. The equation is written as $y'_x y^{-2} + y = x^{-1}$ where the homogeneous equation $u'_x u^{-3} = -1$ has the solution $u(x) = (a + 2x)^{-\frac{1}{2}}$ with a constant a. The inhomogeneous equation is solved by replacing the constant with a function $a(x)$ such that $y(x) = \{a(x) + 2x\}^{-\frac{1}{2}}$ has the derivative

$$y'_x = -\frac{y^3}{2}(a'_x + 2)$$

and the differential equation is equivalent to

$$-y(a'_x + 2) + 2y = \frac{2}{x},$$

$$\frac{a'_x}{\{a(x) + 2x\}^{\frac{1}{2}}} = -\frac{2}{x},$$

hence $\{a(x) + 2x\}^{\frac{1}{2}} = a_0^{\frac{1}{2}} + \log\left(\frac{x_0}{x}\right)^2$ with the initial value $a_0 = y_0^{-2}$ and $y(x) = \left\{a_0 + \log\left(\frac{x_0}{x}\right)^2\right\}^{-1}$.

4.7.25. $y(x) = x \tan x$ with derivatives $y' = x(1 + \tan^2 x) + \tan x$ and $y'' = 2 + 2x^{-1}yy'$.

4.7.26. Denoting $y_x = a_x^{-1}$, the equation becomes

$$a_x^2 y_x^2 + a_x y'_x + bx^{-1} = 1 - \frac{a'_x}{a_x} + bx^{-1} = 0$$

and it has the solution $a_x = cx^b e^{x-x_0}$ with $c = a_0 x_0^{-b}$.

4.7.27. The equation $xu'_x = u^2 - u$ implies

$$\frac{1}{x} = \frac{u'_x}{u^2 - u} = -\frac{(u^2 - u)'_x}{u^2 - u} + \frac{2}{u - 1} u'_x,$$

$$\log(xx_0^{-1}) = -\log \frac{u(u-1)}{u_0(u_0 - 1)} + \log \frac{(u-1)^2}{(u_0 - 1)^2} = \log \frac{(u-1)u_0}{(u_0 - 1)u},$$

$$x = x_0 \frac{(u-1)u_0}{(u_0 - 1)u}.$$

4.7.28. For every $x \neq 0$, the equation is written as $(y_x'^2)'_x - 2(x^{-1}y)'_x = 0$. There exists a constant c such that $2c + 1 > 0$ and

$$y'_x = \sqrt{2}\Big(\frac{y}{x} + c\Big)^{\frac{1}{2}}.$$

Let $u(x) = (y + cx)^{\frac{1}{2}}$, its first order derivative is $u'_x = \frac{1}{2}(y'_x + c)u(x)^{-1}$ and

$$y'_x = 2u'_x u_x - c = \sqrt{2} u x^{-\frac{1}{2}},$$

equivalently $\sqrt{2x}u'_x u_x - u_x = c\sqrt{\frac{x}{2}}$. The homogeneous equation $\sqrt{2x}v'_x = 1$ has the solution $v_x = u_0 + \sqrt{2x}$, with $v(0) = u_0$. The solution of the inhomogeneous equation is $u_x = v_x + h_x$ such that $h_x h'_x + \sqrt{2x}h'_x = \frac{c}{\sqrt{2}}$ with solution h such that $h_x^2 + 2\int_0^x \sqrt{2t}h'_t\,dt = \sqrt{2}cx$.

4.7.29. The function $f = xy'_x - y$ has the derivative $f'_x = xy''_x$ such that $x^3 y''_x - (y - xy'_x)^2 = x^2 f'_x - f_x^2 = 0$ then

$$\frac{1}{f(x)} = -\frac{1}{x} + c,$$

$$f(x) = \frac{x}{1 + cx} = xy'_x - y,$$

with an arbitrary constant c. Let $y(x) = axv(x)$ with an arbitrary constant a be a solution, then $av'(x) = (1 + cx)^{-1}$ and

$$av(x) = \log(1 + cx), \quad y(x) = x\log(1 + cx).$$

4.7.30. This is a nonlinear first order equation for $z = y'_x$, multiplying it by z and integrating yields

$$\frac{2x}{a^2} = (1+z^2)^{-\frac{3}{2}} z'_x,$$
$$(1+z^2)^{-\frac{1}{2}} = 1 + \int_0^x \frac{2s}{a^2} y'_s\, ds,$$

multiplying both sides by zz'_x and integrating provides an implicit equation for (x, y_x, y'_x)

$$(1+z^2)^{\frac{1}{2}} = \frac{z^2}{2} + \frac{2}{a^2}\int_0^x zz'_t \int_0^t sy'_s\, ds\, dt,$$
$$(1+y'^2_x)^{\frac{1}{2}} = \frac{y'^2_x}{2} + \frac{1}{a^2}\Big(y'^2_x \int_0^x sy'_s\, ds - \int_0^x sy'^3_s\, ds\Big).$$

4.7.31. The equation is written as

$$\frac{y''_{xx}}{y'_x} = \frac{1}{x} + \frac{1}{(a^2+x^2)^{\frac{1}{2}}}$$

therefore

$$\log\frac{x_0 y'_x}{y'_0 x} = \int \frac{ds}{(a^2+s^2)^{\frac{1}{2}}} = \frac{1}{a}\operatorname{arg\,sinh}\frac{x}{a}$$

and y_x is deduced by integration of y'_x.

4.7.32. Since the derivative of xyy'_x is $yy'_x + xy'^2 + xyy''_x$, the equation is equivalent to

$$xyy'_x + nx(a^2-x^2)^{\frac{1}{2}} = n\int_0^x (a^2-s^2)^{\frac{1}{2}}\, ds$$

where the last integral equals $I = a^2 \int \cos^2 u\, du$ with $x = a\sin u$, hence $I = a^2\{\arcsin(a^{-1}x) + a^{-1}x(\frac{\pi}{2} - a^{-1}x)\}$. Then

$$yy'_x = na^2x^{-1}\arcsin(a^{-1}x) + n(a\frac{\pi}{2} - x) - n(a^2-x^2)^{\frac{1}{2}},$$
$$n^{-1}y^2(x) = a^2\int_0^x s^{-1}\arcsin(a^{-1}s)\, ds + \frac{a}{2}x(\pi - x) - I.$$

4.7.33. The function $z(x) = y'_x(x)$ satisfies a Bernoulli equation with coefficients $a(x) = (x+a)^{-1}$ and $b(x) = x(x+a)^{-1}$, the primitive of a is $A(x) = \log(x+a) - \log a$ and $be^A = a^{-1}x$. Let $k = 2a(y'_0)^{-1}$

$$y'_x(x) = 2\frac{x+a}{x^2+k},$$
$$y(x) = y_0 + \int_0^x \frac{t+a}{t^2+k}\, dt = y_0 + \log(x^2+k) + 2\int_0^x \frac{a}{t^2+k}\, dt$$
$$= y_0 + \log(x^2+k) + \frac{2a}{\sqrt{k}}\arctan\frac{x}{\sqrt{k}}.$$

4.7.34. The function $y(x) = x(x+1)^{-\frac{1}{2}}$ has the derivatives

$$y' = (x+1)^{-\frac{1}{2}} - \frac{1}{2}y(x+1)^{-1},$$
$$y'' = (x+1)^{-\frac{3}{2}} - 3y\{4(x+1)^2\}^{-1}$$

such that

$$yy''_x + \frac{y'^2_x}{2} = -\frac{y^2}{4(x+1)^2}\Big\{3 - \frac{1}{(x+1)^2}\Big\} + \frac{1}{x+1}.$$

4.7.35. Let $u = y^{\frac{1}{2}}$ with derivative $u' = \frac{1}{2}y'y^{-\frac{1}{2}}$ so $y' = 2uu'$, the equation becomes

$$2u' + \frac{xy}{1-x^2} - x = 0.$$

The homogeneous equation has the solution $y = y_0(1-x^2)^{\frac{1}{2}}$ and the inhomogeneous equation has the solution $u^2(x) = v(x)(1-x^2)^{\frac{1}{2}}$ where $v'_x = xv$ hence $y = y_0(1-x^2)^{\frac{1}{2}}(1+e^{\frac{x^2}{2}})$.

4.7.36. Let $u = y^2$, the equation is equivalent to $u' + \frac{2au}{x^2} = \frac{2b}{x^2}$ and its solutions have the form $u(x) = v(x)e^{-\frac{2a}{x}}$ with a functions v such that

$$dv_x = \frac{2b}{x^2}e^{\frac{2a}{x}}\,dx = -e^{\frac{ay}{b}}\,dy$$

where $y = \frac{2b}{x}$, hence $v(x) = \frac{b}{a}(1-e^{\frac{2a}{x}})$.

8.5 Linear differential equations in $\mathbb{R}^p$

5.7.1. From (2.10), the function P_m satisfies $(1-x^2)y''_{m,x} - 2xy'_{m,x} + m(m+1)y_m = 0$ and the differential equation $y''_{m,x} = 0$ is equivalent to $2xy'_{m,x} - m(m+1)y_m = 0$, its solution is $y_m(x) = \exp\{x^{\frac{1}{2}m(m+1)}\}$.

The equation $y''_{m,x} = f$ has a solution $y_m(x) = u_m(x)\exp\{x^{\frac{1}{2}m(m+1)}\}$ with a function u_m such that $u_m^{-1}u'_m y_m = f$, this is equivalent to

$$\begin{aligned} u_m(x) &= \int_0^x f(s)y_m^{-1}(s)\,ds = F(x)y_m^{-1}(x) + \frac{m(m+1)}{2}\int_0^x F(s)\,\frac{ds}{s}, \\ &= F(x)y_m^{-1}(x) + \frac{m(m+1)}{2}\Big\{F(x)\log x - \int_0^x f(s)\log s\,ds\Big\}. \end{aligned}$$

5.7.2. Let $u = \sum_{n\geq 0} a_n T_n$ and $f = \sum_{n\geq 0} b_n T_n$ be solutions of the differential equation $\Delta u(x) = f$, this entail $\sum_{n\geq 2}(a_n T''_n - b_n T_n) = 0$. The differential equations of Chebyshev's polynomials $(1-x^2)T_n''(x)-xT_n'(x)+n^2T_n(x)=0$ and $T_n'(\cos\theta)=-n\sin(n\theta)$, at $x=\cos\theta$, imply $b_0=4a_2$, $b_1=24a_3$ and

$$\begin{aligned}0&=\sum_{n\geq 2}[a_n\{xT_{n,x}'-n^2T_n\}-(1-x^2)b_nT_n]\\&=\sum_{n\geq 2}[na_n\{\cos(\theta)\sin(n\theta)+n\cos(n\theta)\}+b_n\sin^2(\theta)\cos(n\theta)]\\0&=\sum_{n\geq 2}[2na_n\{\sin((n+1)\theta)+\sin((n-1)\theta)+2n\cos(n\theta)\}\\&\qquad+b_n\{\cos((n-2)\theta)-\cos((n+2)\theta)\}],\end{aligned}$$

this is equivalent to

$$\begin{aligned}0&=\sum_{n\geq 1}[2\{(n-1)a_{n-1}+(n+1)a_{n+1}\}\sin(n\theta)\\&\quad+\sum_{n\geq 2}\{4n^2a_n+b_{n+2}-b_{n-2}\}\cos(n\theta)],\\0&=(n-1)a_{n-1}+(n+1)a_{n+1},\ n\geq 1,\\0&=4n^2a_n+b_{n+2}-b_{n-2},\ n\geq 2,\end{aligned}$$

the last equations are restrictive for the function f.

5.7.3. Let $u(x)=\sum_{n\geq 0}\{a_n\cos(w_nx)+b_n\sin(w_nx)\}$ be the Fourier series of u and, assuming that the coefficients have expansions with the same frequencies, let $p(x)=\sum_{n\geq 0}\{A_n\cos(w_nx)+B_n\sin(w_nx)\}$ and $q(x)=\sum_{n\geq 0}\{C_n\cos(w_nx)+D_n\sin(w_nx)\}$. Writting the differential equation $Lu_t=0$ defined by (4.17) for these series implies

$$\begin{aligned}0&=a_k(w_k^2A_l-C_l)-b_kw_kw_lB_l,\\0&=w_kw_la_kA_l-b_k(w_k^2B_l-D_l),\\0&=a_kC_l+b_kD_l-w_k(w_k+w_l)(a_kB_l+b_kA_l).\end{aligned}$$

There exists a solution if and only the three equations uniquely define a_k and b_k from the other coefficients, in particular

$$\frac{b_k}{a_l}=\frac{w_k^2A_l-C_l}{w_kw_lB_l}=\frac{w_kw_lA_l}{w_k^2B_l-D_l}=\frac{w_k(w_k+w_l)B_l-C_l}{D_l-w_k(w_k+w_l)A_l}.$$

If the frequences of the functions p or q differ from those of u leads to other conditions.

5.7.4. It is equivalent to $(xy'_x)'_x - y = 0$ and, denoting u the primitive of y, it is solution of $xu''_x - u = k$, an arbitrary constant k. The solution of $xu''_x - u = 0$ is the Airy function A so we assume $u = Av$, with a function v satisfying

$$xv''_x A + 2xv'_x A'_x = k.$$

The derivative of v, $w = v'_x$ satisfies the first order differential equation $xw'_x A + 2xwA'_x = k$ the solution of which is

$$v'_x = v'_0 + \frac{k}{A_x^2} \int_{x_0}^{x} \frac{A_s}{s}\, ds,$$
$$v_x = v_0 + v'_0 x + \int_{x_0}^{x} \frac{k}{A_t^2} \int_{x_0}^{t} \frac{A_s}{s}\, ds\, dt.$$

Since the Taylor expansion of the Airy function is known and given by (7.11), the integrals are calculated and they provide an expression of v as a series. Finally, the function y is $y = u'_x = A'_x v + Av'_x$ where

$$A'_x = \frac{1}{3x^2}\Big(2\int_0^x s\, dA_s - \int_0^x sA_s\, ds\Big).$$

5.7.5. By an expansion $y(x) = \sum_{k\geq 0} a_k x^k$, a_0 is arbitrary, the odd coefficients are zero

$$\sum_{k\geq 1} x^{2k-1}\{a_{2k} 2k(2k-1-m) - a_{2k-2}\} = 0$$

and a_{2k+1} is deduced from a_{2k-1} for every $k \geq 1$. If m is even, all coefficients of this sum depend on a_0. If m is odd, a_{m+1} is undetermined and a_{m+n} is expressed according to a_{m+1} for every even $n \geq 2$, all previous coefficients $a_0, \ldots, a_{m-1}$ are zero. In both cases, the series is convergent, let u denote its limit. Another solution is uv with $uv''_x + v'_x(2u'_x - x^{-1}mu) = 0$ therefore $v'_x = ku^{-2}x^m$ with an arbitrary constant k.

5.7.6. Let $u(x,t) = \sum_{j\geq 0}\sum_{k\geq 0} u_{jk} L_j(x) L_k(t)$, the differential equation is equivalent to

$$f(t,x) = \sum_{j\geq 0}\sum_{k\geq 0}\{u_{jk} L''_{j,xx}(x) L_k(t) - a u_{jk} L_j(x) L''_{k,tt}(t)$$
$$+ b u_{jk} L_j(x) L_k(t) + c u_{jk} L'_{j,x}(x) L_k(t) + d u_{jk} L_j(x) L'_{k,t}(t)\}.$$

The expression of the derivatives of the polynomials imply

$$\begin{aligned}
&x^2t^2f(t,x) = \\
&\sum_{j\geq 0}\sum_{k\geq 0}\Big[u_{j+2,k}t^2L_k(t)\{(j+2)(j+1)L_{j+2}(x) - (j+2)(2j+3)L_{j+1}(x) \\
&\quad - (j+2)^2L_j(x)\} - au_{j,k+2}x^2L_j(x)\{(k+2)(k+1)L_{k+2}(t) \\
&\quad - (k+2)(2k+3)L_{k+1}(t) - (k+2)^2L_k(t)\} + bu_{j,k}L_j(x)L_k(t) \\
&\quad - cu_{j+1,k}xt^2\{(j+1)L_{j+1}(x) + (j+1)L_j(x)\}L_k(t) \\
&\quad - du_{j,k+1}x^2tL_j(x)\{(k+1)L_{k+1}(t) + (k+1)L_k(t)\}\Big]
\end{aligned}$$

and the coefficients are identified to those of an expansion of the function $x^2t^2f(t,x)$.

8.6 Partial differential equations

6.8.1. Integrating $u^{-1}u'_y = -ax$ implies $u(x,y) = k(x)e^{-axy}$ with an arbitrary function k depending only on x.

6.8.2. Integrating the right-hand term with respect to x then y gives $u(x,y) = ax^2y + bxy^2 + k_1(x) + k_2(y)$ with arbitrary functions k_1 depending only on x and k_2 depending only on y.

6.8.3. $u(x,y) = \varphi(x^2 - y^2)$ constant, with φ in $C_1(\mathbb{R})$.

6.8.5. $u(x,y) = \varphi(x^2 + y^2)$ constant, with φ in $C_1(\mathbb{R})$.

6.8.6. $u(x,y) = \varphi(ax - by)$ constant, with φ in $C_1(\mathbb{R})$.

6.8.7. $u(x,y) = \frac{x-y}{b-a}f(ax - by)$, with f in $C_1(\mathbb{R})$.

6.8.8. Let $u'_x = az$ and $bu'_y + u'_z = ax$, solving these equations separately yields $u(x,y,z) = axz + k_1(y,z)$ by an integration of u'_x with respect to x, then $u'_x + bu'_y + u'_z = a(x+z) + k'_{1z}(y,z) + bk'_{1y}(y,z)$. Using the solution of Exercise (6.8.6), $u(x,y,z) = w(y - bz) + axz$, by identity of the partial derivatives.

6.8.9. The integration of $u^{-1}u'_x = \frac{-y}{(x+y)^2}$ implies

$$u(x,y) = v(x,y)e^{\frac{k(x,y)}{x+y}}$$

with functions k and v having the derivatives $v'_x - v'_y = 0$ and $k'_x - k'_y = \frac{x-y}{x+y}$. From Exercise (6.8.6), $v(x,y) = \varphi(x+y)$ with φ in $C_1(\mathbb{R})$, and from Exercise (6.8.7), $2k(x,y) = (x-y)f(x+y)$ and the derivatives imply

$$k'_x - k'_y = f(x,y) = \frac{x-y}{x+y}, \quad k(x,y) = \frac{(x-y)^2}{4(x+y)}.$$

6.8.10. Let $u(x,y) = e^{\varphi(xy^{-1})}$ with φ in $C_1(\mathbb{R})$, its derivatives are

$$u'_x = y^{-1}\varphi'(xy^{-1})u \quad u'_y = -xy^{-2}\varphi'(xy^{-1})u$$

and the differential equation becomes homogeneous in x an y. Denoting $t = xy^{-1}$, implies $\varphi'(t) = (1-t^2)^{-1}$ therefore

$$\varphi(t) = \frac{1}{2}\log\frac{1+t}{1-t}$$

and $u(x,y) = (1+t)^{\frac{1}{2}}(1-t)^{-\frac{1}{2}}$.

6.8.11. Integrating separately $x^2u'_x = x^4y^{-1}$ and $y^2u'_y = x^4y^{-1}$ provides solutions depending on an arbitrary function φ of $C_1(\mathbb{R})$

$$u(x,y) = \frac{x^3}{3y} + \frac{x^2}{6} + \varphi(\frac{1}{x} - \frac{1}{y}).$$

6.8.12. Integrating the partial derivative $u'_y = f(x,z,u'_x,u'_z)$ where $u'_x = v(ay^{-1})$ and $u'_z = h(yz^{-1})$, a solution is $u(x,y,z) = axy^{-1} + byz^{-1}$. Another solution depending only on y and z is deduced from $yu'_y + zu'_z = 0$ as $u(y,z) = \varphi(yz^{-1}) + k$, with an arbitrary constant k.

6.8.13. The equation $x(y-1)\,dx + y(x-1)\,dy = 0$ has the solution y such that $x + \log(x-1) + y + \log(y-1)$ is constant, it follows that $(1-y)e^y = k_x(1-x)^{-1}e^{-x}$ and the solution is y such that $(1-y)e^y = e^{F(x)}(1-x)^{-1}e^{-x}$ with a primitive F of $(x-1)^{-1}f_x$.

6.8.14. The solution is a relationship between functions y_x and z_x such that $y_xz_x = k_x$ satisfies the differential equation $k'_x = -2x^{-1}zk_x$ i.e. $k'_x = -2x^{-1}y_xz_x^2$, its primitive is $y_xz_x = y_0z_0 - 2\int_{x_0}^x s^{-1}y_sz_s^2\,ds$.

6.8.15. Let $x\frac{du}{dx} = -y\frac{du}{dy} = \frac{x^2}{y}$, hence

$$y = cx^{-1},$$

$$du = \frac{x}{y}\,dx = \frac{x^2}{c}\,dx,$$

$$u = u_0 + \frac{x^3}{3c} = u_0 + \frac{x^2}{3y},$$

and we can replace the constant u_0 by an arbitrary homogeneous function $\varphi(xy)$.

6.8.16. Writing the equation as

$$x\frac{\partial u}{\partial x}+y\frac{\partial u}{\partial y}=\frac{xy}{u}$$

and assuming that

$$\frac{2u\partial u}{xy}=\frac{\partial x}{x}=\frac{\partial y}{y}$$

implies $y=c_1x$ and $uu'_x=\frac{y}{2}=\frac{c_1x}{2}$ hence

$$u^2=\frac{c_1x^2}{4}+c_2=\frac{xy}{4}+v(\frac{y}{x})$$

with an arbitrary function v.

6.8.17. Let

$$axu=y^2\frac{\partial u}{\partial y}-xy\frac{\partial u}{\partial x},$$

$$\frac{2\partial u}{axu}=-\frac{\partial x}{xy}=\frac{\partial y}{y^2}$$

then $xy=c_1$ and

$$\frac{u'_y}{u}=\frac{ax}{2y^2}=\frac{ac_1}{2y^3},$$

$$\log\frac{u}{u_0}=-\frac{ac_1}{4y^2}+c_2=-\frac{ax}{4y}+c_2,$$

$$u=v(xy)e^{-\frac{ax}{4y}}$$

with an arbitrary function v.

6.8.18. Writing $u=u_{1x}+u_{2y}$ and solving separately the marginal second order differential equations $x^2u''_{xx}=xu'_x-4x^{-4}u_x$ gives the function $u(x,y)=x^2\sin x^{-2}+ky^2$.

6.8.19. Let $u(x,y)=v(x,y)e^x$, the function v satisfies the equation

$$xv''_{xx}+2(x+1)v'_x+2yv'_y+2v=2.$$

It has the solution $v(x,y)=v_1(x)+v_2(y)$ which implies $v(x,y)=a+w(x)+y^{-1}$, with a function w such that $xw''_{xx}+2(x+1)w'_x+2w=0$ and $w(x)=x^{-1}$ is a solution.

6.8.20. $u(x,y) = A(x) + A(y)$, where A is the Airy function solution of Equation (4.16).

6.8.21. The homogeneous equation $y'' = x' = y$ with initial conditions y_0 and y'_0 at $x = 0$ has the solution $y_x = y_0 e^{\pm x}$. Let $y_x = y_0 e^{-x}$, the general form of the derivative of the solution is $y'_x = v_x e^{-x}$ with a function v such that $v'_x = -e^x f'_x$. Then

$$v_x = y_0 - \int_0^x e^s \, df_s = y'_0 - e^x f_x + f_0 + \int_0^x e^s f_s \, ds$$

and $y'_x = v_x e^{-x}$ implies

$$y_x = y_0 + (y'_0 + f_0)(1 - e^{-x}) - \int_0^x e^{s-x} f_s \, ds + \int_0^x \int_0^t e^{s-x} f_s \, ds \, dt.$$

6.8.22. Deriving the differential equation for y'_t implies $y''_t = a y_t^2$, therefore $y_t = -at + b$ and

$$x_t = \frac{a^2 x^3}{3} - abx^2 + b^2 x + c$$

with arbitrary constants b and c.

6.8.23. Since $y_t = \sin t$, we have $x'_t = \sin^3 t = \frac{1}{4}(3 \sin t - \sin 3t)$ and $x_t = \frac{1}{4}(\frac{1}{3} \cos 3t - 3 \cos t) + \frac{2}{3}$.

6.8.24. The differential equations imply $x y'_t - y x'_t = x^2 + y^2$ and $u(x,y) = x^{-1} y$ with

$$u'_t = x'_t u'_x + y'_t u'_y = x^{-2}(x y'_t - y x'_t) = 1 + u_t^2$$

hence $u = \tan t + c$ and $y_t = x_t(\tan t + c)$, with an arbitrary constant c.

6.8.25. Deriving y'_t, $y''_t = a'_t x_t + a_t^2(l - y_t) = a_t^{-1} a'_t y'_t - a_t^2(l - y_t)$ then x_t and y_t are solutions of the homogeneous equations

$$\begin{aligned} 0 &= x''_t - a_t^{-1} a'_t x'_t + a^2 x_t, \\ 0 &= y''_t - a_t^{-1} a'_t z'_t + a_t^2 z_t. \end{aligned}$$

If a is a constant, x_t and y_t are linear combinations of $\sin(at)$ and $\cos(at)$, otherwise $x_t = l \sin \theta_t$ and $y_t = l(1 - \cos \theta_t)$ with $a_t = \theta'_t$.

6.8.26. Writing

$$\frac{\partial R(r)}{\partial x} = \frac{dR(r)}{dr}\frac{\partial r}{\partial x} = \frac{x}{r}\frac{dR(r)}{dr}, \qquad \frac{\partial R(r)}{\partial y} = \frac{y}{r}\frac{dR(r)}{dr}$$

reduces the differential equations to

$$x'_t = ax_t\frac{dR}{dr},$$
$$y'_t = by_t\frac{dR}{dr},$$

and their solutions are $x_t = x_0 e^{aR_t}$ and $y_t = y_0 e^{bR_t}$, with $R_t = R(r_t)$.

6.8.27. By derivation of the equations we obtain differential equations depending separately on x and respectively y

$$\begin{aligned} x''_t &= a_1 y'_t + b_1 x'_t = x_t(a_1 b_2 + b_1^2) + a_1 y_t(a_2 + b_1) + a_1 c_2 + b_1 c_1 \\ &= x_t(a_1 b_2 + b_1^2) + (a_2 + b_1)(x'_t - b_1 x_t - c_1) + a_1 c_2 + b_1 c_1 \\ &= (a_2 + b_1)x'_t + (a_1 b_2 - a_2 b_1)x_t + a_1 c_2 - a_2 c_1 \\ y''_t &= a_2 y'_t + b_2 x'_t = b_2 x_t(a_1 + b_1) + a_1 y_t(a_2 + b_1) + b_1 c_1 \\ &= (a_2 + b_1)y'_t + (b_2 a_1 - a_2 b_1)y_t + b_2 c_1 - b_1 c_2 \end{aligned}$$

they are solved explicitly by Proposition 4.12.

6.8.28. Let $u_t = \int_0^t x_s y_s z_s \, ds$ then $x_t^2 = 2au_t + \alpha$, $y_t^2 = 2bu_t + \beta$ and $z_t^2 = 2cu_t + \gamma$, with

$$u_t'^2 = x_t^2 y_t^2 z_t^2 = (2au_t + \alpha)(2bu_t + \beta)(2cu_t + \gamma).$$

8.7 Special functions

7.8.1. According to Legendre, by the change of variable $x = \frac{1}{2}(1 + y)$, $B_{x,x} = 2^{1-2x} \int_{-1}^{1} (1 - y^2)^{a-1} \, dy$ then with $z = y^2$

$$B_{x,x} = 2^{1-2x} \int_0^1 x^{-\frac{1}{2}} (1 - x)^{a-1} \, dx.$$

7.8.2. Let $y(1 - x^n) = x^n$, then $I_{p,q} = n^{-1} \int_0^\infty y^{\frac{p}{n}-1}(1 + y)^{-\frac{p+q}{n}} \, dy$.

7.8.3. $B_{x,k-x} B_{x,1-x}^{-1}$ is written as

$$\frac{\Gamma_{k-x}}{\Gamma_{1-x}\Gamma_k} = \frac{(1 - x)\cdots(k - 1 - x)}{1.2\cdots(k - 1)} = \prod_{n=1}^{k-1}\left(1 - \frac{x}{n}\right) < (1 - x)^{k-1}$$

and it converges to zero as k tends to infinity.

7.8.4. Let $J_k(x) = B^{(k)}_{x,1-x}$, for $k \geq 1$, an expansion of $(1+s)^{-1}$ and the change of variable $s = e^{-y}$ yield

$$J_k(x) = \sum_{n \geq 0} (-1)^n \int_0^\infty s^{x+n-1} \log^k s\, ds,$$

$$\begin{aligned}\int_0^\infty s^{x+n-1} \log^k s\, ds &= (-1)^k \int_0^1 y^k e^{-(x+n)y}\, dy \\ &= \frac{(-1)^k}{(x+n)^{k+1}} \int_0^{x+n} t^k e^{-t}\, dt\end{aligned}$$

and

$$\int_0^{x+n} t^k e^{-t}\, dt = \Gamma_{k+1}\{1 - e^{-(x+n)}\} - (x+n)e^{-(x+n)} \frac{1-(x+n)^k}{1-(x+n)},$$

this is a sum of terms asymptotically equivalent to n^{-2} therefore $J_k(x)$ is finite for every x in $]0,1[$.

7.8.5. Writing $x + y = y(x+1) + (1-y)x$, from the convexity of φ we have $\varphi(x+y) \geq y\varphi(x+1) + (1-y)\varphi(x)$ for every y in $]0,1[$ and the results follow.

7.8.6. Let $0 \geq x \geq 1$ and $n \geq 1$, and let $\varphi(x) = \log \Gamma_x$, we have

$$\begin{aligned}g_n(x) &= \log \frac{x!(n-1)!}{(n+x-1)!} - \log x + x \log n \\ &= \varphi(x) + \varphi(n) - \varphi(x+n) + x \log n, \\ \varphi(x) - g_n(x) &= \varphi(x+n) - \varphi(n) - x \log n\end{aligned}$$

and from Exercice (7.8.5), $\varphi(x+n) - \varphi(x) \leq x\{\varphi(n+1) - \varphi(n)\}$ hence

$$\varphi(x) - g_n(x) \leq x\{\varphi(n+1) - \varphi(n) - \log n\} = 0.$$

Applying Stirling's formula $x! = x^x e^{-x}\sqrt{2\pi x}$, for every $x > 0$, we obtain

$$\begin{aligned}\varphi(x) - g_n(x) &= \Big(x + n - \frac{1}{2}\Big) \log(n+x-1) - \Big(n - \frac{1}{2}\Big) \log(n-1) \\ &\quad - x - x \log n \\ &= \Big(n - \frac{1}{2}\Big) \log\{1 + (n-1)^{-1}x\} + x \log\{1 + n^{-1}(x-1)\} - x\end{aligned}$$

as n tends to infinity, this leads to the equivalence

$$\varphi(x) - g_n(x) = (n-1)^{-1}\Big(n - \frac{1}{2}\Big)x + n^{-1}x(x-1) - x + o(1) = o(1).$$

The approximation of Γ_{x+n} is a direct consequence of this convergence.

7.8.7. Using (7.9) and the equality $\Gamma_{n+x} = x(x+1)\cdots(x+n-1)\Gamma_x$ for every n

$$\frac{1}{\Gamma_x} = \lim_{n\to\infty} \frac{x(x+1)\cdots(x+n-1)}{\Gamma_{n+x}} = x \lim_{n\to\infty} n^{-x}\frac{x(x+1)\cdots(x+n-1)}{\Gamma_n}$$

$$= x \lim_{n\to\infty} n^{-x} \prod_{k=1}^{n-1}\Big(1+\frac{x}{k}\Big)$$

$$= x \lim_{n\to\infty} \exp\Big\{-x\log n + \sum_{k=1}^{n-1} \log\Big(1+\frac{x}{k}\Big)\Big\}$$

introducing γ_n which converges to γ

$$\frac{1}{\Gamma_x} = xe^{\gamma x}\prod_{k=1}^{\infty} \exp\Big\{\log\Big(1+\frac{x}{k}\Big) - \frac{x}{k}\Big\}$$

$$= xe^{\gamma x}\prod_{k=1}^{\infty}\Big(1+\frac{x}{k}\Big)e^{-\frac{x}{k}}.$$

The product converges since its logarithm $\log(1+\frac{x}{k}) - \frac{x}{k}$ develops as $\sum_{k=1}^{\infty} O(x^2k^{-2})\}$.

7.8.8. Let $\varphi = \log\Gamma$, from Exercice (7.8.7)

$$\varphi(x) = \log x^{-1} - \gamma x + \sum_{k\geq 1}\frac{x}{k} + \sum_{k\geq 1}\log\frac{k}{k+x},$$

$$\varphi'(x) = -x^{-1} - \gamma + \sum_{k\geq 1}\frac{x}{k(k+x)},$$

and from the recurrence formula it is a finite sum $\varphi'(x) = \sum_{n=0}^{k-1}(a+n)^{-1}$ where $x-k=a$ belongs to $]0,1[$.

7.8.9. Since $L_{a,b} = a^{-(b+1)}\Gamma_{b+1}$

$$L_{a,b} = \frac{b}{a^{b+1}}\Gamma_b,$$

$$L_{a,b+k} = \frac{b+k}{a^{b+k+1}}\Gamma_{b+k} = \frac{1}{a}L_{a,b+k-1},$$

for $k\geq 1$. Let $\alpha>0$, $L_{\alpha,b} = \frac{a^{b+1}}{\alpha^{b+1}}L_{a,b}$.

7.8.10. The first integral has a series expansion

$$\frac{1}{a}+\Big(\frac{1}{n+a}-\frac{1}{n-a}\Big)+\cdots+\Big(\frac{1}{kn+a}-\frac{1}{kn-a}\Big)+\cdots$$
$$=\frac{1}{a}-\frac{2a}{n^2-a^2}-\cdots-\frac{2a}{k^2n^2-a^2}-\cdots$$

and this is an expansion of $n^{-1}\pi\cot(n^{-1}a\pi)$. Its derivatives imply, for $\omega=n^{-1}\pi$

$$I_2=\frac{2\omega^3}{\sin^3(a\omega)}\cos(a\omega),$$
$$I_3=\frac{3!\omega^4}{\sin^4(a\omega)}\Big(\frac{1}{3}+\frac{2}{3}\cos^2(a\omega)\Big).$$

7.8.11. The solution is $u_x=u_0\int_{x_0}^x e^{F_s}\,ds$ where F is the primitive of $\lambda-x\sin x$, $F_x=\lambda(x-x_0)+x_0\cos x_0-\sin x_0-x\cos x+\sin x$. The solution is u such that $u'_x=u'_0+u_xF_x-\int_{x_0}^x F_su_s\,ds$ and

$$u_x=u_0+u'_0(x-x_0)+\int_{x_0}^x u_sF_s\,ds-\int_{x_0}^x (x-s)u_sF_s\,ds.$$

On a δ-grid $(x_n)_{n\geq 0}$, $u(x_{n+1})$ approximated by

$$u_{n+1}=u_n+\delta u'_0+u_n\Big\{\int_{x_n}^{x_{n+1}} F_s\,ds-\int_{x_n}^{x_{n+1}}(x-s)F_s\,ds\Big\}.$$

Another expression of the solution is obtained using expansions of the sine function and u as $u_x=\sum_{k\geq 0}a_kx^k$, they imply $2a_2+\lambda a_0=0$, $6a_3+\lambda a_1=0$ and for every $k\geq 2$

$$(k+1)(k+2)a_{k+1}+\lambda a_k=\sum_{m,n\geq 0,2m+n+2=k}\frac{(-1)^m}{(2m+1)!}a_n.$$

These equations include an increasing number of parameters and they must be solved iteratively, with arbitrary constants a_0 and a_1.

7.8.12. By subtraction of the differential equations at a and b

$$0=\{(x+a)-(x+b)\}y''_x-\{(x+a)^{\frac{1}{2}}-(x+b)^{\frac{1}{2}}\}y'_x,$$
$$0=\{(x+a)^{\frac{1}{2}}+(x+b)^{\frac{1}{2}}\}y''_x-y'_x,$$

Let $y'_x=u\{(x+a)^{\frac{1}{2}}-(x+b)^{\frac{1}{2}}\}$, then

$$y''_x=\frac{u}{2}\{(x+a)^{-\frac{1}{2}}-(x+b)^{-\frac{1}{2}}\}+\frac{u'_x}{u}y'_x,$$
$$0=\frac{u}{2}\Big\{\frac{(x+b)^{\frac{1}{2}}}{(x+a)^{\frac{1}{2}}}-\frac{(x+a)^{\frac{1}{2}}}{(x+b)^{\frac{1}{2}}}\Big\}+\{(x+a)-(x+b)\}u'_x$$
$$-\{(x+a)^{\frac{1}{2}}-(x+b)^{\frac{1}{2}}\}u$$
$$0=\{(x+a)^{\frac{1}{2}}+(x+b)^{\frac{1}{2}}\}u'_x-u-\frac{u}{2}\{(x+a)^{-\frac{1}{2}}+(x+b)^{-\frac{1}{2}}\}.$$

Since the above equalities are valid for all a and b, the last equality is also true for b and a third constant c, and by subtraction

$$0 = \{(x+a)^{\frac{1}{2}} - (x+c)^{\frac{1}{2}}\}u_x' - \frac{u}{2}\{(x+a)^{-\frac{1}{2}} - (x+c)^{-\frac{1}{2}}\},$$
$$0 = 2\{(x+a)^{\frac{1}{2}}(x+c)^{\frac{1}{2}}\}u_x' - u,$$
$$u_x = (x+a)^{\frac{1}{2}} - (x+c)^{\frac{1}{2}}.$$

Choosing $b = c$

$$y_x' = (x+a) + (x+b) - 2(x+a)^{\frac{1}{2}}(x+b)^{\frac{1}{2}},$$
$$y_x = \frac{1}{2}\{(x+a)^2 + (x+b)^2\} - 2I,$$

where

$$I = \int_0^x (s+a)^{\frac{1}{2}}(s+b)^{\frac{1}{2}}\,ds = \int_0^x \left\{\left(s + \frac{a+b}{2}\right)^2 - k^2\right\}^{\frac{1}{2}} ds,$$

with a constant k such that $2k = a - b < 2s + a + b$, then

$$I = k\int_0^{y_x} (y^2-1)^{\frac{1}{2}}\,dy = k\int_0^{u_x} \sinh^2 u\,du = \frac{k}{2}\int_0^{u_x} \{\cosh(2u) - 1\}\,du$$
$$= \frac{k}{4}\{\sinh(2\operatorname{arg\,cosh} y_x) - 2\operatorname{arg\,cosh} y_x\},$$

where $y_x = k^{-1}\{x + \frac{1}{2}(a+b)\} = \cosh u_x$.

7.8.13. Let $u_x = \sum_{m\geq 0} a_m\{\cos\omega_m(x) + b_m \sin\omega_m(x)\}$, it has the derivatives

$$u_x' = -\sum_{m\geq 0} a_m\omega_m'(x)\sin\omega_m(x) + \sum_{m\geq 0} b_m\omega_m'(x)\cos\omega_m(x),$$
$$u_x'' = -\sum_{m\geq 0} a_m\{\omega_m''(x)\sin\omega_m(x) + \omega_m'^2(x)\cos(\omega_m x)\}$$
$$+\sum_{m\geq 0} b_m\{\omega_m''(x)\cos\omega_m(x) - \omega_m'^2(x)\sin\omega_m(x)\},$$

and it satisfies the equation

$$0 = \sum_{m\geq 0}[(1-x^2)\{b_m\omega_m''(x) - a_m\omega_m'^2(x)\} - b_m x\omega_m'(x) + \alpha^2 a_m]\cos(\omega_m x)$$
$$-\sum_{m\geq 0}[(1-x^2)\{a_m\omega_m''(x) + b_m\omega_m'^2(x)\} + a_m x\omega_m'(x) - \alpha^2 b_m]\sin\omega_m(x)$$

then solving these equations for all m and x defines the coefficients of u_x in the expansion and provides another form of the solution.

Let $A_m = b_m^{-1} a_m + a_m^{-1} b_m$ and $B_m = b_m^{-1} a_m - a_m^{-1} b_m$, assuming that the coefficients of the sine and cosine functions are zero we get a differential equation

$$0 = A_m \frac{\omega''_m(x)}{\omega'_m(x)} - B_m \frac{x}{1-x^2},$$
$$\omega'_m(x) = k(1-x^2)^{-\frac{B_m}{2A_m}}$$

with arbitrary constants a_m, b_m and k. It primitive is

$$\omega_m(x) = I_{1,2-\frac{B_m}{A_m}}(x) + c$$

with an arbitrary constant c where I is the Eulerian function I defined with $n = 2$ (Section 7.1) with an integral from zero to $x < 1$.

8.8 Programs

The figures have been drawn with continuous lines on thin grids using the R graphics functions plot for the real functions and perspect for the spatial graph of bivariate functions. The values of the bivariate functions are written in matrices by the R function outer which requires the definition of new R functions f(x,y).

Graph of the additive wave function of Figure (1.2)

```
x=seq(0,15,by=.3)
y=seq(0,8,by=.2)
z=matrix(0,length(x),length(y))
f<- function(x,y) {u=sin(pi*x/3); v=sin(pi*y/3); (u+v)/2}
z=outer(x,y,f)
persp(x,y,z,theta=30,phi=20)
```

Graph of the multiplicative wave function of Figure (3.2)

```
x=seq(0,15,by=.3); y=seq(0,8,by=.2)
z=matrix(0,length(x),length(y))
f<- function(x,y) {u=sin(pi*x/3); v=sin(pi*y/3); u*v/2}
z=outer(x,y,f)
persp(x,y,z,theta=10,phi=20)
```

Function $u_1(x,t) = x^{-1}e^{at}$ (Chap. 1)

```
x=seq(1,15,by=.3); y=seq(0,8,by=.3)
z=matrix(0,length(x),length(y))
```

```
f<- function(x,y) {u=sqrt(2)*y; exp(u)/x}
z=outer(x,y,f)
```

Function $u_2(x,t) = u_0 \exp(-2t^2(\lambda x)^{-1}\}$ (Chap. 1)

```
x=seq(1,15,by=.6); y=seq(0,15,by=.6)
z=matrix(0,length(x),length(y))
g<- function(x,y) {u=sqrt(2)*y**2; exp(-u)/(.2*x)}
z2=outer(x,y,g)
```

Graph of the complex curve $p(z) = (z - ki)^2$ of Figure (1.1)

```
a=seq(-pi,pi,by=.01)
x=cos(a); y=sin(a)
z= x+ 1i*y
zz=(z-.3*1i)**2+1
z1=Re(zz);z2=Im(zz)
plot(z1,z2,type="l")
```

Graph of the elliptic function of Figure (4.1)

```
a=seq(-pi,pi,by=.01)
r=.25;x=r*cos(a);y=r*sin(a)
x1=x+rep(.4,n);xm1=x-rep(.4,n)
y1=y+rep(.4,n);ym1=y-rep(.4,n)
```

Graphs of Figures (1.6) and (1.7) defined by Equation (1.16)

```
k=6; t=seq(-pi,pi,by=.02)
x= -k*(cos(t))**k+k*(k-1)*((sin(t))**2)*(cos(t))**(k-2)
y= -k*(sin(t))**k+k*(k-1)*((cos(t))**2)*(sin(t))**(k-2)
plot(x,y,type="l")
```

Graph of Figure (3.7)

```
k=4; t=seq(-pi,pi,by=.02)
x= k*sin(t)*cos(t)**(k-1); y= -k*cos(t)*sin(t)**(k-2)
```

Double folium

```
k=10; t=seq(-pi,pi,by=.02)
r=4*k*(cos(t))*((sin(t))**2)
x=r*cos(t); y= r*sin(t)
plot(x,y,type="l")
lines(-x,y)
```

Graph of Figure (3.8)

```
r=1; k=6
t=seq(.01,2*pi,by=.001)
r=sin(k*t)*k
x=r*cos(t); y= r*sin(t)
```

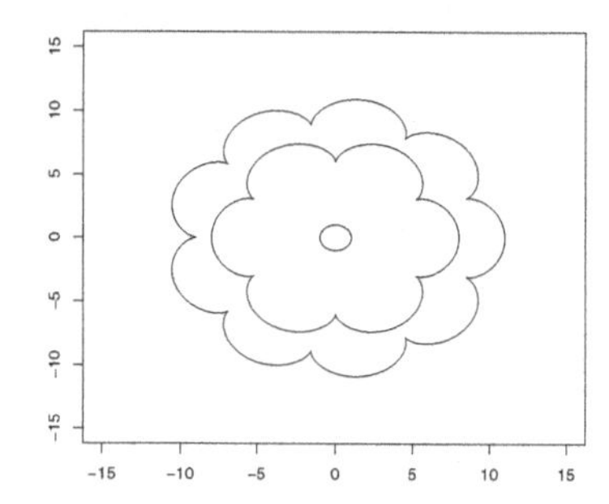

Fig. 8.2 Graph of cycloids.

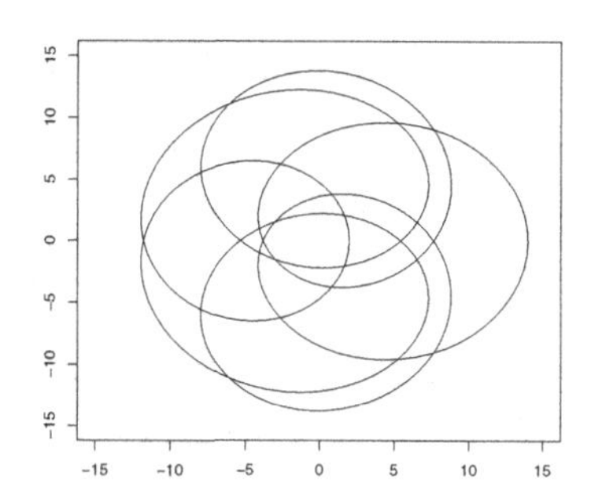

Fig. 8.3 Graph of cycloids.

Hypocycloïd, Figure (8.2)

```
t=seq(-pi,pi,by=.02)
x=7*cos(t)+cos(7*t)
y=7*sin(t)+sin(7*t)
x2=10*cos(t)+cos(10*t)
y2=10*sin(t)+sin(10*t)
a=cos(t); b=sin(t)
plot(x,y,type="l")
lines(x2,y2)
lines(a,b)
```

Disymmetric hypocycloïd, Figure (8.3)

```
t=seq(-pi,pi,by=.02)
x=6*(cos(t))**3+8*cos(6*t)
y=6*sin(t)+8*sin(6*t)
```

Graph of Figure (8.1)

```
t=seq(-pi,pi,by=.02)
x=5*cos(t)+6*cos(5*t)
y=5*sin(t)+6*sin(5*t)
```

Graph of a deltoïd, Figures (1.8) with $k = 4$ and (1.9) with $k = 2.5$. With an integer k, the deltoïd has $k + 1$ edges and with a fractional number, the graph shows deltoïds turning around a central point.

```
t=seq(0,10*pi,by=.02)
x=k*cos(t)+cos(k*t)
y=k*sin(t)-sin(k*t)
```

Graph of an astroïd, Figure (1.4)

```
k=7; t=seq(-pi,pi,by=.02)
x=(cos(t))**k; y=(sin(t))**k
```

Graph of a torus

```
k=50; t=seq(-pi,pi,by=.001)
x=k*cos(t)+(k+1)*cos(k*t)
y=-k*sin(t) -(k+1)*sin(k*t)
```

Graph of a hypocycloïd, Figure (8.1)

```
x2=-3*(cos(t))**3+6*cos(t)*(sin(t))**2
y2=6*sin(t)*(cos(t))**2-3*(sin(t))**3
```

Graph of a star, Figure (1.5)

```
t=seq(-pi,pi,by=.02)
x2=-5*(cos(t))**5+20*((sin(t))**2)*(cos(t))**3
y2=20*((cos(t))**2)*(sin(t))**3-5*(sin(t))**5
```

Graph of the function Gamma on $]1, 2]$, Figure (3.1)

```
x=seq(0,30,by=.0001)
n=length(x)
y=0;g=0
f=function(x,y){u=(x**(y-1))*exp(-x);.0001*cumsum(u)}
for(i in 1:21){s=1+(i-1)*.05
z=outer(x,s,f)
g[i]=z[n]; y[i]=s}
f1=function(x,y){u=(x**(k*y-1))*exp(-x**k/k)
.0001*cumsum(u)}
y1=0;g1=0;k=10
for(i in 1:1){s=i*.1
```

```
z=outer(x,s,f1)
g1[i]=z[n]*2**(1-s)
y1[i]=s}
plot(y,g,xlim=c(0,2),ylim=c(0,10),frame.plot=TRUE,type=''l'')
lines(y1,g1,type=''l'')
```

Fourier'wave function (1822), Figure (1.3)

```
x=seq(0,2*pi,by=.01)
y=cos(x)-cos(3*x)/3+cos(5*x)/5-cos(7*x)/7+cos(9*x)/9
-cos(11*x)/11+cos(13*x)/13-cos(15*x)/15+cos(17*x)/17
-cos(19*x)/19+cos(21*x)/21
```

Linear wave, Figure (3.3)

```
a=20; r=2; t=seq(.2,2*pi,by=.01)
x=t-a*sin(t*r/a)/r
y=a-a*cos(t*a/r)/r
```

Waves increasing in an angle

```
t=seq(0,k*pi,by=.01)
x=t*sin(a*t)
y=t*cos(a*t)
```

Waves increasing in a domain between a curve $f(t)$, calculated in ft, and its symmetric $-f(t)$

```
t=seq(0,k*pi,by=.01)
x=ft*sin(a*t); y=ft*cos(a*t)
```

Exponential spiral, Figure (3.2)

```
k=4; t=seq(.2,k*pi,by=.01)
r=k/(t**.2)
x=cos(k*t) +tan(t)
y=r*sin(k*t)+tan(t)
```

Regular spiral, Figure (3.5)

```
k=40; a=.2; t=seq(0,k*pi,by=.01)
x=t*sin(t*a)
y=t*cos(t*a)
```

Spring, Figure (3.6)

```
k=100; a=15; t=seq(0,k*pi,by=.01)
x=cos(t) +sin(t*a)
y=cos(t)+cos(t*a)
```

Cylinder

```
k=195; a=.01; t=seq(0,k*pi,by=.01)
x=sin(t)+sin(t*a)
y=sin(t)+cos(t*a)
```

Epicycloid, Figure (3.4)

```
r=3; a=.3; b=7; t=seq(.2,20*pi,by=.01)
y=a*r*sin(t/r)-r*sin(3*t/r)/b
x=a*r*cos(t/r)+r*cos(3*t/r)/b
```

Bibliography

Atkinson, F. V. (1955). On second order non-linear oscillation, *Pacif. J. Math* **5**, pp. 643–647.

Bertrand, J. L. (1870). *Traité de calcul différentiel et de calcul intégral* (Gauthier-Villard, Paris).

Boccardo, L., Gallouët., T. and F., M. (1992). Unicité de la solution de certaines équations elliptiques non linéaires, *C. R. Acad. Sci. Paris, série 1* **315**, pp. 1159–1164.

Byerly, E. B. (1893). *An elementary treatise on Fourier's series and spherical, cylindrical, and ellipsoidal harmonics, with applications to problems in mathematical physics* (Ginn and Company, Boston, New York, Chicago London).

Cartan, H. (1971). *Calcul différentiel* (Hermann, Paris).

Didon, F. (1868). *Etude de certaines fonctions analogues aux fonctions X_n de Legendre, etc. Thesis* (Gauthier-Villard, Paris).

Everitt, W. N. (2005). A catalogue of Sturm-Liouville differential equations. In *Sturm-Liouville theory* (W. O. Amrein, A. M. Hinz and D. P. Pearson (editors). Birkhauser Verlag, Basel).

Fourier, J. (1822). *Théorie analytique de la chaleur* (Firmin Didot, Paris).

Gottlieb, D. and Orszag, S. A. (1977). *Numerical Analysis of Spectral Methods: Theory and Applications* (SIAM, Philadelphia).

Hadamard, M. (1907). Les problèmes aux limites des équations aux dérivées partielles, *J. Phys. Theor. Appl.* **6**, pp. 202–241.

Heywood, H. B. and Fréchet, M. (1912). *L'équation de Fredholm et ses applications á la physique mathématique* (Hermann, Paris).

Lagrange, J. L. (1853). *Mécanique analytique* (Mallet-Bachelier, Paris).

Legendre, A. M. (1805). *Nouvelles méthodes pour la détermination des orbites des comètes* (Firmin Didot, Paris).

Legendre, A. M. (1826). *Traité des fonctions elliptiques et des intégrales Eulériennes, avec des tables pour en faciliter le calcul numérique* (Huzard-Courcier, Paris).

Leray, J. and Lions, J.-L. (1965). Quelques résultats de Visik sur les problèmes elliptiques non linéaires par les méthodes de Minty-Browder, *Bull. Soc. Math. France* **93**, pp. 97–107.

Lions, J.-L. (1965a). Sur certaines équations paraboliques non linéaires, *Bull. Soc. Math. France* **93**, pp. 155–175.

Lions, J.-L. (1965b). Sur un nouveau type de problème non linéaire pour opérateurs hyperboliques du 2e ordre, *Sém. J. Leray* **2**, pp. 17–33.

Lions, J.-L. (1969). Quelques remarques sur les inégalités variationnelles, *Sém. J. Leray* **3**, pp. 23–30.

Lions, J.-L. and Strauss, W. A. (1965). Some non-linear evolution equations, *Bull. Soc. Math. France* **93**, pp. 43–96.

Liouville, J. (1839). Sur l'intégration des équations linéaires aux différentielles partielles, *J. Math. Pure Appl.* **4**, pp. 1–6.

Poincaré, H. (1890). Sur les équations aux dérivées partielles de la physique mathématique, *Amer. J. Math.* **12**, pp. 211–294.

Poisson, S. D. (1838). *Traité de mécanique* (Garnier, Paris).

Pons, O. M.-T. (2012). *Inequalities in Analysis and Probability* (World Sci. Publi. Co., Singapore).

Sturm, C. (1836). Sur les équations différentielles linéaires du second ordre, *J. Math. Pures Appl.* **1**, pp. 106–186.

Sturm, C. (1861). *Cours de mécanique de l'école polytechnique* (Mallet-Bachelier, Paris).

Sturm, C. (1868). *Cours d'analyse de l'école polytechnique, 3e éd.* (Gauthier-Villars, Paris).

Timmermans, A. (1854). *Traité de calcul différentiel et de calcul intégral* (Académie Royales de Belgique, Bruxelles).

Trench, W. (2001). *Elementary differential equations* (Brooks, Cole Thomson learning).

Index

Printed in the United States
by Baker & Taylor Publisher Services